WATER TREATMENT IN DEVELOPED AND DEVELOPING NATIONS

An International Perspective

WATER TREATMENT IN DEVELOPED AND DEVELOPING NATIONS

An International Perspective

Edited by

Victor Monsalvo, PhD

AAP APPLE ACADEMIC PRESS

Apple Academic Press Inc.
3333 Mistwell Crescent
Oakville, ON L6L 0A2
Canada

Apple Academic Press Inc.
9 Spinnaker Way
Waretown, NJ 08758
USA

Exclusive worldwide distribution by CRC Press, a member of Taylor & Francis Group

Printed in the United States of America on acid-free paper

International Standard Book Number-13: 978-1-77188-241-5 (Hardcover)

Library and Archives Canada Cataloguing in Publication

Water treatment in developed and developing nations: an international perspective / edited by Victor Monsalvo, PhD.

Includes bibliographical references and index.
ISBN 978-1-77188-241-5 (bound)
1. Sewage--Purification--Developing countries. 2. Sewage--Purification--Developed countries. I. Monsalvo, Victor (Victor M.), editor

TD627.W28 2015 628.309172'4 C2015-901598-7

Library of Congress Cataloging-in-Publication Data

Water treatment in developed and developing nations : an international perspective / editor: Victor Monsalvo, PhD. -- 1st ed.

pages cm
Other title: Green enology in practice
Includes bibliographical references and index.
ISBN 978-1-77188-241-5 (alk. paper)
1. Sewage--Purification--Developed countries. 2. Sewage--Purification--Developing countries. 3. Water treatment plants--Developed countries. 4. Water treatment plants--Developing countries. I. Monsalvo, Victor M., editor.

TD745.W366 2015 628.1--dc23 2015007775

Apple Academic Press also publishes its books in a variety of electronic formats. Some content that appears in print may not be available in electronic format. For information about Apple Academic Press products, visit our website at **www.appleacademicpress.com** and the CRC Press website at **www.crcpress.com**

Cover photos:
Top: A sewage treatment plant in northern Portugal. Photo by Dantadd. Used with permission per the Creative Commons Attribution-Share Alike 2.5 Generic license. http://en.wikipedia.org/wiki/Water_treatment#mediaviewer/File:Bragan%C3%A7a43.jpg.
Bottom left: Kenya - Ruai treatment plant (ponds) and Dandora dumping site. Photo by SuSanA Secretariat - https://www.flickr.com/photos/gtzecosan/7044489061/. Used with permission per the Creative Commons Attribution 2.0 Generic license.
Bottom right: Water purification systems a Bret lake, Switzerland. Photo by Rama. Used with permission per the Creative Commons Attribution-Share Alike 2.0 France license. http://commons.wikimedia.org/wiki/File:Usine_Bret_MG_1648.jpg.

About the Editor

VICTOR MONSALVO, PhD

Professor Victor Monsalvo is an environmental scientist with a PhD in chemical engineering from the University Autonoma de Madrid, where he later became a professor in the chemical engineering section. As a researcher, he has worked with the following universities: Leeds, Cranfield, Sydney, and Aachen. He took part of an active research team working in areas of environmental technologies, water recycling, and advanced water treatment systems. He has been involved in sixteen research projects sponsored by various entities. He has led nine research projects with private companies and an R&D national project, coauthored two patents (national and international) and a book, edited two books, and written around fifty journal and referred conference papers. He has given two key notes in international conferences and has been a member of the organizing committee of five national and international conferences, workshops, and summer schools. He is currently working as senior researcher in the Chemical Processes Department at Abengoa Research, Abengoa.

Contents

Acknowledgment and How to Cite

The editor and publisher thank each of the authors who contributed to this book. The chapters in this book were previously published elsewhere. To cite the work contained in this book and to view the individual permissions, please refer to the citation at the beginning of each chapter. Each chapter was read individually and carefully selected by the editor; the result is a book that provides a nuanced look at the the treatment of wastewater around the world. The chapters included are broken into two sections.

The articles in the first section were chosen to cover topics related to wastewater treatment in developing countries (according to the United Nations' categories). Topics include:

- Reclaimed water for irrigation reuse in developing countries (chapter 1).
- Sludge-handling practices in Micronesia (chapter 2).
- The removal of phthalate esters from Chinese water sources (chapter 3).
- Disposal of domestic wastewater in Nigeria (chapter 4).
- Ameba-enrichment in a South African wastewater treatment plant (chapter 5).
- Bioenergy from wastewater produced by a Brazilian meat-processing plant (chapter 6).

In the second section we turn our attention to wastewater treatment in developed nations, focusing on the following topics:

- The presence of various pharmaceutical contaminants in the River Thames (chapter 7).
- Wastewater recycling in Greece (chapter 8).
- The impact on surface water from contaminants released from German wastewater treatment plants (chapter 9).
- A Canadian constructed wetland's effectiveness for the removal of various contaminants from wastewater (chapter 10).
- Accessing irrigation from treated wastewater in the United States (chapter 11).

- The spacial distribution of fecal indicator bacteria in the groundwater beneath two American wastewater treatment plants (chapter 12).
- Detection of contamination from retinoid acid reception agonists in Japanese wastewater treatment plants (chapter 13).

By looking at a variety of water treatment methods and technologies, within the context of developing and developed nations' differing resources, we gain a better perspective on the effectiveness of techniques being used around the world. Selecting the right wastewater treatment technology for each circumstance requires an understanding of what are the most effective alternatives.

List of Contributors

Jimme M. Abba
Department of Geography, University of Maiduguri, Nigeria

Pradip Adhikari
Postdoctoral Fellow, Dept. of Soil Environment and Atmospheric Sciences, Univ. of Missouri, Columbia, Missouri, USA

Julie C. Anderson
Richardson College for the Environment, Department of Environmental Studies and Sciences and Department of Chemistry, The University of Winnipeg, Winnipeg, MB, R3B 2E9, Canada

Silvia L. F. Andersen
Federal University of Santa Catarina, Brazil

Andreas Angelakis
Institute of Iraklion, Hellenic Agricultural Organization DEMETER-N.AG.RE.F., 71307 Iraklion and Hellenic Water Supply and Sewerage Services Association, 41222 Larissa, Greece

Hauwa Lawan Badawi
National Commission for Museums and Monuments, Maiduguri, Nigeria

T. G. Barnard
Water and Health Research Centre, University of Johannesburg, P.O. Box 17011, Doornfontein 2028, South Africa

C. Bartie
Water and Health Research Centre, University of Johannesburg, P.O. Box 17011, Doornfontein 2028, South Africa and National Institute for Occupational Health, P.O. Box 4788, Johannesburg 2000, South Africa

Alexandra Botzat
Department of Ecology – Conservation Ecology, Faculty of Biology, Philipps University of Marburg, Marburg, Germany

Michael J. Bowes
Natural Environment Research Council, Centre for Ecology and Hydrology, Wallingford, United Kingdom

Jules C. Carlson
Richardson College for the Environment, Department of Environmental Studies and Sciences and Department of Chemistry, The University of Winnipeg, Winnipeg, MB, R3B 2E9, Canada and Department of Environment and Geography, University of Manitoba, Winnipeg, MB, R3T 2N2, Canada

Jonathan K. Challis
Richardson College for the Environment, Department of Environmental Studies and Sciences and Department of Chemistry, The University of Winnipeg, Winnipeg, MB, R3B 2E9, Canada and Department of Chemistry, University of Manitoba, Winnipeg, MB, R3T 2N2, Canada

Zhonglin Chen
State Key Laboratory of Urban Water Resource and Environment, School of Municipal and Environmental Engineering, Harbin Institute of Technology, Harbin 150090, China

David Daniel
Dept. of Economics and International Business, New Mexico State University, Las Cruces, New Mexico 88003-8003, USA

Rennio F. de Sena
Federal University of Santa Catarina, Brazil and Federal University of Paraíba, Brazil

Ganna Fedorova
Department of Chemistry, Umeå University, Umeå, Sweden and University of South Bohemia in Ceske Budejovice, Faculty of Fisheries and Protection of Waters, South Bohemian Research Center of Aquaculture and Biodiversity of Hydrocenoses, Vodnany, Czech Republic

Jerker Fick
Department of Chemistry, Umeå University, Umeå, Sweden

Aziza K. Genena
Federal University of Santa Catarina, Brazil and Federal Technological University of Paraná, Brazil

Roman Grabic
Department of Chemistry, Umeå University, Umeå, Sweden and University of South Bohemia in Ceske Budejovice, Faculty of Fisheries and Protection of Waters, South Bohemian Research Center of Aquaculture and Biodiversity of Hydrocenoses, Vodnany, Czech Republic

Mark L. Hanson
Department of Environment and Geography, University of Manitoba, Winnipeg, MB, R3T 2N2, Canada

Jonathan Harris
Environmental Health Sciences Program, East Carolina University, 3400 suite Carol Belk Building, Greenville, NC 27858, USA

Charles Humphrey
Environmental Health Sciences Program, East Carolina University, 3400 suite Carol Belk Building, Greenville, NC 27858, USA

Michihiko Ike
Division of Sustainable Energy and Environmental Engineering, Osaka University, Suita, Osaka, Japan

Andreas Ilias
Land Reclamation Institute of Thessaloniki, Hellenic Agricultural Organization DEMETER-N. AG.RE.F., 57400 Sindos Thessaloniki, Greece

Daisuke Inoue
Division of Sustainable Energy and Environmental Engineering, Osaka University, Suita, Osaka, Japan

Josef D. Järhult
Section of Infectious Diseases, Department of Medical Sciences, Uppsala University, Uppsala, Sweden

Humberto J. José
Federal University of Santa Catarina, Brazil

Abba Kagu
Department of Geography, University of Maiduguri, Nigeria

Ghazanfar A. Khan
Department of Chemistry, Umeå University, Umeå, Sweden

Charles W. Knapp
David Livingstone Centre for Sustainability, Department of Civil & Environmental Engineering, University of Strathclyde, Glasgow, Scotland, G1 1XN, UK

Richard H. Lindberg
Department of Chemistry, Umeå University, Umeå, Sweden

Yu Liu
State Key Laboratory of Urban Water Resource and Environment, School of Municipal and Environmental Engineering, Harbin Institute of Technology, Harbin 150090, China

Jennifer E. Low
Richardson College for the Environment, Department of Environmental Studies and Sciences and Department of Chemistry, The University of Winnipeg, Winnipeg, MB, R3B 2E9, Canada

Danielle B. Luiz
Federal University of Santa Catarina, Brazil and EMBRAPA Fishery and Aquaculture, Brazil

Axel Magdeburg
Department Aquatic Ecotoxicology, Biological Sciences Division, Goethe University Frankfurt am Main, Frankfurt am Main, Germany

John G. Mexal
Dept. of Plant and Environmental Sciences, New Mexico State University, New Mexico 88003-8003, USA

Regina F. P. M. Moreira
Federal University of Santa Catarina, Brazil

P. Muchesa
Water and Health Research Centre, University of Johannesburg, P.O. Box 17011, Doornfontein 2028, South Africa

O. Mwamba
Water and Health Research Centre, University of Johannesburg, P.O. Box 17011, Doornfontein 2028, South Africa

Tsuyoshi Nakanishi
Laboratory of Hygienic Chemistry and Molecular Toxicology, Gifu Pharmaceutical University, Gifu, Japan

Jörg Oehlmann
Department Aquatic Ecotoxicology, Biological Sciences Division, Goethe University Frankfurt am Main, Frankfurt am Main, Germany

Michael O'Driscoll
Department of Geological Sciences, East Carolina University, 204 Graham, Greenville, NC 27858, USA

Björn Olsen
Section of Infectious Diseases, Department of Medical Sciences, Uppsala University, Uppsala, Sweden and Section for Zoonotic Ecology and Epidemiology, School of Natural Sciences, Linnaeus University, Kalmar, Sweden

Athanasios Panoras
Land Reclamation Institute of Thessaloniki, Hellenic Agricultural Organization DEMETER-N. AG.RE.F., 57400 Sindos Thessaloniki, Greece

Kristin Quednow
Analytical Environmental Chemistry, Institute for Atmospheric and Environmental Sciences, Goethe University Frankfurt am Main, Frankfurt am Main, Germany

Joseph D. Rouse
Water and Environmental Research Institute of the Western Pacific, University of Guam, UOG Station, Mangilao 96923, Guam

Kazuko Sawada
Division of Sustainable Energy and Environmental Engineering, Osaka University, Suita, Osaka, Japan

Horst Fr. Schröder
RWTH Aachen University, Germany

Kazunari Sei
Division of Sustainable Energy and Environmental Engineering, Osaka University, Suita, Osaka, Japan

Jimin Shen
State Key Laboratory of Urban Water Resource and Environment, School of Municipal and Environmental Engineering, Harbin Institute of Technology, Harbin 150090, China

Manoj K. Shukla
Dept. of Plant and Environmental Sciences, New Mexico State University, New Mexico 88003-8003, USA

Andrew C. Singer
Natural Environment Research Council, Centre for Ecology and Hydrology, Wallingford, United Kingdom

Hanna Söderström
Department of Chemistry, Umeå University, Umeå, Sweden

Daniel Stalter
Department Aquatic Ecotoxicology, Biological Sciences Division, Goethe University Frankfurt am Main, Frankfurt am Main, Germany

Elaine Virmond
Federal University of Santa Catarina, Brazil and EMBRAPA Agroenergy, Brazil

Yuichiro Wada
Division of Sustainable Energy and Environmental Engineering, Osaka University, Suita, Osaka, Japan

John R. White
Wetland & Aquatic Biogeochemistry Laboratory, Department of Oceanography and Coastal Sciences, Louisiana State University, Baton Rouge, LA 70803, USA

Charles S. Wong
Richardson College for the Environment, Department of Environmental Studies and Sciences and Department of Chemistry, The University of Winnipeg, Winnipeg, MB, R3B 2E9, Canada and Department of Chemistry, University of Manitoba, Winnipeg, MB, R3T 2N2, Canada

Florentina Zurita
Environmental Quality Laboratory, Centro Universitario de la Ciénega, University of Guadalajara, Ocotlán, Jalisco 47820, Mexico

Introduction

Inadequate wastewater treatment has serious consequences for both human health and the environment. Poorly treated wastewater increases the risk of infectious diseases, spreads antibiotic-resistant genes, and exposes humans, animals, and the environment to a wide range of potentially dangerous chemicals.

Unfortunately, wastewater treatment is often expensive. Municipal governments, particularly those in developing nations, often lack the necessary economic and workforce resources to provide adequate treatment. Water supply and available space impose other limitations on a region's wastewater options.

UN studies show that high-income, developed nations treat about 70 percent of their wastewater, while only 8 percent of wastewater undergoes any kind of treatment in low-income developing nations. At the same time, particularly in water-scarce developing nations, wastewater volumes have increased substantially in recent years, due in part to urban migration.

Meanwhile, a growing world population, the increasing scarcity of water resources, and the rise in fertilizer prices all contribute to the trend to reuse wastewater for agricultural purposes. Many farmers in water-scarce developing nations irrigate with wastewater because it is the only year-round water source they have, and it reduces the need for purchasing fertilizer. The health risks associated with this practice merit ongoing monitoring and investigation.

The articles in this compendium by no means offer a complete perspective on the current research and developing technologies related to wastewater treatment in regions around the world. They do, however, provide a representative cross-sample of both developing and developed nations' wastewater treatment facilities. Given the growing importance of wastewater management to international health, ongoing research on treatment and reuse technologies is essential.

Victor Monsalvo, PhD

In Chapter 1, Zurita and White evaluate three different two-stage hybrid ecological wastewater treatment systems (HEWTS) with combinations of horizontal flow (HF) constructed wetlands (CWs), vertical flow (VF) CWs and stabilization ponds (SP) for the removal of Organic-N, NH_4^+, NO_3^-, Total N, Total P, Total Coliforms (TCol) and *Escherichia Coli*, BOD, COD and TSS. The overall goal of the study was novel in comparison to most other studies in that the authors sought to evaluate and compare the efficiency of the three HEWTSs for water quality improvements, while minimizing nutrient removal from the wastewater in order to generate high quality reclaimed water for reuse for irrigation of crops. The most effective systems were those systems containing a vertical flow component, either HF-VF or VF-HF. In these two HEWTS, NH_4^+ was reduced by 85.5% and 85.0% respectively, while NO_3^- was increased to 91.4 ± 17.6 mg/L and to 82.5 ± 17.2 mg/L, respectively, an artifact of nitrification. At the same time, *E. coli* was reduced by 99.93% and 99.99%, respectively. While the goal of most wastewater treatment is focused on reducing nutrients, the results here demonstrate that two-stage HEWTSs containing VF components can be used to produce a high quality effluent while retaining inorganic nutrients, thereby conserving this valuable resource for reuse as irrigation water for agriculture in subtropical developing countries where water and fertilizer resources are scarce or expensive.

In Chapter 2, by Rouse, a survey of wastewater treatment facilities in the Federated States of Micronesia revealed a lack of fully functional treatment systems and conditions that potentially could lead to adverse environmental impacts and public health concerns. Due to inadequate facilities, the amount and composition of wastewater entering the plants as well as the degree of treatment being achieved is largely unknown. In some cases raw sewage is being discharged directly into the ocean and waste sludge is regularly taken by local residents for agricultural purposes without adequate treatment. In addition, the need to establish best management practices for placement and maintenance of septic tanks is urgent. Furthermore, development of eco-friendly solutions is needed to more effectively treat wastewater from industrial and agricultural sources in an effort to abate current pollution problems. Comparisons of treatment methods being used and problems encountered at different locations in the islands would provide valuable information to aid in the development of sustainable treatment practices throughout Micronesia.

The presence of phthalate esters (PAEs) in the environment has gained a considerable attention due to their potential impacts on public health. Chapter 3, by Liu and colleagues, reports the first data on the occurrence of 15 PAEs in the water near the Mopanshan Reservoir—the new and important water source of Harbin city in Northeast China. As drinking water is a major source for human exposure to PAEs, the fate of target PAEs in the two waterworks (Mopanshan Waterworks and Seven Waterworks) was also analyzed. The results demonstrated that the total concentrations of 15 PAEs in the water near the Mopanshan Reservoir were relatively moderate, ranging from 355.8 to 9226.5 ng/L, with the mean value of 2943.1 ng/L. DBP and DEHP dominated the PAE concentrations, which ranged from 52.5 to 4498.2 ng/L and 128.9 to 6570.9 ng/L, respectively. The occurrence and concentrations of these compounds were heavily spatially dependent. Meanwhile, the results on the waterworks samples suggested no significant differences in PAE levels with the input of the raw waters. Without effective and stable removal of PAEs after the conventional drinking water treatment in the waterworks (25.8% to 76.5%), the risks posed by PAEs through drinking water ingestion were still existing, which should be paid special attention to the source control in the Mopanshan Reservoir and some advanced treatment processes for drinking water supplies.

Chapter 4, by Kagu and colleagues, evaluates domestic wastewater disposal in some selected wards of urban Maiduguri. The unprecedented population growth in Maiduguri generally is directly proportional to its demand for water supply for both domestic and industrial needs. Consequently the more water is consumed, the more its waste is generated. This in turn encourages its indiscriminate disposal particularly in an environment with poor drainage system such as most parts of Maiduguri urban area. It is against this background that the study examined how domestic wastewater is disposed in Maiduguri using the following objectives: to quantify the amount of domestic wastewater generated in the sampled areas, to study and describe the methods used in disposing wastewater, and to highlight the issues due to poor drainage in the town. The scope therefore, covers only domestic wastewater disposal in the selected wards with reference to individual household wastewater disposal system. This was based on planned and unplanned drainage system areas. New G R A and Bulumkutu were selected as the combination of the planned and unplanned areas, Hausari as unplanned and Gwange as planned areas respectively. Purposive, stratified and accidental sampling methods were employed at different stages in the sampling of the wards and the respondents. Three hundred and sixty copies of questionnaire designed to capture the objectives of the study were administered in the three selected wards (120 for each ward). The data obtained were analyzed using simple descriptive statistic and ANOVA. From the results obtained, the mean responses of the residents with regard to the wastewater generated from bathing, washing, food preparation and others, Maisandari has the highest values: 2.1583a, 1.3417a, 1.1250a and 1.1250a at 0.05 significant levels. This implies that most of the wastewater generated in the study area came from Maisandari ward. As a result of the serious environmental and health implication of this indiscriminate wastewater disposal in the study area, it is therefore recommended, that Borno State government should ensure effective/efficient maintenance of the existing waste disposal system and those areas without adequate drainage system should be provided by the government. The State government should provide drainages in the urban which are channelized in to central outlet system where it can be easily treated before final disposal for other uses.

Free-living amoebae pose a potential health risk in water systems as they may be pathogenic and harbor potential pathogenic bacteria known as amoebae resistant bacteria. In Chapter 5, Muchesa and colleagues observed free-living amoebae in 150 (87.2%) of the environmental water samples. In particular, *Acanthamoeba* sp. was identified in 22 (12.8%) using amoebal enrichment and confirmed by molecular analysis. FLA were isolated in all 8 stages of the wastewater treatment plant using the amoebal enrichment technique. A total of 16 (9.3%) samples were positive for FLA from influent, 20 (11.6%) from bioreactor feed, 16 (9.3%) from anaerobic zone, 16 (9.3%) from anoxic zone, 32 (18.6%) from aerators, 16 (9.3%) from bioreactor effluent, 11 (6.4%) from bioreactor final effluent, and 45 (26.2%) from maturation pond. This study provides baseline information on the occurrence of amoebae in wastewater treatment plant. This has health implications on receiving water bodies as some FLA are pathogenic and are also involved in the transmission and dissemination of pathogenic bacteria.

Chapter 6, by José and colleagues, looks at the wastewater produced in meat processing plants. Meat processing plants worldwide use approximately 62 Mm^3 per year of water. Only a small amount of this quantity becomes a component of the final product. The remaining part becomes wastewater with high biological and chemical oxygen demands, high fat content and high concentrations of dry residue, sedimentary and total suspended matter as well as nitrogen and chloride compounds (Sroka et al., 2004). Of the components usually found in these effluents, blood can be considered as the most problematic due to its capacity to inhibit floc formation during physicochemical wastewater treatment and its high biochemical (BOD5, biochemical oxygen demand during decomposition over a 5-day period) and chemical oxygen demand (COD). In fact, even with correct handling during meat processing, this activity generates 2.0 and 0.5 liters of blood as effluent for each bovine animal and pig, respectively (Tritt & Schuchardt, 1992). The treatment of both the solid wastes and the wastewater from the meat processing industry represents one of the greatest concerns associated with the agro-industrial sector globally, mainly due to the restrictions that international trade regulations have imposed over their use and the related environmental issues.

In Chapter 7, by Singer and colleagues, the concentration of eleven antibiotics (trimethoprim, oxytetracycline, ciprofloxacin, azithromycin, cefotaxime, doxycycline, sulfamethoxazole, erythromycin, clarithromycin, ofloxacin, norfloxacin), three decongestants (naphazoline, oxymetazoline, xylometazoline) and the antiviral drug oseltamivir's active metabolite, oseltamivir carboxylate (OC), were measured weekly at 21 locations within the River Thames catchment in England during the month of November 2009, the autumnal peak of the influenza A[H1N1]pdm09 pandemic. The aim was to quantify the pharmaceutical response to the pandemic and compare this to drug use during the late pandemic (March 2010) and the inter-pandemic periods (May 2011). A large and small wastewater treatment plant (WWTP) were sampled in November 2009 to understand the differential fate of the analytes in the two WWTPs prior to their entry in the receiving river and to estimate drug users using a wastewater epidemiology approach. Mean hourly OC concentrations in the small and large WWTP's influent were 208 and 350 ng/L (max, 2070 and 550 ng/L, respectively). Erythromycin was the most concentrated antibiotic measured in Benson and Oxford WWTPs influent (max = 6,870 and 2,930 ng/L, respectively). Napthazoline and oxymetazoline were the most frequently detected and concentrated decongestant in the Benson WWTP influent (1650 and 67 ng/L) and effluent (696 and 307 ng/L), respectively, but were below detection in the Oxford WWTP. OC was found in 73% of November 2009's weekly river samples (max = 193 ng/L), but only in 5% and 0% of the late- and inter-pandemic river samples, respectively. The mean river concentration of each antibiotic during the pandemic largely fell between 17–74 ng/L, with clarithromycin (max = 292 ng/L) and erythromycin (max = 448 ng/L) yielding the highest single measure. In general, the concentration and frequency of detecting antibiotics in the river increased during the pandemic. OC was uniquely well-suited for the wastewater epidemiology approach owing to its nature as a prodrug, recalcitrance and temporally- and spatially-resolved prescription statistics.

In Greece, and particularly in many southeastern and island areas, there is severe pressure on water resources, further exacerbated by the high demand of water for tourism and irrigation in summertime. The integration of treated wastewater into water resources management is of paramount importance to meet future demands. Despite this need, only a few projects of effluent reuse have been implemented, most of them being pilot projects of crop or landscape irrigation. The most important projects which are currently in practice are those of Thessaloniki, Chalkida, Malia, Livadia, Amfisa, Kalikratia, and Chersonissos. Chapter 8, by Ilias and colleagues, examines the project in Thessaloniki, at the most important wastewater reuse site, the secondary effluent of the city's Waste Water Treatment Plant (WWTP) (165,000 m^3/day) is used for agricultural irrigation after mixing with freshwater at a 1:5 ratio. The main crops irrigated are rice, corn, alfalfa and cotton. A few other projects are under planning, such as that at Iraklion, Agios Nikolaos and several island regions. Finally, it should be mentioned that there are several cases of indirect reuse, especially in central Greece. However, the reuse potential in Greece is limited, since effluent from Athens's WWTP, serving approximately half of the country's population, is not economically feasible due to the location of the plant.

Since the 1980s, advances in wastewater treatment technology have led to considerably improved surface water quality in the urban areas of many high income countries. However, trace concentrations of organic wastewater-associated contaminants may still pose a key environmental hazard impairing the ecological quality of surface waters. To identify key impact factors, Stalter and colleagues analyzed the effects of a wide range of anthropogenic and environmental variables on the aquatic macroinvertebrate community in Chapter 9. The authors assessed ecological water quality at 26 sampling sites in four urban German lowland river systems with a 0–100% load of state-of-the-art biological activated sludge treated wastewater. The chemical analysis suite comprised 12 organic contaminants (five phosphor organic flame retardants, two musk fragrances, bisphenol A, nonylphenol, octylphenol, diethyltoluamide, terbutryn), 16 polycyclic aromatic hydrocarbons, and 12 heavy metals. Non-metric multidimensional scaling identified organic contaminants that are mainly wastewater-associated (i.e., phosphor organic flame retardants, musk fragrances, and diethyltoluamide) as a major impact variable on macroinvertebrate species composition. The structural degradation of streams was also identified as a significant factor. Multiple linear regression models revealed a significant impact of organic contaminants on invertebrate populations, in particular on *Ephemeroptera, Plecoptera*, and *Trichoptera* species. Spearman rank correlation analyses confirmed wastewater-associated organic contaminants as the most significant variable negatively impacting the biodiversity of sensitive macroinvertebrate species. In addition to increased aquatic pollution with organic contaminants, a greater wastewater fraction was accompanied by a slight decrease in oxygen concentration and an increase in salinity. This study highlights the importance of reducing the wastewater-associated impact on surface waters. For aquatic ecosystems in urban areas this would lead to: (i) improvement of the ecological integrity, (ii) reduction of biodiversity loss, and (iii) faster achievement of objectives of legislative requirements, e.g., the European Water Framework Directive.

The discharge of complex mixtures of nutrients, organic micropollutants, and antibiotic resistance genes from treated municipal wastewater into freshwater systems are global concerns for human health and aquatic organisms. Antibiotic resistance genes (ARGs) are genes that have the ability to impart resistance to antibiotics and reduce the efficacy of antibiotics in the systems in which they are found. In the rural community of Grand Marais, Manitoba, Canada, wastewater is treated passively in a sewage lagoon prior to passage through a treatment wetland and subsequent release into surface waters. Using this facility as a model system for the Canadian Prairies, the two aims of Chapter 10, by Anderson and colleagues, were to assess: (a) the presence of nutrients, micropollutants (i.e., pesticides, pharmaceuticals), and ARGs in lagoon outputs, and (b) their potential removal by the treatment wetland prior to release to surface waters in 2012. As expected, concentrations of nitrogen and phosphorus species were greatest in the lagoon and declined with movement through the wetland treatment system. Pharmaceutical and agricultural chemicals were detected at concentrations in the ng/L range. Concentrations of these compounds spiked downstream of the lagoon following discharge and attenuation was observed as the effluent migrated through the wetland system. Hazard quotients calculated for micropollutants of interest indicated minimal toxicological risk to aquatic biota, and results suggest that the wetland attenuated atrazine and carbamazepine significantly. There was no significant targeted removal of ARGs in the wetland and the data suggest that the bacterial population in this system may have genes imparting antibiotic resistance. The results of this study indicate that while the treatment wetland may effectively attenuate excess nutrients and remove some micropollutants and bacteria, it does not specifically target ARGs for removal. Additional studies would be beneficial to determine whether upgrades to extend retention time or alter plant community structure within the wetland would optimize removal of micropollutants and ARGs to fully characterize the utility of these systems on the Canadian Prairies.

Land application of treated wastewater is increasing particularly in areas where water stress is a major concern. The primary objective Adhikari and colleagues in Chapter 11 was to quantify the effect of irrigation with aerated lagoon treated wastewater on soil properties. Core and bulk soil samples were collected from areas under the canopies of mesquite and creosote and intercanopy areas from each of the three plots. Irrigation water quality from 2006 to 2008 showed that average sodium adsorption ratio (SAR), electrical conductivity (EC) and pH of irrigation water were 37.16, 5.32 dS m-1 and 9.7, respectively. The sprinkler uniformity coefficients of irrigated plot-I was 49.34 ± 2.23 % and irrigated plot-II was 61.57 ± 2.11 %. Within irrigated and between irrigated and un-irrigated plots, most soil physical properties remained similar except saturated hydraulic conductivity (K_s) which was significantly higher under mesquite canopies than in the intercanopy areas. Chloride (Cl^-) concentrations below 60 cm depth were higher under creosote than mesquite canopies in irrigated plots indicating deeper leaching of Cl^-. Nitrate (NO_3^-) concentrations below 20 cm depth under canopy and intercanopy areas were low indicating no leaching of NO_3^-.The average SAR to 100 cm depth under shrub canopies was 18.46 ± 2.56 in irrigated plots compared to 2.94 ± 0.79 in the un-irrigated plot. The Na^+ content of creosote was eleven times higher un-irrigated than un-irrigated plot and Na^+ content of herbaceous vegetation was three times higher in the irrigated than unirrigated. Thus irrigation with high sodium wastewater has exacerbated the soil sodicity and plant Na^+ content. Since the majority of mesquite roots are found within 100 cm, and creosote and herbaceous vegetation roots are found within 25 cm from soil surface, a further increase in sodicity may threaten the survival of woody and perennial herbaceous vegetation of the study site.

On-site wastewater treatment systems (OWS) are a common means of wastewater treatment in coastal North Carolina, where the soils are sandy and groundwater is relatively close to the surface (<5 m). Wastewater contains elevated concentrations of pathogenic microorganisms that can contaminate groundwater and surface water if OWS are not operating efficiently and distributing wastewater equally to all drainfield trenches. The objectives of Humphrey and colleagues in Chapter 12 were to compare the distribution of fecal indicator bacteria (FIB) in groundwater beneath a large low-pressure pipe (LPP) OWS and a large pump to distribution box system, and to determine the effectiveness of the systems in reducing FIB including total coliform, *E. coli,* and enterococci. Monitoring wells were installed at the fronts and ends of the drainfields for sample collection. Groundwater beneath the LPP had a more homogeneous spatial distribution of *E. coli* and enterococci concentrations and the specific conductivity of groundwater was also more uniform relative to groundwater beneath the distribution box system. Both systems were effective (>99%) at reducing FIB concentrations before discharge to groundwater. Results indicate that the LPP did enhance the distribution of FIB in groundwater beneath the drainfield area relative to the pump to distribution box system. Although the LPP system had a vadose zone over 2 m thinner than the pump to distribution box system, FIB treatment was similar. Enterococci was the most resilient FIB of the three tested.

Retinoic acid (RA) receptor (RAR) agonists are potential toxicants that can cause teratogenesis in vertebrates. To determine the occurrence of RAR agonists in municipal wastewater treatment plants (WWTPs), in Chapter 13 Sawada and colleagues examined the RARα agonistic activities of influent and effluent samples from several municipal WWTPs in Osaka, Japan, using a yeast two-hybrid assay. Significant RARα agonistic activity was detected in all the influent samples investigated, suggesting that municipal wastewater consistently contains RAR agonists. Fractionations using high-performance liquid chromatography, directed by the bioassay, found several bioactive peaks from influent samples. The RAR agonists, all-trans RA (atRA), 13-cis RA (13cRA), 4-oxo-atRA, and 4-oxo-13cRA, possibly arising from human urine, were identified by liquid chromatography ion trap time-of-flight mass spectrometry. Quantification of the identified compounds in municipal WWTPs confirmed that they were responsible for the majority of RARα agonistic activity in WWTP influents, and also revealed they were readily removed from wastewater by activated sludge treatment. Simultaneous measurement of the RARα agonistic activity revealed that although total activity typically declined concomitant with the reduction of the four identified compounds, it remained high after the decline of RAs and 4-oxo-RAs in one WWTP, suggesting the occurrence of unidentified RAR agonists during the activated sludge treatment. Environ. Toxicol. Chem. 2012;31:307–315. © 2011 SETAC

PART I

DEVELOPING COUNTRIES

CHAPTER 1

Comparative Study of Three Two-Stage Hybrid Ecological Wastewater Treatment Systems for Producing High Nutrient, Reclaimed Water for Irrigation Reuse in Developing Countries

FLORENTINA ZURITA AND JOHN R. WHITE

1.1 INTRODUCTION

Irrigation of crops with raw, municipal wastewater has been a common practice for many decades in developing countries such as China, Mexico, Peru, Egypt, Lebanon, Morocco, India and Vietnam, mainly due to its nutrient value recognized by farmers [1]. Moreover, in some poor areas of developing countries like Mexico, wastewater reuse represents a critical opportunity of improving living standards by increasing income and ensuring food supplies [2]. Unfortunately, the use of untreated municipal wastewater in an agricultural setting poses risks to human health due to the

Comparative Study of Three Two-Stage Hybrid Ecological Wastewater Treatment Systems for Producing High Nutrient, Reclaimed Water for Irrigation Reuse in Developing Countries. © *Zurita F and White JR.* Water ***6****,2 (2014), doi:10.3390/w6020213. Licensed under Creative Commons Attribution 3.0 Unported License, http://creativecommons.org/licenses/by/3.0.*

potential presence of excreta-related pathogens (viruses, bacteria, protozoan and multicellular parasites), skin irritants and toxic chemicals including heavy metals; although it is uncommon to find unsafe levels of heavy metals in municipal wastewater [3]. Consequently, it is important to both treat the wastewater and select wastewater treatment processes that reduce pathogen while retaining nutrients if the water is to be applied for irrigation purposes [4]. Reuse of treated, high-quality reclaimed wastewater for agriculture not only protects human health but is also a good conservation strategy by reducing the consumption of limited drinking water for irrigation and reducing fertilizer costs to the agricultural sector in low-income countries.

Constructed wetlands and waste stabilization ponds are the most widely used ecological wastewater treatment systems in use in the world [5]; although they require significantly more land area than other treatment options. These technologies have proven to be effective treatment alternatives, using natural processes for treating wastewater in small and medium communities, mainly, worldwide. These systems are capable of reaching nearly 100% removal of parasitic eggs due to longer retention times in comparison to more expensive and energy-intensive conventional technologies [6]. In general, a one-stage system is usually not sufficient to effect pathogen reduction to safe target levels [7]. Nitrogen removal in constructed wetlands (CWs) and in particular, stabilization ponds (SPs) is often limited due to the lack of a sophisticated, controlled series of environmental conditions that promote settling, then oxidation followed by reduction which is required for organics N removal, and for promotion of coupled nitrification -denitrification [8]. The same is true for phosphorus removal.

Constructed wetlands have been extensively evaluated mostly in temperate climate prevalent in developed countries; in contrast, the experiences are less abundant under tropical and subtropical areas. In rich countries, the design criteria and guidelines have emphasized nutrient removal. However, the limited capacity of natural systems for nutrient removal is an advantage when the treatment goal is to produce a reclaimed wastewater for irrigation to promote plant growth. A low pathogen concentration, high nutrient content (in particular N) and low presence of heavy metals or other toxic pollutant in reuse water are very desirable in reclaimed wastewater for agricultural irrigation [9]. Hybrid ecological wastewater treatment systems take advantages of the strengths of each different type of CWs and SPs for improving water quality and combine them in order to produce high quality reclaimed water, such as a partially or fully nitrified effluent with low concentrations of indicator organisms and BOD [10,11]. Therefore, in this study three, relatively low cost hybrid ecological wastewater treatment systems (HEWTSs) were evaluated for the treatment of a high-ammonium concentration wastewater generated at a university over one year of operation. The goal was to compare the efficiency of the three HEWTSs arranged as two-stage systems for pollutant removal in order to produce high-quality reclaimed water appropriate for agricultural irrigation claiming the coupled benefits of water and nutrient recycling.

1.2 MATERIALS AND METHODS

1.2.1 DESCRIPTION OF THE WETLAND SYSTEMS

The study was carried out at the Centro Universitario de la Ciénega in Ocotlán, Jalisco, México from September 2009 to August 2010. The climate in the area is classified as warm and wet with rainfall in summer (ACw). The altitude is between 1530 and 1600 m above sea level. A 1100 L tank was used to store the wastewater which was pumped daily from the sewer line located on the campus. The tank was adapted to provide some opportunity for sedimentation of solids. The wastewater was a mixture of gray water (from a cafeteria), and sewage and wastewater from teaching and research laboratories. A total flow rate of ~200 L/d of wastewater was treated and distributed equally among the HEWTSs. The design hydraulic loading rate for the HF-CW, VF-CW and SP were 6.9 cm/d, 14.5 cm/d and 6.8 cm/d.

Three two-stage HEWTSs were evaluated in duplicate (Figure 1). System I consisted of a horizontal flow CW followed by a stabilization pond (HF-SP). The CWs were continuously fed with a theoretical hydraulic retention time of 3 days. The effluent from the CWs flowed by gravity to the stabilization ponds. System II was also configured with a horizontal flow CW as a first stage which was then followed by a vertical flow CW as a second stage (HF-VF). The horizontal flow CW operated in the same way as in the system I but the effluent was collected in a tank and pumped intermittently every 2 h on to the substrate of the vertical flow CW. System III was configured with a vertical flow CW followed by a horizontal flow CW (VF-HF). The vertical flow CW was intermittently fed by a pump programmed to discharge 2.8 L every 2 h on to the surface, specifically over the plant without a distribution system. The effluent flowed by gravity to the next stage.

The dimensions of the horizontal flow CWs, stabilization ponds and vertical flow CWs were 120 cm × 40 cm × 50 cm (L × W × H); 70 cm × 70 cm × 70 cm (L × W × H) and 48 cm × 48 cm × 110 cm (L × W × H), respectively. All the units were constructed of fiberglass and reinforced with external iron bars for supports. The horizontal flow CWs were planted with 6 (25–30 cm-height) individual of *Zantedeschia aethiopica* plants and the vertical flow CWs were planted with 1 individual, adult plant of *Strelitzia reginae*. These species were previously used in CWs for domestic wastewater treatment with successful results [12]. After six months of experimentation, the *Z. aethiopica* plants were replaced with *Canna indica* (a well-known wetland plant) due to the fact that the former plants desiccated during the dry season characterized by low air humidity and high ambient temperatures. Ground tezontle rock was used as the media in all the CWs after first being sieved through a 0.5-mm-opening sieve to remove finer particles which would typically clog subsurface-flow systems. The media sieve analysis revealed a d_{10} of 0.645 mm, d_{60} of 2.3 mm and a uniformity coefficient (UC) of 3.6.

1.2.2 WATER QUALITY PARAMETERS

The systems were fed with wastewater since the beginning but allowed to stabilize for four months and then monitored weekly for the following eight months. Organic-N, Ammonia, Nitrate, total N, BOD, COD, TSS, total P, TCol, *E. Coli*, pH, OD and Conductivity were measured at the influent and effluent of each system. Chemical and biological water quality parameters were determined as described in the Standard Methods for the examination of Water and Wastewater [13]. Total coliforms and *E. Coli* were quantified by the Colilert method. Samples were analyzed immediately after they were taken, in the Quality Environmental Laboratory at the university. When this was not possible, samples were preserved at 4 °C and analyzed within 24 h. A potentiometer (Thermo Scientific 3 Star) was used to measure pH and conductivity.

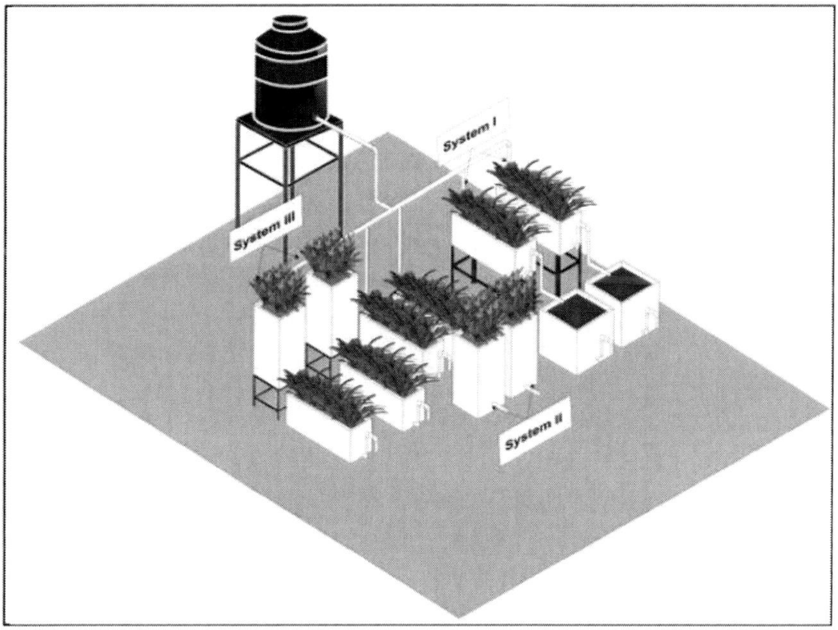

FIGURE 1: Hybrid ecological wastewater treatment systems. System I: HF-SP, System II: HF-VF and System III: VF-HF.

1.2.3 DATA ANALYSIS

A randomized block design was used to analyze the data in this study. Multifactor analysis of variance (ANOVA) was carried out using the Statgraphics Centurion XVI software package to check differences amongst treatments (influent and the three HEWTSs or influent and the two stages in each individual HEWTS). A significance level of $p = 0.05$ was used for all statistical tests, and values reported are the mean (average) ± standard error of the mean. When a significant difference was observed between treatments in the ANOVA procedure, multiple comparisons were made using the Least Significant Difference (LSD) test for differences between means.

1.3 RESULTS AND DISCUSSION

The mean characteristics of both the influent water and the treated water for each of the three HEWTSs are presented in Table 1, Table 2 and Table 3. Overall, the influent was dominated by ammonium-N, comprising almost 92% of total N with a mean TP value of 12.4 mg/L. Other water quality parameters of the influent include 140 mg/L BOD, 273 mg/L COD and 61.8 mg/L TSS (Table 1, Table 2 and Table 3).

1.3.1 ORGANIC NITROGEN

Organic-N in the output of the HEWTSs was significantly different in comparison to the input ($p < 0.05$). In the (HF-SP) an increase to a mean of 10.5 mg/L was observed due to the presence of algae in stage II which assimilated ammonia and nitrate producing Organic-N [14]. In contrast, Organic N was reduced in the other two hybrid HEWTS (HF-VF; VF-HF) reaching statistically similar values of 1.6 and 1.2 mg/L, respectively in the effluents (Table 1). The increase in Org-N in the first system and the reduction in the other two hybrid constructed wetlands demonstrate the superior performance of these two systems for the reduction of organic N.

TABLE 1: Performance summary for the three hybrid ecological wastewater treatment systems (HEWTSs) with respect to nutrient removal. Average ± standard error of the mean. Entire system removal percentages are in parentheses with bold letter.

Parameter	Influent	System I: HF-SP		System II: HF-VF		System III: VF-HF	
		1st stage HF-CW	2nd stage SP	1st stage HF-CW	2nd stage VF-CW	1st stage VF-CW	2nd stage HF-CW
Org-N (mg/L)	7.1 ± 1.2	3.2± 1.0	10.5± 1.0	2.8 ± 0.7	1.6 ± 0.7	1.4 ± 0.8	1.2 ± 0.8
Org-N Removal (%)		54.9	−228.1 (−47.9)	60.6	42.9 (77.5)	80.3	* (83.1)
NH_4^+-N (mg/L)	128.2 ± 11.4	103.1± 11.4	35.8 ± 12.2	103.4 ± 12.4	18.6 ± 5.6	25.1 ± 6.7	19.2 ± 6.2
NH_4^+-N Removal (%)		19.6	65.3 (72.0)	19.3	82.0 (85.5)	80.4	* (85.0)
NO_3^--N (mg/L)	4.2 ± 1.4	1.95± 0.3	14.1 ± 1.8	1.97 ± 0.6	91.4 ± 17.6	108 ± 16.3	82.5 ± 17.2
NO_3^--N Removal (%)		53.6	−623 (−236)	53.1	−4540 (−2076)	−2471	23.6 (−1864)
TN (mg/L)	139.5 ± 12.1	108.3 ± 12.1	60.4 ± 9.9	108.2 ± 22.2	111.6 ± 13.2	134.5 ± 21.1	102.9 ± 13.1
TN Removal (%)		22.4	44.2 (56.7)	22.4	* (20)	*	23.5 (26.2)
TP (mg/L)	12.4 ± 1.1	11.8± 1.0	11.4 ± 1.0	12.1 ± 1.1	12.2 ± 1.0	11.3 ± 1.1	12.4 ± 1.1
TP removal (%)		*	*	*	*	*	*

*Note: * No significant difference with regard to the concentration in the previous stage.*

1.3.2 AMMONIUM

Ammonium concentration in the raw wastewater was unexpectedly high (Table 1), perhaps due to the use of liquid culture medium employed in the Biology cell and Microbiology labs and subsequent disposal down the sinks. The NH_4^+-N was reduced significantly in all three systems ($p < 0.05$) with a higher similar decrease in the HF-VF and VF-HF systems. These results support the findings by other authors with respect to the important role that vertical flow CWs plays for treating high-ammonia wastewater

[15]. The reduction of ammonium concentrations was achieved mainly in the VF unit (>80%), regardless of position in the HEWTs. The presence of vertical flow CWs in the HEWTSs significantly increased the efficiency of the treatment by more than 13% with respect to the (HF-SP) system containing no VF component. The aerobic conditions present in VF systems likely contributed to this ammonium reduction due to the oxidation of NH_4^+ to NO_3^- (nitrification) under aerobic condition [16]. This result is in line with other studies using a similar VF system with gravel as a substrate, such as that in which a reduction of ammonium of 47.5% was achieved under similar influent concentrations [17] and another one under Mediterranean weather, where a reduction of 74% was reached with an influent concentration of 43.1 mg/L [18]. In addition, the ammonium-reduction performance of these two HEWTS containing VF components were similar to 88% of ammonia-N removal found in other study [9] in a hybrid VF-HF system as the mean reduction in our systems averaged 85.3%.

1.3.3 NITRATE

As expected, nitrate increased significantly in the three HEWTSs ($p < 0.05$). In concert with the high reduction rates of ammonium, there were higher and analogous increases in nitrate concentration in both systems containing VF treatment components (Table 1, Figure 2a–c). This result was driven by the greater nitrification capacity in the VF components in comparison to the stabilization ponds which anchored the other system. In the three different systems, there was a significant difference between the two stages in the nitrate concentration ($p < 0.05$). The system composed of VF followed by HF had a 23.6% nitrate reduction from the first stage to the second stage. The HF wetland systems typically have greater reduction rates for nitrate [19,20]. However, our system may have lacked sufficient organic carbon to stimulate denitrification [21]. The BOD was almost entirely removed in the first stage leaving little bioavailable carbon for denitrification in the second stage, having as a consequence a nitrate accumulation [8]. According to several researchers, a source of dissolved organic matter should be provided to improve nitrate removal by recirculation or by adding an external organic source such as methanol to provide the needed electron donors for denitrification [22,23]. However, in the case where the goal of treatment is to provide a nitrified effluent, these results indicate that the two systems containing VF systems are preferred and are equally effective in maintaining bioavailable N which can be used for crop production. Other authors have arrived at the same conclusions when the treated wastewater is to be reused in irrigation [24].

1.3.4 TOTAL N

Total N was significantly reduced in the three HEWTSs ($p < 0.05$) with greater removal in the HF-SP system in comparison to systems containing VF components (Table 1, Figure 3). It is likely that ammonia volatilization, common in algae dominated, aerobic stabilization ponds was responsible for this higher removal of TN. Photosynthetic activity in surface waters can drive up pH during the day leading to ammonium volatilization [25]. The pH (measured from 8–10 am each day) of the HF-SP system, averaged above 8 during all stages of treatment. Since TN was dominated by ammonium in the influent, a greater removal of TN by the system suggests removal of N from the system by either ammonia volatilization or coupled nitrification-denitrification versus simply converting ammonium to nitrate (denitrification) as was the case in the two VF systems. Although aerobic conditions generally predominate in aerobic stabilization ponds during the day driven by photosynthetic activity of algae, dissolved oxygen can drop as a result of diurnal variations to a very low level during night time permitting denitrification to take place [26]. Recall in the two systems containing VF components, it is noticeable (Figure 3) that nitrogen in the effluent is composed primarily of nitrate while in contrast, in the system containing the stabilization pond, nitrogen is primarily in the ammonium form. As previously stated, this nitrate accumulation was due to the depletion of biodegradable organic compounds in the previous stage that inhibited denitrification, as has been reported by others [8]. Moreover, the average total N removal in the HF-VF and VF-HF systems were similar to the 29% removal, achieved in HF-VF systems under warm climate in Mediterranean region. An increase to 66% total N removal was only possible through 100% wastewater recirculation through the system [15].

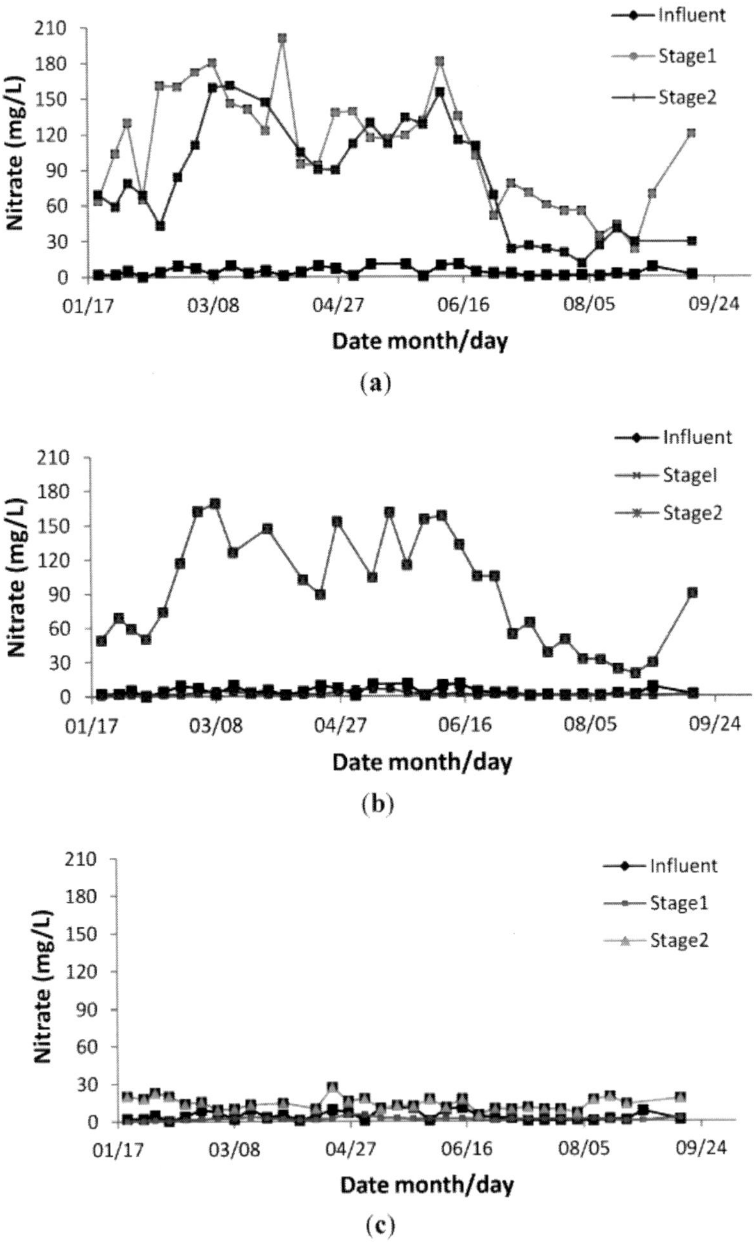

FIGURE 2: Nitrate concentration in the three HEWTSs along the monitoring period. (a) HF-SP; (b) HF-VF; (c) VF-HF.

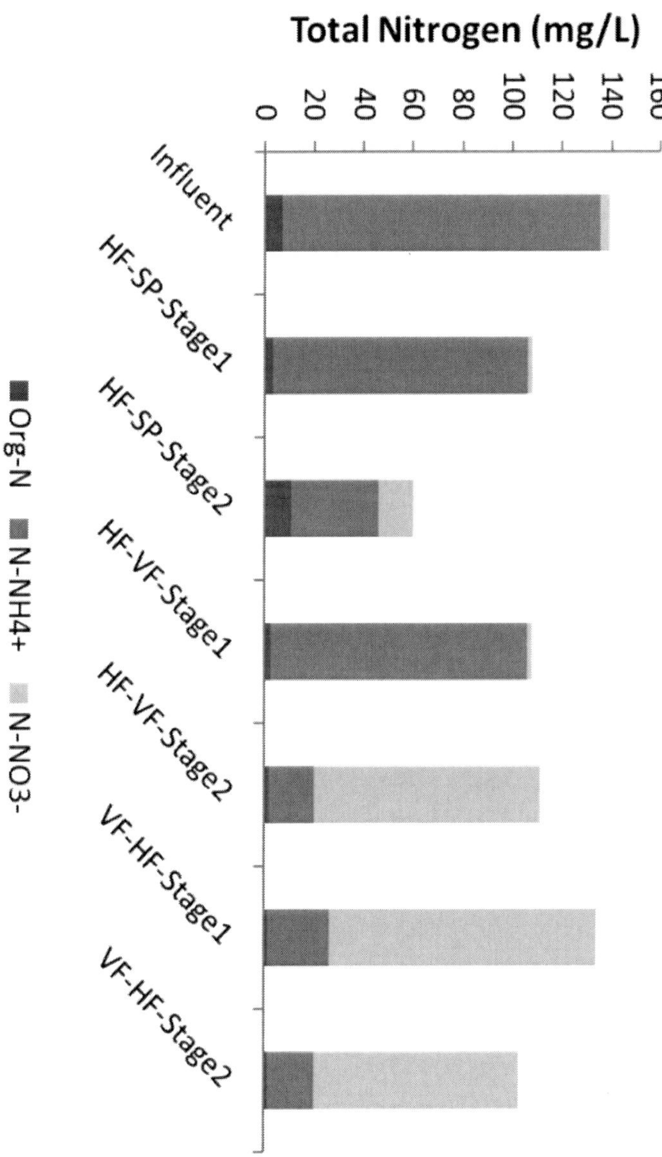

FIGURE 3: Average TN concentrations in the influent and effluents of the HEWTSs.

1.3.5 TOTAL P

It is well known that TP removal in constructed wetlands can be complicated due to the coupling of both organic and inorganic removal mechanisms [27] available for P in wetland systems as well as regeneration of P from organic matter back into the water column [28,29]. As a result many single treatment systems and hybrid systems have reported low TP removal rates [30,31]; in a VF-HF system with gravel, a removal of only 21% was achieved [11]. In our case, the decrease in TP concentration at the effluent of the three HEWTSs was not significant ($p > 0.05$). The same finding was obtained between the two stages in each system ($p > 0.05$). In previous studies, with the same substrate, ~45% and ~50% of phosphorus removal was achieved in the horizontal flow CWs and vertical flow CWs, respectively [12]. The lack of TP removal in the horizontal flow (HF) system in this work was likely due to the fact that six months after beginning the experiment (two months after beginning the monitoring period), the ornamental emergent plants (*Z. aethiopica*) began to desiccate in the horizontal flow CWs in all the three systems. As the plants senesced and decayed, they are essentially releasing P to the systems. In the vertical flow components, there was non-uniform distribution of the influent as there was no distribution system that evenly distributed the wastewater over the surface of the bed, but instead the wastewater was fed over the plant. This uneven distribution pattern may have contributed to the low P removal. In vertical flow CWs, adsorption and precipitation of phosphorus is effective in systems where wastewater continuously comes into contact with the filtration substrate [32]. While the substrate provides sites for P removal by sorption initially, the sorption sites can become saturated, over time [33]. It is at that point that the expansion of the root systems from the developing plants (plant uptake) would replace sorption to the substrate as the major P removal mechanism over time. However, if the goal of the water reuse is for agricultural purposes, the low removal rate of P in this case is preferable, much as it was for N, as the nutrients will be available for whatever crop receives the reclaimed water [11].

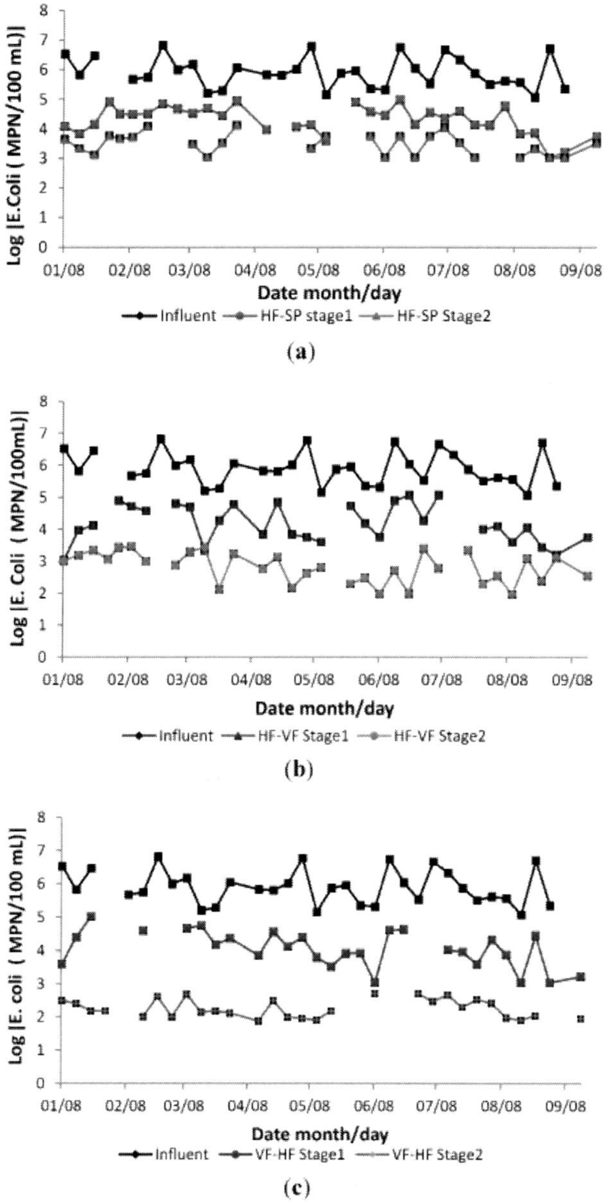

FIGURE 4: *Escherichia coli* concentration in the three EWTSs along the monitoring period. (a) HF-SP; (b) HF-VF; (c) VF-HF.

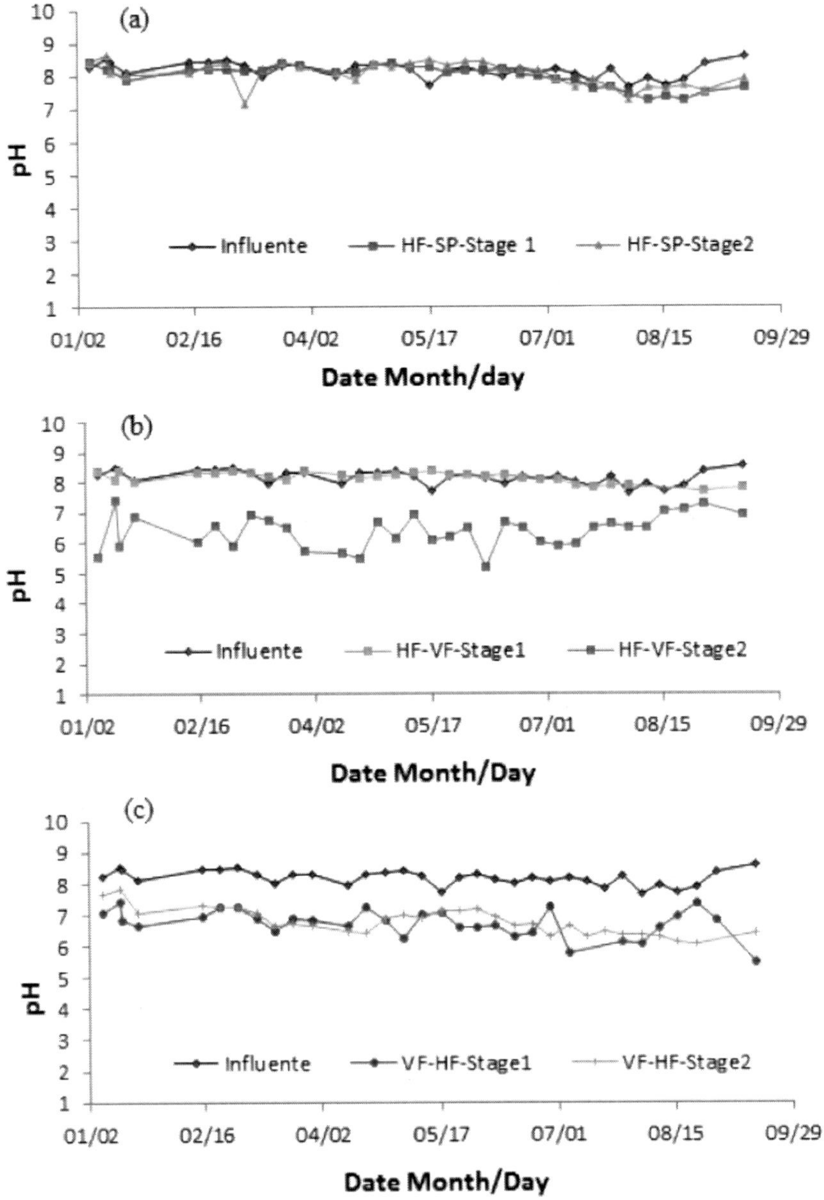

FIGURE 5: PH in the three EWTSs along the monitoring period. (a) HF-SP; (b) HF-VF; (c) VF-HF.

1.3.6 DISINFECTION PERFORMANCE

Total coliform removal was significantly lower in the HF-SP system in comparison to the other two systems which contained VF components ($p < 0.05$) and whose removals were similar. An increase in TCol was observed in the stabilization ponds (Table 2) perhaps related to feces of birds frequently observed in the ponds. In addition, it is well known that TCol can reproduce in surface water when stimulated by high nutrient availability [34].

TABLE 2: Performance summary for the three HEWTSs with respect to indicator organisms. Average ± standard error of the mean. Entire system removal percentages are in parentheses with bold letter.

Parameter	Influent	System I: HF-SP		System II: HF-VF		System III: VF-HF	
		1st stage HF CW	2nd stage SP	1st stage HF-CW	2nd stage VF-CW	1st stage VF-CW	2nd stage HF-CW
Tot.Coliform (MPN/100mL)	$2.5 \times 10^6 \pm 9.9 \times 10^5$	$2.0 \times 10^5 \pm 6.3 \times 10^4$	$3.8 \times 10^5 \pm 1.4 \times 10^5$	$2.1 \times 10^5 \pm 6.6 \times 10^4$	$1.6 \times 10^4 \pm 5.7 \times 10^3$	$1.1 \times 10^5 \pm 4.1 \times 10^4$	$7.7 \times 10^4 \pm 2.5 \times 10^4$
Tot.Coliform Removal (%)		92.0	−90.0 (84.8)	91.6	92.38 (99.36)	95.6	30.00 (96.92)
E. coli (MPN/100 mL)	$1.6 \times 10^6 \pm 6.8 \times 10^5$	$3.1 \times 10^4 \pm 9.6 \times 10^3$	4210 ± 1457	$3.1 \times 10^4 \pm 1.3 \times 10^4$	1060 ± 326	$3.8 \times 10^4 \pm 6.2 \times 10^3$	213.1 ± 59.0
E. coli Removal (%)		98.06	86.42 (99.74)	98.06	96.58 (99.93)	97.63	99.44 (99.99)

With regard to *E. coli* reductions, findings were similar to the results of TCol. The HF-VF and VF-HF configured systems were more effective (3.2 log unit reduction and 3.9 log unit reduction, respectively) ($p < 0.05$) than the HF-SP system (2.6 log unit reduction) (Table 2, Figure 4a–c). The predominant aerobic conditions in VF systems represent the main factor responsible for their higher efficiency in indicator organism removal; the aerobic conditions are not optimal for their success and allow higher predator abundance [8,12]. In this way, the HF-VF and VF-HF systems evaluated in this work in subtropical climate were highly efficient and achieved a reduction of *E. coli* that fulfills the levels established in the Mexico's current national guideline for reclaimed water reuse for agricultural irrigation purposes [35]. Similar findings were obtained in VF-HF systems in a study performed in a tropical climate [7] and in a HF-VF system which treated wastewater produced by a hotel under hot Mediterranean climate, where the authors report a 99.93%–99.99% removal of indicator organisms [24]. In addition, these results satisfy the 2006 WHO guidelines that require a 3–4 log unit pathogen reduction by wastewater treatment in order to protect the health of those working in wastewater-irrigated field and those consuming wastewater-irrigated food crops [36].

TABLE 3: Performance summary for the three HEWTSs with respect to organic load and control parameters. Average ± standard error of the mean. Entire system removal percentages are in parentheses with bold letter.

Parameter	Influent	System I: HF-SP		System II: HF-VF		System: VF-HF	
		1st stage HF-CW	2nd stage SP	1st stage HF-CW	2nd stage VF-CW	1st stage VF-CW	2nd stage HF-CW
BOD (mg/L)	140.6 ± 28.2	27.3 ± 5.1	55.5 ± 9.6	24.3 ± 3.8	6.1 ± 0.9	6.7 ± 1.5	4.8 ± 0.72
BOD Removal (%)		80.6	−103.3 (60.5)	82.7	74.9 (95.7)	95.2	* (96.6)
COD (mg/L)	273.5 ± +50.0	96.9 ± 11.8	277.8 ± 43.5	95.4 ± 17.5	56.9 ± 7.4	66.7 ± 8.9	55.8 ± 8.2
COD Removal (%)		64.6	−186.7 (0.0)	65.1	40.4 (79.2)	75.6	* (79.6)
TSS (mg/L)	61.8 ± 11.7	12.3 ± 2.4	138.3 ± 31.0	8.3 ± 2.0	4.6 ± 1.0	10.6 ± 3.6	4.5 ± 1.0
TSS Removal (%)		80.1	−1024.4 (−123.8)	84.9	49.5 (92.4)	82.8	57.4 (92.7)
Conductivity (μS/cm)	1797 ± 359	1774 ± 119	1387 ± 119	1693 ± 375.5	1369 ± 291	1381 ± 285	1457 ± 409
pH	8.2 ± 0.08	8.0 ± 0.08	8.1 ± 0.08	8.2 ± 0.14	6.4 ± 0.13	6.7 ± 0.12	6.8 ± 0.12
DO (mg/L)	1.5 ± 0.5	5.5 ± 0.6	8.8 ± 1.7	4.7 ± 0.7	4.3 ± 0.6	6.9 ± 0.5	5.2 ± 0.5

*Note: * No significant difference with regard to the concentration in the previous stage.*

With respect to the individual components that made up these hybrid in-series treatment systems, the removal rate of TCol in each of the components were similar to those obtained in previous studies with either one-stage horizontal flow CWs and vertical flow CWs treatment systems [12]. Therefore, these results demonstrate that at least two stages of treatment in HEWTs are required to achieve the disinfection level in order to qualify the reuse of reclaimed water in a safe approach for irrigation to agricultural fields.

1.3.7 BOD AND COD

The HF-VF and VF-HF systems were more effective for BOD removal as well as for COD removal with significant greater reductions compared to the HF-SP system ($p < 0.05$) (Table 3). In the latter system, both BOD and COD increased in the stabilization ponds after a decrease in the first-stage due to presence of algae, which essentially converted sewage-BOD to algal-BOD [37]. In the VF-HF system, BOD was almost completely reduced (>95%) in the VF first stage, so that no significant reduction was observed in the second stage ($p > 0.05$); analogous results were reported in VF-HF systems where the authors found a removal of 89% and 90% of BOD by using gravel and lapilli as substrate, respectively, and no additional BOD removal in the HF stage [11]. A similar pattern was observed with respect to the COD (Table 3).

1.3.8 TOTAL SUSPENDED SOLIDS

The concentration of TSS in the influent was low due to the partial sedimentation process which took place in the holding tank. The TSS increased for the HF-SP system due to the presence of algae in the stabilization pond while TSS was significantly reduced in the other two systems ($p < 0.05$), averaging 92% for both the HF-VF and VF-HF systems.

1.3.9 PH

The pH is an important characteristic of water when considering reuse of reclaimed water for irrigation [38]. It is well documented that irrigation water should have a pH ranging between 5 and 6.5 in order to maximize nutrient uptake by plants [39]. The values of pH averaged above 8 for the HF-SP system likely driven by photosynthetic activity in the stabilization pond (Figure 5a). In contrast, the HF-VF and VF-HF systems effectively modified the pH to an average value of 6.6 (Table 3, Figure 5b,c) due mainly to the nitrification process in the vertical flow CWs.

1.3.10 ELECTRICAL CONDUCTIVITY (EC)

Although an increase in EC might be expected in the effluent of CWs, some studies have found that this does not frequently happen with a HRT < 3 days but for a HRT > 10 days [11,40]. In this study, EC was reduced significantly in the three systems ($p < 0.05$) reaching similar values. According to some authors [41], a decrease in EC despite significant water losses is explained by uptake of micro and macroelements and ions by plants and their removal through adsorption to plant roots, litter and settled suspended particles. Moreover, the ground tezontle rock used as the media in the CWs has demonstrated a high capacity for total dissolved solid removal in previous studies [42]. The final values in the three effluents averaged less than 1.5 dS/m, which is considered as the threshold value from which a reduction in crop yield potential due to salinity, can be expected in salt-sensitive species [43,44].

1.4 CONCLUSIONS

Two-stage hybrid ecological wastewater treatment systems comprised of constructed wetlands and stabilization ponds can be configured and combined in order to optimize the inorganic nutrient retention and the removal of pathogens and other detrimental water quality parameters. In this study, the most effective systems capable of providing high quality, nutrient-rich water for agriculture, for the water quality parameters we evaluated, were those systems containing a vertical flow CW component, either HF-VF or VF-HF. A high-nitrified effluent, desirable in reclaimed water reuse for crop irrigation, was produced in the two systems due to efficient nitrification in the vertical flow components. There was no significant difference amongst the three systems with regard to TP removal. Regarding the indicator organisms, the two most effective systems achieved a >3 log unit *E. coli* reduction, fulfilling the WHO guidelines for wastewater treatment systems and complying the <1000 MPN/100 mL Mexican standard for treated wastewater reuse in agriculture. In this way, we have demonstrated that it is possible to remove harmful pathogenic organisms by using at least two-stage systems, essentially disinfecting the reclaimed water using natural processes and thereby negating the need to use expensive conventional disinfectants. The production and reuse of high-quality, reclaimed water is a recommended practice in order to protect human health, maximize recycling of nutrients and reduce demand on high quality drinking water sources that are used for irrigation in many parts of the world, conserving this vital resource for human consumption. Moreover, in the developing world, where fertilizers costs are relatively high, the coupling of low-cost wastewater treatment while maximizing reuse of nutrients makes both economic sense and improves long-term sustainability of the coupled human-agricultural system.

REFERENCES

1. Jiménez, B.; Drechsel, P.; Koné, D.; Bahri, A.; Raschid-Sally, L.; Qadir, M. Wastewater, sludge and excreta use in developing countries: An overview. In Wastewater Irrigation and Health. Assessing and Mitigating Risk in Low-Income Countries; Drechsel, P., Scott, C.A., Raschid-Sally, L, Redwood, M., Bahri, A., Eds.; International Water Management Institute and International Development Research Centre (IDRC): London, UK, 2010; pp. 3–27.
2. Jiménez, B. Irrigation in developing countries using wastewater. Int. Rev. Environ. Strateg. 2006, 2, 229–250.
3. Bos, R.; Carr, R.; Keraita, B. Assessing and mitigating wastewater-related health risks in low-income countries: An introduction. In Wastewater Irrigation and Health. Assessing and Mitigating Risk in Low-Income Countries; Drechsel, P., Scott, C.A., Raschid-Sally, L, Redwood, M., Bahri, A., Eds.; International Water Management Institute and International Development Research Centre (IDRC): London, UK, 2010; pp. 29–47.
4. Jiménez, B.; Mara, D.; Carr, R.; Brissaud, F. Wastewater treatment for pathogen removal and nutrient conservation: Suitable systems for use in developing countries. In Wastewater Irrigation and Health. Assessing and Mitigating Risk in Low-Income Countries; Drechsel, P., Scott, C.A., Raschid-Sally, L., Redwood, M., Bahri,

A., Eds.; International Water Management Institute and International Development Research Centre (IDRC): London, UK, 2010; pp. 149–169.

5. Zurita, F.; Roy, E.D.; White, J.R. Municipal wastewater treatment in Mexico: Current status and opportunities for employing ecological treatment systems. Environ. Technol. 2012, 33, 1151–1158.
6. Sharafi, K.; Fazlzadehdavil, M.; Pirsaheb, M.; Derayat, J.; Hazrati, S. The comparison of parasite eggs and protozoan cysts of urban raw wastewater and efficiency of various wastewater treatment systems to remove them. Ecol. Eng. 2012, 44, 244–248.
7. García, J.A.; Paredes, D.; Cubillos, J.A. Effect of plants and the combination of wetland treatment type systems on pathogen removal in tropical climate conditions. Ecol. Eng. 2013, 58, 57–62.
8. Saeed, T.; Sun, G. Enhanced denitrification and organics removal in hybrid wetland columns: Comparative experiments. Bioresour. Technol. 2011, 102, 967–974.
9. Marecos Do Monte, H.; Albuquerque, A. Analysis of constructed wetland performance for irrigation reuse. Water Sci. Technol. 2010, 61, 1699–1705.
10. Vymazal, J. Horizontal sub-surface flow and hybrid constructed wetlands systems for wastewater treatment. Ecol. Eng. 2005, 25, 478–490.
11. Herrera-Melián, J.A.; Martín-Rodríguez, A.J.; Araña, J.; Gonzalez-Díaz, O.; González-Enríquez, J.J. Hybrid constructed wetlands for wastewater treatment and reuse in the Canary Islands. Ecol. Eng. 2010, 36, 891–899.
12. Zurita, F.; De Anda, J.; Belmont, M.A. Treatment of domestic wastewater and production of commercial flowers in vertical and horizontal subsurface-flow constructed wetlands. Ecol. Eng. 2009, 35, 861–869.
13. American Public Health Association; American Water Works Association; Water Environment Federation. Standard Methods for the Examination of Water and Wastewater; APHA: Washington, DC, USA, 2005.
14. Park, J.B.K.; Craggs, R.J.; Shilton, A.N. Wastewater treatment high rate algal ponds for biofuel production. Bioresour. Technol. 2011, 102, 35–42.
15. Ayas, S.C.; Aktas, Ô.; Findik, N.; Akca, L.; Kinaci, C. Effect of recirculation on nitrogen removal in a hybrid constructed wetland system. Ecol. Eng. 2012, 40, 1–5.
16. White, J.R.; Reddy, K.R. Potential nitrification and denitrification rates in a phosphorus-impacted subtropical peatland. J. Environ. Qual. 2003, 32, 2436–2443.
17. Gikas, G.D.; Tsihrintzis, V.A. A small-size vertical flow constructed wetland for onsite treatment of household wastewater. Ecol. Eng. 2012, 44, 337–343.
18. Ávila, C.; Salas, J.J.; Martín, I.; Aragón, C.; García, J. Integrated treatment of combined sewer wastewater and stormwater in a hybrid constructed wetland system in southern Spain and its further reuse. Ecol. Eng. 2013, 50, 13–20.
19. Gardner, L.M.; White, J.R. Denitrification enzyme activity as a potential spatial indicator of nitrate loading in a Mississippi River diversion wetland soil. Soil Sci. Soc. Am. J. 2010, 74, 1037–1047.
20. VanZomeren, C.; White, J.R.; DeLaune, R.D. Ammonification and denitrification rates in coastal louisiana bayou sediment and marsh soil: Implications for Mississippi River diversion management. Ecol. Eng. 2013, 54, 77–81.
21. White, J.R.; Reddy, K.R. The influence of nitrate and phosphorus loading on denitrifying enzyme activity in Everglades wetland soils. Soil Sci. Soc. Am. J. 1999, 63, 1945–1954.

22. Vymazal, J.; Kropfelová, L. A three-stage experimental constructed wetland for treatment of domestic sewage: First 2 years of operation. Ecol. Eng. 2011, 37, 90–98.
23. Tanner, C.C.; Sukias, J.P.S.; Headley, T.R.; Yates, C.R.; Stott, R. Constructed wetlands and denitrifying bioreactors for on-site and decentralized wastewater treatment: Comparison of five alternative configurations. Ecol. Eng. 2012, 42, 112–123.
24. Masi, F.; Martinuzzi, N. Constructed wetlands for the Mediterranean countries: Hybrid systems for water reuse and sustainable sanitation. Desalination 2007, 215, 44–55.
25. Reddy, K.R.; DeLaune, R.D. Biogeochemistry of Wetlands: Science and Applications; CRC Press, Taylor & Francis Group: Boca Raton, FL, USA, 2008.
26. Lai, P.C.C.; Lam, P.K.S. Major pathways for nitrogen removal in wastewater stabilization ponds. Water Air Soil Poll. 1997, 94, 125–136.
27. Moustafa, M.Z.; White, J.R.; Coghlan, C.C.; Reddy, K.R. Influence of hydropattern and vegetation on P reduction in a constructed wetland under high and low mass loading rates. Ecol. Eng. 2012, 42, 134–145.
28. Bostic, E.M.; White, J.R.; Reddy, K.R.; Corstanje, R. Evidence of phosphorus distribution in wetland soil after the termination of nutrient loading. Soil Sci. Soc. Am. J. 2010, 74, 1808–1815.
29. Zhang, W.; White, J.R.; De Laune, R.D. Diverted Mississippi River sediment as a potential phosphorus source affecting louisiana water quality. J. Freshwater Ecol. 2012, 27, 575–586.
30. Vohla, C.; Kõiva, M.; Bavorb, H.J.; Chazarencc, F.; Mandera, U. Filter materials for phosphorus removal from wastewater in treatment wetlands—A review. Ecol. Eng. 2011, 37, 70–89.
31. Wu, H.; Zhang, J.; Li, P.; Zhang, J.; Xie, H.; Zhang, B. Nutrient removal in constructed microcosm wetlands for treating polluted river water in northern China. Ecol. Eng. 2011, 37, 560–568.
32. Zhao, Y.J.; Hui, Z.; Chao, X.; Nie, E.; Li, H.J.; He, J.; Zheng, Z. Efficiency of two-stage combinations of subsurface vertical down-flow and up-flow constructed wetland systems for treating variation in influent C/N ratios of domestic wastewater. Ecol. Eng. 2011, 37, 1546–1554.
33. Belmont, M.A.; White, J.R.; Reddy, K.R. Phosphorus sorption characteristics of sediments in Lake Istokpoga and the upper chair of lakes. J. Environ. Quality 2009, 38, 987–996.
34. Tyagi, V.K.; Chopra, A.K.; Kazmi, A.A.; Kumar, A. Alternative microbial indicators of faecal pollution: Current perspective. Iran J. Environ. Health. Sci. Eng. 2006, 3, 205–216.
35. Secretaría de Medio Ambiente y Recursos Naturales (SEMARNAT). Que Establece los Límites Máximos Permisibles de Contaminantes en las Descargas de Aguas Residuales en Aguas y Bienes Nacionales; Norma Oficial Mexicana NOM-001-SEMARNAT-1996; OfficialGazette of the Federation, Secretary of Government: Mexico City, Mexico, 1996.
36. Mara, D.; Bos, R. Risk analysis and epidemiology: The 2006 WHO guidelines for the safe use of wastewater in agriculture. In Wastewater Irrigation and Health. Assessing and Mitigating Risk in Low-Income Countries; Drechsel, P., Scott, C.A., Raschid-Sally, L, Redwood, M., Bahri, A., Eds.; International Water Management

Institute and International Development Research Centre (IDRC): London, UK, 2010; pp. 51–62.
37. Shilton, N.; Walmsley, A. Solids and Organics in Pond Treatment Technology; Shilton, A., Ed.; IWA Publishing: London, UK, 2005.
38. Carr, G.; Potter, R.B.; Nortcliff, S. Water reuse for irrigation in Jordan: Perceptions of water quality among farmers. Agr. Water Manag. 2011, 98, 847–854.
39. Ghehsareh, A.M.; Samadi, N. Effect of soil acidification on growth indices and microelements uptake by greenhouse cucumber. Afr. J. Agr. Res. 2012, 7, 1659–1665.
40. Díaz, F.J.; O'Geen, A.T.; Dahlgren, R.A. Agricultural pollutant removal by constructed wetlands: Implications for water management and design. Agr. Water Manag. 2012, 104, 171–183.
41. Kyambadde, J.; Kansiime, F.; Dalhammar, G. Nitrogen and phosphorus removal in substrate-free pilot constructed wetlands with horizontal surface flow in Uganda. Water Air Soil Poll. 2005, 165, 37–59.
42. Zurita, F.; Del Toro-Sánchez, C.L.; Gutierrez-Lomelí, M.; Rodríguez-Sahagún, A.; Castellanos-Hernández, O.A; Ramírez-Martínez, G.; White, J.R. Preliminary study on the potential of arsenic removal by subsurface flow constructed mesocosms. Ecol. Eng. 2012, 47, 101–104.
43. Pedrero, F.; Alarcón, J.J. Effects of treated wastewater irrigation on lemon. Desalination 2009, 246, 631–639.
44. De Miguel, A.; Martínez-Hernández, V.; Leal, M.; González-Naranjo, V.; de Bustamante, I.; Lillo, J.; Salas, J.J.; Palacios-Díaz, M.P. Short-term effects of reclaimed water irrigation: Jatropha curcas L. cultivation. Ecol. Eng. 2013, 50, 44–51.

CHAPTER 2

Sustainability of Wastewater Treatment and Excess Sludge Handling Practices in the Federated States of Micronesia

JOSEPH D. ROUSE

2.1 INTRODUCTION

Vast distances separating the dispersed populations of island countries in the Pacific Ocean have traditionally hindered their development due to the high cost for providing services, including the basic infrastructure required to address sanitation issues. The Federated States of Micronesia (FSM) alone consists of 607 islands scattered over more than one million square miles (three million square kilometers) of ocean [1]. In addition to the geographical distances, cultural and linguistic heterogeneities among the island communities of the FSM, while serving as a hallmark of national identity, also potentially add to isolation effects. Located in the Caroline archipelago of the western Pacific (Figure 1), the islands of the FSM are grouped into the four geopolitical states of (from east to west) Kosrae, Pohnpei, Chuuk, and Yap. The nation's capital is located in Palikir on the island of Pohnpei. As of July 2013, the estimated population of the FSM is 106,104, with a reported distribution of Kosraean 6.2%, Pohnpeian 24.2%, Chuukese 48.8%, Yapese 5.2%, Yap outer islanders 4.5%, and others 11.1% [2].

Effective management of wastewater treatment plays a crucial role in protecting the health of people and safeguarding the fragile environments of the Pacific islands [4]. The existing treatment facilities in the FSM, though, are not adequately inventoried and the limited information that is available is largely esoteric in nature, thus making it difficult to evaluate their effectiveness. Accordingly, a knowledge base is needed to allow for comparisons of methods being used at different locations with the goal of building upon success and, where applicable, avoiding the duplication of mistakes in our quest for development of sustainable infrastructures throughout the region.

The objective of this study is to compile an inventory of the wastewater-related infrastructure in the FSM that would be of use in identifying where potential weaknesses exist and prioritizing for future water-sector projects. This objective is met by conducting field surveys to document the existing wastewater treatment systems in the population centers of Kosrae,

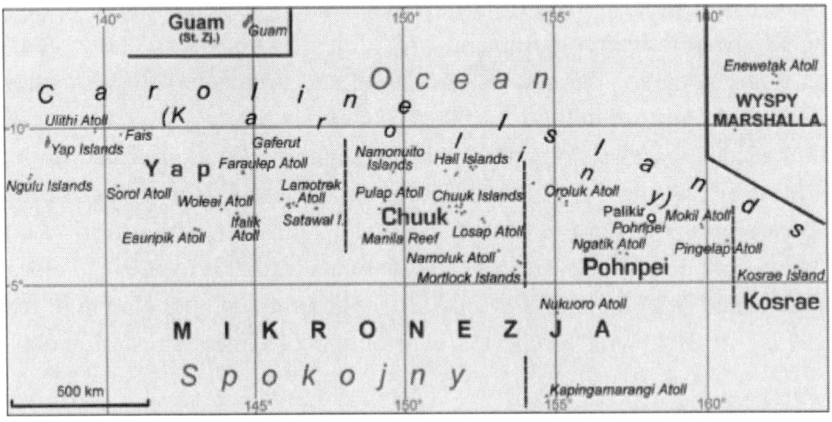

FIGURE 1: Map of the Federated States of Micronesia located in the Caroline Islands of the western Pacific. Image source: Aotearoa at pl.wikipedia [3].

Pohnpei, Chuuk, and Yap. Items to be targeted include the types of unit processes being used, degree of treatment being targeted, design capacities, degrees of treatment being achieved (removal efficiency, regulatory compliance), and locations where treated effluents are being discharged. Of equal concern is the handling of excess sludge generated at wastewater treatment plants, including the methods being used for treatment and disposal of waste sludge, or reuse of biosolids. In addition, characterization of industrial wastewater sources is also a critical issue.

2.2 METHODOLOGY

The methods employed over the course of this project involved civil and environmental engineering fieldwork. Prior to commencing fieldwork, points of contact were established in each state to clarify the purpose of the investigation to local authorities. Preliminary inquiries were then conducted to gather as much relevant information as possible by correspondence prior to commencing with visits.

Site visits started with meeting pertinent local officials on utility boards and other government agencies to review previously received preliminary information and gather additional information concerning locations and specific details of local facilities. Subsequently, visits were conducted to all known treatment facilities to document all pertinent factors and take photographs.

A concerted effort was made to glean a complete overview of the regional wastewater treatment and disposal practices including plans for future work from interviews with local officials and discussions with plant operators. Prior to leaving a particular area, every effort was made to synthesize the findings and see if any outstanding questions remained to be answered.

2.3 HIGHLIGHTS OF FIELDWORK FINDINGS

2.3.1 KOSRAE

In Kosrae State, matters related to water and wastewater are managed under the Department of Transportation and Infrastructure; it is hoped, though, that a Kosrae Utilities Authority (KUA) will be formed in the near future to focus solely on issues related to water and wastewater. There is one centralized wastewater treatment system on the island of Kosrae constructed in 1983 called the Tofol oxidation pond, which is a system of three ponds functioning in series (Figure 2). Each pond has a footprint of approximately 150 by 50 ft (45 by 15 m); and with an assumed depth of 4 ft (1 m), this gives a total volume for the system of 90,000 ft^3 (700,000 gal, or 2000 m^3). Input to the system is from a sewer line servicing the nearby hospital, community college, and high school. The average inflow is estimated to be less than 1000 gal/day (4000 L/day) and the effluent (or treated outflow) discharges to the nearby Tofol River. By these numbers, the hydraulic retention time (HRT) would be approximately 700 days, or nearly 2 years; thus, the system does appear to have a luxuriously long HRT. Accordingly, the hydraulics of the system would be largely controlled by rainfall, evaporation, and infiltration, rather than wastewater flow. Water quality analyses are not being conducted, though the system does appear to be working well with evidence of fish flourishing in all three ponds; however, recently there had been some clogging in the 8-in (diameter) line between the first two ponds resulting in a brief overflow. As a natural safety-factor, the pond system is buffered by a mangrove bordering the site that would offer some assimilative capacity in the event of system failure.

Around 15 years ago, by government directive, 1165 septic tanks were installed on Kosrae, which brought the total coverage for wastewater treatment services in the state to 94% (Table 1). Included in this total is an additional undocumented number of previously installed septic tanks. For a standard single-family unit, the new septic tank consists of a large first holding tank of 600 gal (2200 L) connected to a smaller second tank of 300 gal (1100 L), from which the effluent overflows into a leaching well of 400 gal (1500 L) containing a gravel bed (for a double-family unit, the volumes are scaled up by a factor of approximately 1.5). It is said that in some cases leaching fields were installed in place of the standard leaching well, but this could not be confirmed. Furthermore, in the densely populated Lelu Village—a residential bed-town—there is insufficient space for use of either leaching fields or wells, thus the basic design for septic tanks consists of only the first two compartments (i.e., no leaching well), from which the effluent discharges to a sewer collection network servicing the whole island village. This entire effluent flow is then directed to a concrete collection pit with an estimated capacity of 1500 gal (5700 L), which discharges by gravity flow through a half-mile (ca. one kilometer) long outfall line into the ocean.

Installation of septic tanks has not been governed by any specific design standards for setbacks or density (i.e., number of tanks per unit area). Typically, they have been constructed on a per-family basis and included an adjoining "out-house" consisting of a partially enclosed concrete structure with shower, sink, and toilet (Figure 3). About twice a month a pump truck draws excess sludge from one septic tank and transports it to the oxidation ponds, thus the vast majority have not been serviced during the 15 or more years since their installation. Nonetheless, the septic-tank culture does appear to be functioning well. However, septic tanks only provide a primary level of treatment, i.e., effluent is essentially raw sewage minus the solid material that has settled out in the tank, thus the impact of discharges on receiving water bodies (both the ocean and groundwater) would be a point of concern.

FIGURE 2: Wastewater treatment system in Kosrae consisting of three oxidation ponds (only one pond shown). Image source: author.

FIGURE 3: Typical outhouse with adjoining septic tank in Lelu Village, Kosrae. Image source: author.

TABLE 1: Numbers of septic tanks installed on Kosrae by government directive. Source: Department of Transportation and Infrastructure [5].

Village	Septic tanks installed	Approximate date of installation
Lelu	350	1994
Tafunsak	340	1997
Utwa	200	1998
Malem	275	1999
Total	1,165	— -

A solid-waste landfill (Fukuoka type—semi-areobic landfill technology) was installed on Kosrae in 2010, which includes a leachate collection system. While the landfill, per se, would not be a part of this study, the leachate generated at the landfill would be of concern as an industrial wastewater. The leachate flow is mostly a function of the amount of rainfall over the exposed landfill and is collected in a catch pond with an impervious lining (see Figure 4). Out-flow from the pond passes through a gravel-sand bed and is discharged into the coastal mangrove. The leachate is tested for chemical oxidation demand (COD—an aggregate indicator of organic compounds) and pH on a monthly basis with each testing event consisting of multiple measurements taken from the leachate and nearby environmental samples. Test results have consistently been below the imposed regulatory limits of 100 mg COD/L and pH of 10, which is attributable to diligent segregation of waste materials prior to deposition. In the event of a violation of effluent quality requirements, the public is to be alerted to avoid the discharge area. If discharge quality does become an issue in the future, some form of treatment process would have to be considered.

FIGURE 4: Lined leachate catch pond at the "Fukuoka-type" landfill on Kosrae. Image source: author.

FIGURE 5: Contact-stabilization type activated-sludge wastewater treatment plant on Pohnpei. Image source: author.

FIGURE 6: Inoperative industrial wastewater treatment plant of tuna-packaging plant on Pohnpei. Image source: author.

2.3.2 POHNPEI

In Pohnpei State, matters related to water and wastewater are managed by the Pohnpei Utilities Corporation (PUC). In Kolonia Village, the main business center on the island of Pohnpei, there is a municipal wastewater treatment plant consisting of a contact-stabilization type activated-sludge process, which is apparently about 60 years old (Figure 5). The plant was designed for 700 house-hold connections, but is thought to be serving 2 to 3 times that number now. However, data analyses for wastewater have not been conducted for over a decade, thus the actual flow rate and treatment efficiency are not known. While the existing plant does appear to be functioning properly by visual inspection, the activated sludge in the aeration basin appears to be much too thin, which could be due to the hydraulic overload on the system in its current state. A new plant, though, of the same type with approximately the same capacity is currently under construction, which will run in parallel with the old plant, and should effectively double the current capacity. The outfall line discharges effluent into the adjoining crowded bay approximately 1000 ft (300 m) from the shore; however, there is a plan in effect to install a new outfall in the near future that would be about twice as long. From the site of the existing plant, though, construction of an outfall capable of discharging beyond the mouth of the bay into the open sea would not be practicable.

Influent to the wastewater treatment plant does not include industrial wastewater, though it does receive inputs from restaurants and kitchens and thus would include some fats, oils, and grease (FOG). However, there is a tuna-packaging plant in the industrial zone near the airport with its own wastewater treatment plant that is currently inoperative (Figure 6). Ownership of that treatment plant is said to have been transferred to PUC about a decade ago, though the tuna-packaging plant is still in business and, apparently, discharging untreated industrial wastewater directly to the bay. Though not actively investigated in this study, it is known that there are a lot of piggeries on Pohnpei, from which runoff of swine waste into the bay (or elsewhere) is considered to be a significant problem. While there are said to be a couple of digesters for larger operations, the vast majority of this waste discharge appears to be going untreated.

Additionally, there is a solid-waste dump near the airport, concerning which there has been some discussion about the need to construct a proper landfill with leachate collection and the ability to generate energy from waste. Furthermore, the existing municipal wastewater treatment plant, discussed above, yields 500 to 600 gal (ca. 2000 L) of excess sludge per month, which is applied to a drying bed near the airport. After drying, the sludge is supposed to be transferred to the dump, though apparently it is always taken directly from the bed by local residents and used for agricultural purposes.

Septic tanks on Pohnpei are typically constructed with an adjoining seepage well rather than a leaching field and it is said that drawing excess sludge from the tanks is almost non-existent—only one tank is serviced every 2 or 3 years—even though PUC does have a pump truck for that purpose. The capital area in Pilikir Village is served by a very large septic tank of approximately 120,000 gal (450 m^3) with a footprint of approximately 20 by 40 ft (6 by 12 m) and a leaching field said to extend 100 yards (90 m). This facility appears to be intended for use by the government complex, though a lift-station has been installed to tie in nearby residences.

2.3.3 CHUUK

In Chuuk State, matters related to water and wastewater are managed by the Chuuk Public Utilities Corporation (CPUC). On the island of Weno (traditionally named Moen, the main population center located in the Chuuk Islands cluster, see Figure 1) there is a contact-stabilization type activated-sludge wastewater treatment plant located near the airport. The plant appears to be of the same construction as the one on Pohnpei and is about 40 years old (commissioned in 1973); however, it has only been operated intermittently over the years and is currently shut down (Figure 7). Untreated sewage thus bypasses the plant and is being discharged in the Chuuk lagoon, a world-famous diving site. The amount being discharged, though, may not be very great because much of the sewer collection network has been out of operation due to grinder-pump stations and lift stations being out of service, and currently under repair. CPUC is actively making progress towards getting the network back on line in preparation for the anticipated refurbishing of the old treatment plant. The Weno sewer collection system is the most extensive network in the FSM, covering the full extents of the northern and western shores of Weno Island.

There are, apparently, no industrial wastewater sources on Weno that would contribute to the soon-to-be-refurbished treatment plant other than FOG from restaurants and possibly discharges from laundromats and automotive shops. However, there are various farming activities, including household pigpens, which could constitute sources for agricultural wastewater that would require some form of treatment if environmental discharges become an issue.

For solid waste disposal, there is only a waste dump on Weno, without leachate collection. Additionally, CPUC is called to draw excess sludge from a septic tank only about once every three months; though they are currently making an effort to teach proper operation and maintenance of septic tanks and grease traps to local residents and businesses.

FIGURE 7: Inoperative wastewater treatment plant on Weno Island in Chuuk state. Image source: author.

FIGURE 8: Imhoff tank wastewater treatment plant in Colonia Village, Yap, with a map of the sewer network painted on wall. Image source: author.

2.3.4 YAP

In Yap State, matters related to water and wastewater are managed by the Yap State Public Service Corporation (YSPSC). In Colonia, the main business center on the island of Yap, the municipal wastewater treatment plant consists of an Imhoff tank system, with dual units operating in parallel, which is about 40 years old (commissioned around 1974, see Figure 8). While Imhoff tanks are relatively easy to operate and have very low energy requirements due to the absence of aeration and internal recycle, they provide only a primary level of treatment consisting of a limited removal of suspended solids. Data analyses of influent and effluent are not being performed; thus, treatment efficiency is not known. Furthermore, the amount of intermittent flow entering the plant on a daily basis is not known, making it difficult to estimate loading rates. In addition, the outfall line from the plant is known to be broken and is discharging the partially treated effluent into a shallow part of the bay about 500 ft (150 m) from shore.

There are no known industrial wastewater sources on Yap Island now; however, there are some activities at the petroleum refinery in Colonia that could be of concern. In addition, there are some relatively small fish-cleaning operations and possibly FOG discharges from older establishments (including restaurants) that are not equipped with grease traps, all of which could be contributing to the treatment demand of the Imhoff tank system. Furthermore, various farming activities, including household pigpens, could constitute sources for agricultural industrial wastewater that would require some form of treatment if environmental discharges become an issue.

For solid waste disposal, there is only a waste dump on Yap, which does not employ leachate collection; though a new landfill is under construction (at the time of this report), which will include a leachate collection system as employed on Kosrae. The landfill leachate will constitute an industrial wastewater, which will require monitoring and would require treatment if discharge standards are not met.

At the municipal wastewater treatment plant (the Imhoff tank system, discussed above) an unknown amount of excess sludge is periodically drawn from the tanks and applied to a drying bed. After drying, the sludge is supposed to be transferred to the dump, though apparently it is always taken by local residents and used for agricultural purposes.

Yap State has a relatively new pump truck that is used to draw excess sludge from a septic tank about once per month, the contents of which are delivered to the treatment plant (via the sewer). YSPSC and regulatory officials are actively working on evaluating and upgrading various issues including efficiency of wastewater treatment and installation and governance of septic tanks.

2.4 DISCUSSION AND CONCLUSIONS

The distribution of wastewater treatment services across the four states of the FSM is summarized in Table 2. As shown, Kosrae has nearly complete coverage due to the extensive initiative to install septic tanks a few years ago. This apparent advantage over the other states, though, is influenced by the population and business activities in Kosrae being confined to only one island and thus not having inhabitants on outer islands, which would be largely accounted for under the "None" category. However, considering that septic tanks provide only a primary level of treatment, further treatment of the collected effluent from tanks in Lelu Village prior to discharge in the bay would be worth pursuing. In addition, investigating the means of extending the sewer collection line in Tofol Village to take advantage of the greatly under-loaded oxidation pond system would be a valuable avenue of inquiry.

TABLE 2: Wastewater treatment services in the FSM (2010). Source: Office of SBOC [6].

State:	KOSRAE	POHNPEI	CHUUK	YAP
Households *	1,143	6,289	7,024	2,311
Sewer connections	4%	17%	9%	12%
Septic tanks	94%	35%	31%	31%
Other	<1%	18%	12%	8%
None	1.4%	30%	48%	49%

* *One household represents approximately 6 persons.*

In Chuuk State, the relatively low coverage for sewer connections (9%) may increase soon due to ongoing repairs of the sewer network and the pending restoration of the central treatment plant on Weno Island. However, given the history of difficulties in maintaining operation of the activated-sludge based wastewater treatment plant and the numerous grinder and lift stations in the collection network, installation of a few strategically located pre-fabricated package treatment plants might prove to be a more effective pursuit. Relying mostly on attached-growth, or biofilm technology, package plants offer a secondary (or greater) level of treatment in compliance with most regulatory standards with relatively low O&M requirements. Furthermore, with package plants, flexibility in scale would be possible by adding on capacity with modular units on an as-needed basis. This approach could be applicable on other islands too, though feasibilities studies would be needed on a case-by-case basis.

On the island of Yap, it would be beneficial to find a cost-effective means of upgrading the well-functioning Imhoff tank system so as to provide a more advanced level of treatment. One positive feature of the existing system, though, is that no energy input is required to operate the plant (following the last lift-station in the collection network). However, to achieve a secondary level of treatment, aeration of the tank would be necessary to enhance biological activity, which would constitute significant energy consumption and operational cost. Furthermore, for traditional biological treatment, pumping is required to retain and circulate the biomass (activated sludge) in the system, which would be yet an additional energy requirement. However, another method to harness biological treatment power that requires less energy input would be to use an attached-growth process, which eliminates the internal pumping requirement to retain effective biomass [7,8]. As a relatively simple retrofit, a frame with the attachment medium, or fabric, could be fitted to the existing tank. As a point of inquiry, though, some testing would be necessary to see what level of aeration (if any) would be required to achieve adequate results. Fortuitously, the existing Imhoff tank system consists of two parallel units, which would allow for retrofitting only one unit and making a comparison of results. However, as an unavoidable tradeoff for improved effluent quality, some increase in waste sludge production would occur; there are, though, some beneficial uses of waste sludge worth considering (see below).

A repeatedly encountered question during discussions with local regulators and operators addressed a need to define the required treatment levels for sewage sludge so that it can be considered safe for use as a biosolids product. Currently, at locations where settled sludge is collected and put on a drying bed, it is quickly whisked away and used for agricultural purposes despite the lack of adequate treatment, thus stressing the need for regulatory supervision. Though not binding in the FSM, the US EPA offers widely accepted definitions for different classes of biosolids that could serve as a guideline [9]. Exceptional quality (i.e., Class A) biosolids, which have no crop-harvesting restrictions, consist of treated residuals that contain no detectable levels of pathogens and low levels of heavy metals. Technologies that can meet Class A standards must process the biosolids for a sufficient length of time at a temperature high enough to yield a product in keeping with the required pathogen cut. Composting is one such method, which can offer an environmentally friendly method to recycle the nutrients and organic matter found in municipal wastewater solids [10]. A recent pilot project on the island of Guam demonstrated the feasibility of composting manure mixed with other organic materials, which when used as a soil amendment resulted in enhanced agricultural productivity as compared to use of synthetic fertilizers; furthermore, use of the compost helped correct various unfavorable soil properties commonly found on tropical Pacific islands [11].

In addition, the technologies involved in deriving energy from waste sludge (and solid waste in general) are becoming more and more practicable for small scale applications. For remotely located island communities, in particular, where petroleum-based fuels are so costly, the impetus clearly exists for making progress toward energy independence, in addition to protecting the environment. While this approach might not be fully feasible at this time, before long a study on this topic would be in order.

Apart from the inoperative treatment plant at the tuna-packaging plant on Pohnpei, there do not appear to be any pressing issues concerning industrial wastewater treatment. With improvements to local economies, though, the (re)emergence of tuna packing and other industries on the islands may occur, along with the wastewater they yield, which will have to be followed closely. In addition, swine farming operations are known to be a source of agricultural industrial pollution on Pohnpei and, to a lesser degree, on the other islands as well. Apart from a few larger operations, much of the problem consists of runoff from pigpens at the household level, thus collection and transport of the waste to a central treatment facility would not be practicable. In this case, as a low-cost, eco-friendly solution, the use of vetiver grass merits consideration. This robust, yet non-invasive, grass can be planted strategically to intercept waste runoff and serve as a natural treatment system and a protective buffer [12], and has specifically demonstrated effectiveness for treatment of piggery waste in a pond application [13]. Furthermore, vetiver grass has been evaluated for use in watershed management in southern Guam [14].

Wastewater treatment requirements vary for different communities throughout the FSM; however, some underlying similarities among these tropical islands do exist. Accordingly, comparisons of methods being used and the results obtained at different locations will be of value for planning purposes. Such information is needed to assist in decision making for future improvements with a goal of developing sustainable wastewater treatment infrastructures in Micronesia and throughout the communities of the western Pacific.

REFERENCES

1. Countries and Their Cultures. Available online: http://www.everyculture.com/Ma-Ni/Federated-States-of-Micronesia.html (accessed on 16 September 2013).
2. The World Factbook. Available online: https://www.cia.gov/library/publications/the-world-factbook/geos/fm.html (accessed on 16 September 2013).
3. Wikimedia Commons. Available online: http://commons.wikimedia.org/wiki/File:Federated_States_of_Micronesia-map_PL.png (accessed on 16 September 2013).
4. Pacific Islands Applied Geoscience Commission. Pacific Wastewater Policy Statement. Available online: http://pacificwater.org/userfiles/file/water%20publication/WastewaterPolicy.pdf (accessed on 16 September 2013).
5. Department of Transportation and Infrastructure, Tofol, Kosrae State, FSM.
6. Office of Statistics, Budget and Economic Management, Overseas Development Assistance, and Compact Management (SBOC), Palikir, Pohnpei State, FSM.
7. Furukawa, K.; Rouse, J.D.; Yoshida, N.; Hatanaka, H. Mass cultivation of anaerobic ammonium-oxidizing sludge using a novel nonwoven biomass carrier. J. Chem. Eng. Jpn. 2003, 36, 1163–1169.
8. Rouse, J.D.; Yazaki, D.; Cheng, Y.; Koyama, T.; Furukawa, K. Swim-bed technology as an innovative attached-growth process for high-rate wastewater treatment. Jpn. J. Water Treat. Biol. 2004, 40, 115–424.
9. U.S. Environmental Protection Agency. A Plain English Guide to the EPA Part 503 Biosolids Rule. Available online: http://water.epa.gov/scitech/wastetech/biosolids/503pe_index.cfm (accessed on 16 September 2013).
10. Trojak, L. Composting on Maui hits its stride. Available online: http://www.biocycle.net/2008/09/composting-on-maui-hits-its-stride/ (accessed on 16 September 2013).
11. Golabi, M.H.; Marler, T.E.; Smith, E.; Cruz, F.; Lawrence, J.H.; Denney, M.J. Use of Compost as an Alternative to Synthetic Fertilizers for Crop Production and Agricultural Sustainability for the Island of Guam; Food & Fertilizer Technology Center: Taipei, Taiwan, 2003. Available online: http://www.fftc.agnet.org/library.php?func=view&id=20110808092915 (accessed on 16 September 2013).
12. Ash, R.; Truong, P. The Use of Vetiver Grass for Sewerage Treatment. Available online: http://www.vetiver.com/AUS_ekeshire01.pdf (accessed on 16 September 2013).

13. Kong, X.; Weiwen, L.; Biqing, W.; Fuhe, L. Study on Vetiver's Purification from Pig Farm. Proceedings of the Third International Conference on Vetiver, Guangzhou, China, 6–9 October 2003; pp. 181–185. Available online: http://www.vetiver.org/TVN_ICV3_proceedings.htm (accessed on 16 September 2013).
14. Golabi, M.H.; Iyekar, C.; Minton, D.; Raulerson, C.L.; Drake, J.C. Watershed management to meet water quality standards by using the vetiver system in southern Guam. AU J. Tech. 2005, 9, 64–70.

CHAPTER 3

Occurrence and Removal Characteristics of Phthalate Esters from Typical Water Sources in Northeast China

YU LIU, ZHONGLIN CHEN, AND JIMIN SHEN

3.1 INTRODUCTION

Phthalate acid esters, a class of chemical compounds mainly used as plasticizers for polyvinyl chloride (PVC) or to a lesser extent other resins in different industrial activities, are ubiquitous in the environment and have evoked interest in the past decade due to endocrine disrupting effects and their potential impacts on public health [1–3].

Worldwide production of PAEs is approximately 6 million tons per year [4]. As PAEs are not chemically bound to the polymeric matrix in soft plastics, they can enter the environment by losses during manufacturing processes and by leaching or evaporating from final products [5]. Therefore, the occurrence and fate of specific PAEs in natural water environments have been observed, and also there are a lot of considerable controversies with respect to the safety of PAEs in water [5–10].

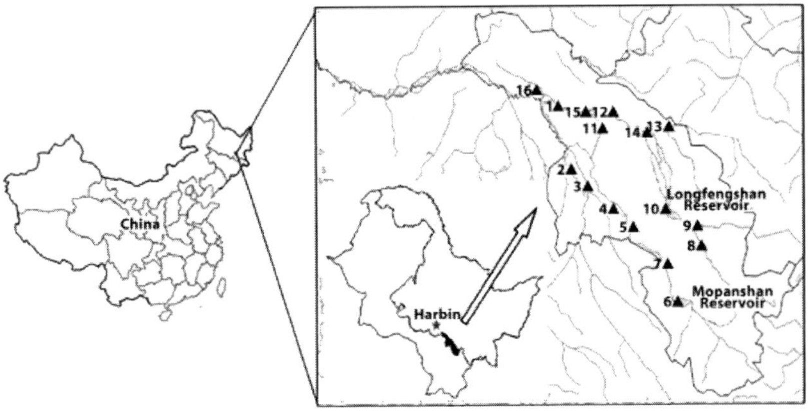

FIGURE 1: Spatial distribution of the 16 sampling sites near the Mopanshan Reservoir in Northeast China.

Six PAE compounds, including dimethyl (DMP), diethyl (DEP), dibutyl (DBP), butylbenzyl (BBP), di(2-ethylhexyl) (DEHP), and di-n-octyl phthalate (DNOP), are classified as priority pollutants by the U.S. Environmental Protection Agency (EPA). Though the toxicity of PAEs to humans has not been well documented, for some years, the Ministry of Environmental Protection in China has regulated phthalates as environmental pollutants. In addition, the standard in China concerning analytical controls on drinking waters does not specifically identify any PAEs as organic pollutant indexes to be determined by the new drinking water standard in 2007 (Standard for drinking water quality; GB5749-2006), which was forced to be monitored for the drinking water supplies in 2012. Consequently, official data about the presence of these pollutants in the aquatic environment of some cities are not available.

Harbin, the capital of Heilongjiang Province, is a typically old industrial base and economically developed city with a population of over 3 million in Northeast China. As the newly enabled water source for Harbin city, Mopanshan waterworks supply the whole city with drinking water

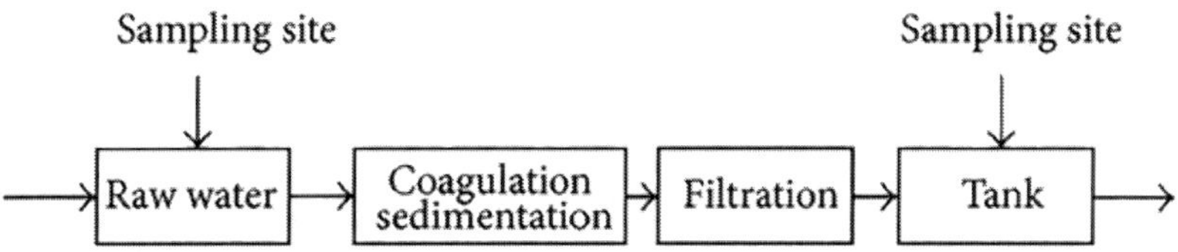

FIGURE 2: Schematic diagram of the waterworks.

through long distance transfer from the Mopanshan Reservoir. To the authors' knowledge, the occurrence and fate of PAEs in the water near this area and its relative waterworks have not previously been examined.

The objectives of this study were (i) to determine the occurrence of PAEs and clarify the fate and distribution of the pollutants in the water source, (ii) to examine the two waterworks, where traditional drinking water treatment step was evaluated on the PAEs removal efficiencies, and (iii) to evaluate the potential for adverse effects of PAEs on human health. Therefore, the investigation of PAEs in the water can provide a valuable record of contamination in Harbin city.

3.2 MATERIALS AND METHODS

3.2.1 CHEMICAL

Fifteen PAEs standard mixture, including dimethyl phthalate, diethyl phthalate, diisobutyl phthalate, di-n-butyl phthalate, di(4-methyl-2-pentyl) phthalate, di(2-ethoxyethyl) phthalate, di-n-amyl phthalate, di-n-hexyl phthalate, butyl benzyl phthalate, di(hexyl-2-ethylhexyl) phthalate, di(2-n-butoxyethyl) phthalate, dicyclohexyl phthalate, di(2-ethylhexyl) phthalate, di-n-nonyl phthalate, di-n-octyl phthalate at 1000 μg/mL each, and surrogate standards, consisting of diisophenyl phthalate, di-n-phenyl phthalate, and di-n-benzyl phthalate, in a mixture solution of 500 μg/mL each, were supplied by AccuStandard Inc. As the internal standard, benzyl benzoate, was also purchased from AccuStandard Inc. All solvents (acetone, hexane, and dichloromethane) used were HPLC-grade and were purchased from J. T. Baker Co. (USA). Anhydrous sodium sulfate (Tianjin Chengguang Chemical Reagent Co., China) was cleaned at 600°C for 6 h and then kept in a desiccator before use.

TABLE 1: Detailed descriptions of the sampling locations.

Number	Sampling site	Name	Latitude	Longitude
1#	Lalin, Jinsan Bridge	Mangniu River	E127°02′53″	N45°5′45″
2#	Xingsheng Town, Gaojiatun	Lalin River	E127°06′11″	N44°51′57″
3#	Dujia Town, Shuguang	Lalin River	E127°10′44″	N44°48′41″
4#	Shanhe Town, Taipingchuan	Lalin River	E127°18′36″	N44°43′50″
5#	Xiaoyang Town, Qichuankou	Lalin River	E127°23′54″	N44°39′46″
6#	Shahezi Town	Mopanshan Reservoir	E127°38′59″	N44°24′17″
7#	Shahezi Town, Gali	Lalin River	E127°35′17″	N44°31′56″
8#	Chonghe Town, Changcuizi	Mangniu River	E127°46′36″	N44°35′46″
9#	Chonghe Town, Xingguo	Mangniu River	E127°35′17″	N44°31′57″
10#	Longfeng Town	Mangniu River	E127°35′19″	N44°43′48″
11#	Changpu Town, Zhonghua	Mangniu River	E127°16′30″	N45°1′42″
12#	Erhe Town, Shuanghe	Mangniu River	E127°18′32″	N45°4′28″
13#	Zhiguang Town, Songjiajie	Mangniu River	E127°34′3″	N45°1′20″
14#	Zhiguang Town, Wuxing	Mangniu River	E127°29′20″	N45°0′43″
15#	Changpu Town, Xingzhuang	Mangniu River	E127°10′42″	N45°4′36″
16#	Yingchang Town, Xingguang	Lalin River	E126°55′8″	N45°9′0′

3.2.2 SAMPLE COLLECTION AND PREPARATION

Harbin, the capital of Heilongjiang province in China, imports nearly all of its drinking water from two sources: the Mopanshan Reservoir (long-distance transport project) and the Songhua River (old water source).

Water samples from the Mopanshan Reservoir and Mopanshan waterwork were collected in 2008. The locations of the sampling sites near the

Mopanshan Reservoir are presented in Figure 1; as a comparison, sampling from another Harbin water supply—Seven waterworks with water source from the Songhua River was performed in 2011. In each waterworks, with considering the hydraulic retention time, three sets of raw water samples (influent) were taken, and then the final finished waters (effluent) after the whole treatment process were carried out for the collections. The flow diagram of the waterworks is shown in Figure 2.

Samples were collected using 2.0 L glassjars from 0.5 m below the water surface. During the whole sampling process, the global position system was used to locate the sampling stations (details in Table 1). All samples were transferred to the laboratory directly after sampling and stored at 4°C prior to extraction within 2 d.

Water samples were filtered under vacuum through glass fiber filters (0.7 μm pore sizes). Prior to extraction, each sample was spiked with surrogate standards. The water samples were extracted based on a classical liquid phase extraction method (USEPA, method 8061) with slight modifications. Briefly, 1 L of water samples was placed in a separating funnel and extracted by means of mechanical shaking with 150 mL dichloromethane, and then, with filtration on sodium sulfate (about 20 g), the organic extracts were concentrated using a rotary evaporator. The exchange of solvent was done by replacing dichloromethane with hexane. Finally, they were reduced to 0.5 mL under gentle nitrogen flow. The internal standard was added to the sample prior to instrumental analysis.

3.2.3 CHEMICAL ANALYSIS

The extracted compounds were determined by gas chromatography coupled to mass spectrometer analysis as described in other publications [11, 12]. Briefly, extracted samples were injected into an Agilent 6890 Series GC equipped with a DB-35MS capillary column (Agilent; 30 m × 0.25 mm i.d.; 0.25 μm film thickness) and an Agilent 5973 MS detector, operating in the selective ion monitoring mode. The column temperature was initially set at 70°C for 1 min, then ramped at 10°C/min to 300°C and held constant for 10 min. The transfer line and the ion source temperature were maintained at 280 and 250°C, respectively. Helium was used as the carrier gas at a flow rate of 1 mL/min. The extracts (2.0 μL) were injected in splitless mode with an inlet temperature of 300°C.

3.2.4 QUALITY ASSURANCE AND QUALITY CONTROL

All glassware was properly cleaned with acetone and dichloromethane before use. Laboratory reagent and instrumental blanks were analyzed with each batch of samples to check for possible contamination and interferences. Only small levels of PAEs were found in procedural blanks in some batches, and the background subtraction was appropriately performed in the quantification of concentration in the water samples. Calibration curves were obtained from at least 3 replicate analyses of each standard solution. The surrogate standards were added to all the samples to monitor matrix effects. Recoveries of PAEs ranged from 62 to 112% in the spiked water and the surrogate recoveries were 75.1 ± 12.7% for diisophenyl phthalate, 72.3 ± 14.3% for di-n-phenyl phthalate, and 102.6 ± 10.4% for di-n-benzyl phthalate in the water samples. The determination limits ranged from 6 to 30 ng/L. A midpoint calibration check standard was injected as a check for instrumental drift in sensitivity after every 10 samples, and a pure solvent (methanol) was injected as a check for carryover of PAEs from sample to sample. All the concentrations were not corrected for the recoveries of the surrogate standards.

3.3 RESULTS AND DISCUSSION

3.3.1 PAES IN WATER SOURCE

The PAEs in the waters from the sampling sites near the Mopanshan Reservoir were investigated and the results are presented in Table 2. The Σ_{15}PAEs concentrations ranged from 355.8 to 9226.5 ng/L, with the geometric mean value of 2943.1 ng/L. Among the 15 PAEs detected in the waters, DIBP, DBP, and DEHP were measured in all the samples, with the average concentrations being 196.6, 801.4, and 1774.1 ng/L, respectively. The three PAE congeners are important and popular addictives in many industrial products including flexible PVC materials and household products, suggesting the main source of PAE contaminants in the water [13]. The correlations of the concentration of DEHP, DBP, and DIBP with the concentrations of total PAEs in the water samples are shown in Figure 3, and a relatively significant correlation between Σ_{15}PAEs and DEHP concentration was found, suggesting the important part played by DEHP in total concentrations of PAEs in the water bodies near the Mopanshan Reservoir. In other words, the contribution of DEHP to total PAEs was higher than that of other PAE congeners in the water samples. In addition, DMP, DEP and DNOP, with the mean value of 14.0, 28.4, and 54.1 ng/L, respectively, only detected at some sampling sites, and have also attracted much attention as the priority pollutants by the China National Environmental Monitoring Center. On the contrary, the concentrations of 6 PAEs (DMEP, BMPP, BBP, DBEP, DCHP, and DNP) were below detection limits in all the water samples near the Mopanshan Reservoir, which is easily explained in terms of much lower quantities of present use in China.

TABLE 2: Concentrations of 15 PAEs in water samples near the Mopanshan Reservoir.

PAEs	Abbreviation	Water (ng/L)		
		Range	Mean	Frequency
Dimethyl phthalate	DMP	nd–42.4	14.0	8/16
Diethyl phthalate	DEP	nd–55.0	28.4	15/16
Diisobutyl phthalate	DIBP	40.0–658.8	196.6	16/16
Dibutyl phthalate	DBP	52.5–4498.2	801.4	16/16
bis(2-methoxyethyl)phthalate	DMEP	nd	nd	0/16
bis(4-methyl-2-pentyl)phthalate	BMPP	nd	nd	0/16
bis(2-ethoxyethyl)phthalate	DEEP	nd–54.6	12.8	5/16
Dipentyl phthalate	DPP	nd–92.5	45.5	10/16
Dihexyl phthalate	DHXP	nd–65.1	16.2	5/16
Benzyl butyl phthalate	BBP	nd	nd	0/16
bis(2-n-butoxyethyl)phthalate	DBEP	nd	nd	0/16
Dicyclohexyl phthalate	DCHP	nd	nd	0/16
bis(2-ethylhexyl)phthalate	DEHP	128.9–6570.9	1774.1	16/16
Di-n-octyl phthalate	DNOP	nd–448.2	54.1	5/16
Dinonyl phthalate	DNP	nd	nd	0/16
PAEs		355.8–9226.5	2943.1	—

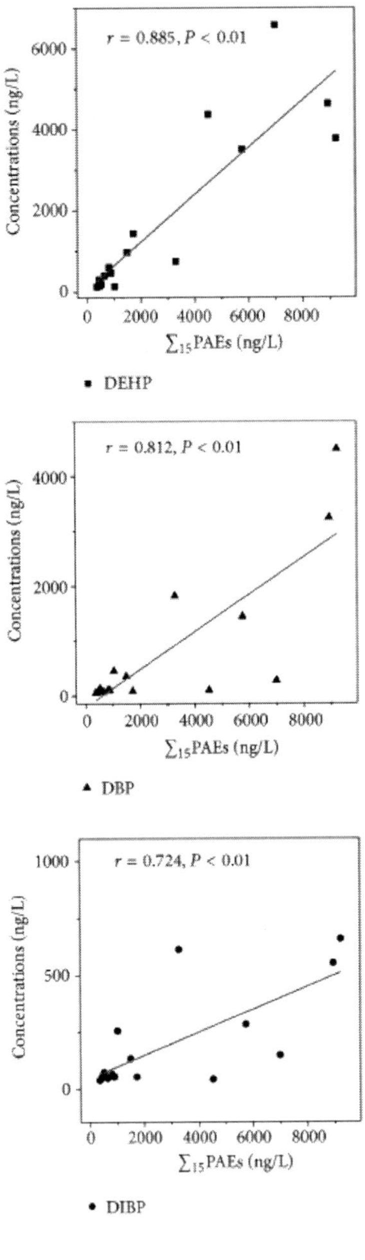

FIGURE 3: Correlations of the concentration of the major PAEs (DEHP, DBP, and DIBP) with the concentrations of total PAEs in the water samples.

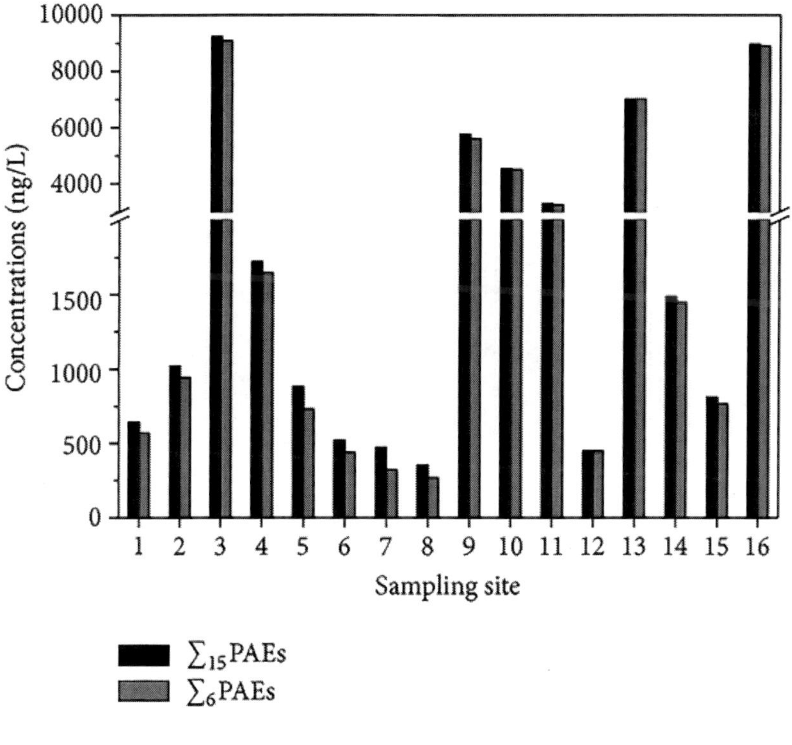

FIGURE 4: Spatial distributions of the 16 sampling sites near the Mopanshan Reservoir.

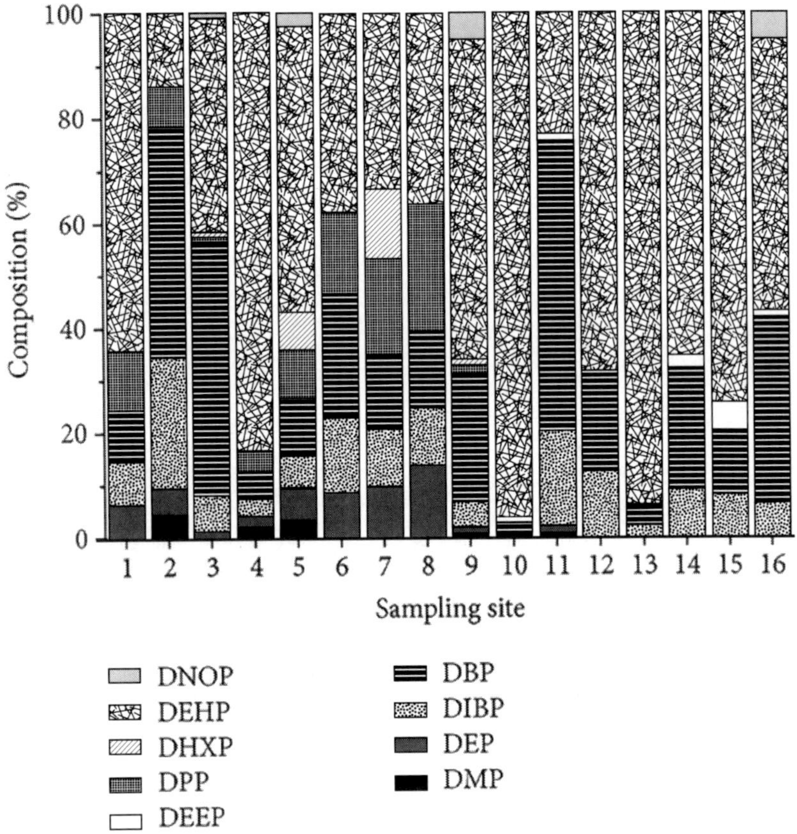

FIGURE 5: PAE composition of the water samples in 16 sampling sites.

The distribution of 15 PAEs (Σ_{15}PAEs) studied and 6 US EPA priority PAEs (Σ_6PAEs; including DMP, DEP, DIBP, DBP, DEHP, and DNOP) in the waters from different sampling sites are shown in Figure 4. There was an obvious variation in the total Σ_{15}PAEs concentrations in water samples near the Mopanshan Reservoir; the concentrations of Σ_6PAEs from the same sampling sites varied from 269.0 to 9086.0 ng/L, with an average of 2868.6 ng/L, the distribution spectra of which observed for all the sampling sites were similar to Σ_{15}PAEs. The highest levels of PAEs contamination were seen on the sampling site 3, followed by some relatively heavily polluted sites (site 16, 13, 9, 10, and 11), indicating that these sites served as important PAE sources, and the spatial distribution of PAEs was site specific. The levels in the waters examined varied over a wide range, especially in the Mangniu River near the Mopanshan Reservoir (in some case over more than one order of magnitude). In general, it should be noted that there might be a relation of PAEs levels with the input of local waste, such as sewage water, food packaging, and scrap material near the sampling point, which were found during the sampling period.

3.3.2 PAE CONGENER PROFILES IN THE WATER

Different PAE patterns may indicate different sources of PAEs. Measurement of the individual PAE composition is helpful to track the contaminant source and demonstrate the transport and fate of these compounds in water [11]. The relative contributions of the 9 detectable PAE congeners Σ_{15}PAEs to the concentrations in the water are presented in Figure 5. It is clear that DEHP was the most abundant in the water samples with the exception of site 2, 3 and 11, contributions ranging from 33.1% to 96.3%, followed by DBP, ranging from 2.1% to 36.3%. The results are consistent with the above data for overall analysis that DBP and DEHP are the dominant components of the PAEs distribution pattern in each sampling site, which reflected the different pattern of plastic contaminant input during the sampling period. Similarly, DIBP and DBP are used in epoxy resins or special adhesive formulations, with the different proportions of these two PAE congeners, which were also the important indicator of the information polluted by PAEs for the sampling locations. Although the limited sample number draws only limited conclusions, there is still reason to note that a leaching from the plastic materials into the runoff water is possible, and that water runoff from the contaminated water is a burden pathway from different sources of PAEs [14].

TABLE 3: Comparison of the concentrations of DEHP and DBP in the water bodies (ng/L).

Location	DEHP			DBP			Reference
	Range	Median	Mean	Range	Median	Mean	
Surface water, Germany	330–97800	2270	—	120–8800	500	—	[14]
Surface water, the Netherlands	nd–5000	320	—	66–3100	250	—	[15]
Seine River estuary, France	160–314	—	—	67–319	—	—	[16]
Tama River, Japan	13–3600	—	—	8–540	—	—	[17]
Velino River, Italy	nd–6400	—	—	nd–44300	—	—	[2]
Surface water, Taiwan	nd–18500	—	9300	1000–13500	—	4900	[18]
Surface water, Jiangsu, China	556–15670.7	—	—	16–5857.5	—	—	[19]
Yangtze River, mainstream, China	3900–54730	—	—	nd–35650	—	—	[20]
Middle and lower Yellow River, China	347–31800	—	—	nd–26000	—	—	[21]
Second Songhua River, China	nd–1752650	370020	—	nd–5616800	—	717240	[22]
Urban lakes, Guangzhou, China	87–630	170	240	940–3600	1990	2030	[11]
Xiangjiang River, China	620–15230	—	—	—	—	—	[23]
This study	128.9–6570.9	671.0	1774.1	52.5–4498.2	110.3	801.4	

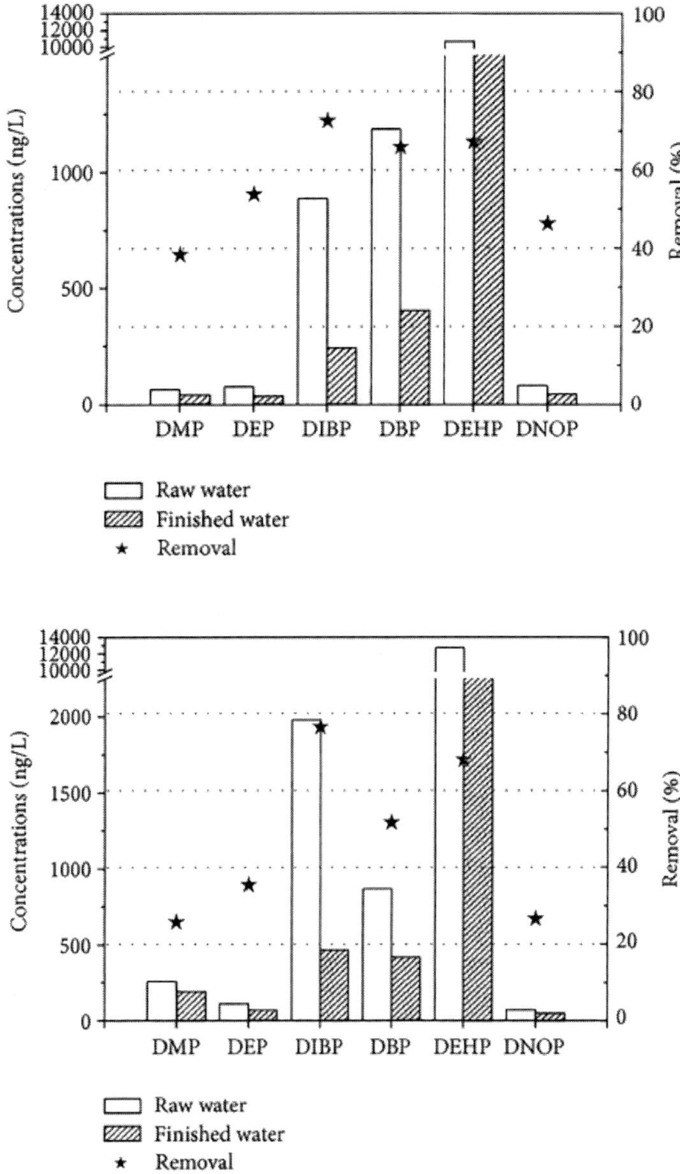

FIGURE 6: PAEs detected in the water samples from the MPSW (a) and SW (b).

3.3.3 COMPARISON WITH OTHER WATER BODIES

Comparison with the total PAEs concentrations is nevertheless very limited due to the different analysis compounds. However, the individual PAE, namely, DBP and DEHP, is by far the most abundant in other researches, and it is possible to make a camparison with our findings. In this case, the results of DBP and DEHP concentrations published in literatures for kinds of water bodies are presented together in Table 3.

The DEHP and DBP concentrations of the present study showed, to some extent, lower concentration levels than those reported in the other water bodies in China. In comparison, the results were comparable or similar to those from the examinations described in other foreign countries. For example, as shown in Table 3, the DEHP concentrations were similar to the surface waters from the Netherlands and Italy described in the literature. Meanwhile, these concentrations in this study were quite lower than the Yangtze River and Second Songhua River in China.

In conclusion, as compared to the results of other studies, the waters near the Mopanshan Reservoir were moderately polluted by PAEs. Therefore, there is a definite need to set up a properly planned and systematic approach to water control near the Mopanshan Reservoir.

3.3.4 PAE LEVELS IN THE WATERWORKS

The Mopanshan Waterworks (MPSW), equipped essentially with the water source of the Mopanshan Reservoir, has been investigated in order to assess the fate of the PAEs during the drinking water treatment process. For comparison, a sampling campaign for the determination of PAEs levels in the Seven Waterworks (SW, waterworks with the old water source of the Songhua River) was carried out. Both waterworks operate coagulation, sedimentation and followed by filtration treatment process, which are typical traditional drinking water treatment.

The measured concentrations in the raw and finished water of the two investigated waterworks are shown in Figure 6. Six out of fifteen PAEs were detected in the finished water from the two waterworks. The detected PAEs were DMP, DEP, DIBP, DBP, DEHP, and DNOP, and the other investigated phthalates are of minor importance, with concentrations all below the limit of detection. As show in Figure 6, the measured concentrations of the analyzed PAEs in the finished water of the two waterworks varied strongly. The most important compound in the finished water was DEHP, with the mean concentration of 3473.7 and 4059.2 ng/L for the MPSW and SW, respectively, suggesting the highest relative composition of total PAE concentrations in the drinking water.

For the raw water from the different water sources, DMP and DIBP concentrations in the MPSW were much lower than the ones in the SW, and the concentrations of DEP, DBP, DEHP, and DEHP were relatively comparable in the two waterworks. The removal of PAEs by these two waterworks ranged from 25.8% to 76.5%, which varied significantly without stable removal efficiencies. The lower removal efficiencies for DMP and DNOP were observed in the SW, with the removal less than 30%. For both the waterworks, no sound removal efficiencies were obtained for the PAEs, indicating that the traditional drinking water treatment cannot show good performance to eliminate these micro pollutants, which has nothing to do with the type of the water source.

Traditional drinking water treatment focuses on dealing with the particles and colloids in terms of physical processes. Many studies of the environmental fates of PAEs have demonstrated that oxidation or microbial action is the principal mechanism for their removal in the aquatic systems [24–26]. Therefore the treatment process should be the combinations with the key techniques for removing PAEs from the water. On the other hand, since the removal efficiencies of PAEs by these advance drinking water treatments in waterworks has not been systematically studied to date, further research in this direction would seem to be required.

3.3.5 EXPOSURE ASSESSMENT OF PAES IN WATER

To evaluate the potential and adverse effects of PAEs, quality guidelines for surface water and drinking water standard were used. The results indicated that the mean concentrations of DBP and DEHP at levels were well below the reference doses (RfD) regarded as unsafe by the EPA for the surface water. The levels were not also above the RfD recommended by China (Environmental Quality Standard for Surface Water of China, GB3838-2002). On the other hand, the amounts of DEHP present in public water supplies should be lower than the drinking water standard (0.006 mg/L for EPA and 0.008 mg/L for China). According to the results, the concentrations of DEHP in drinking water samples from the MPSW were lower than the limited value.

PAEs are also considered to be endocrine disrupting chemicals (EDCs), whose effects may not appear until long-term exposure. According to the results, PAEs were detected in the drinking water constantly ingested in daily life, indicating that drinking water is an important source of human exposure to PAEs contaminants. Assuming a daily water consumption rate of 2 L and an average body weight of 60 kg for adults, the average daily intake of DEHP, DBP, and DIBP by way of drinking water from MPSW was calculated to be 115.8, 13.52, and 8.1 ng/kg/d, respectively. In comparison, the values of SW with the water source of the Songhua River were relatively higher, with the calculated results of 135.3, 14.0, and 15.5 ng/kg/d for DEHP, DBP, and DIBP, respectively. In this study, the estimated daily intake levels of DEHP from the drinking water were quite lower than the RfD of 20000 ng/kg/d released by the EPA. However, some of PAEs are partly metabolised in the organism, and future experiments should be focused on determining the potential effects of the metabolites [27].

Currently treated water from the Songhua River is for nonpotable uses, and MPSW is the exclusive waterworks run for the water supply of Harbin city. However, population growth and drought cycle are limited by the availability of raw water from the Mopanshan Reservoir. To meet the increasing demand, local and regional water authorities have begun a campaign of second water supply project from the Songhua River again, which needs the protection of the Songhua River and advanced water treatment for the source.

3.4 CONCLUSIONS

This study provided the first detailed data on the contamination status of 15 PAEs in the water near the Mopanshan Reservoir. The concentration range of 15 PAEs in the samples was from 355.8 to 9226.5 ng/L, with the mean value of 2943.1 ng/L. DEHP and DBP were the main pollutants among 15 PAEs, accounting for the main watershed pollution. The occurrence and distribution of PAEs from different sampling sites in the water source varied largely, suggesting that the spatial distribution of PAEs was site specific. In addition, the monitoring of PAEs in the waterworks also showed that PAEs can be detected in the drinking water, and certain toxicological risks to drinking water consumers were found. These results reported here contribute to an understanding of how PAE contaminants are distributed in the new water source and the waterworks in Harbin city as well as forming a basis for further modeling, risk assessment, and selection of drinking water treatment technology.

In conclusion, the results implied that no urgent remediation measures were required with respect to PAEs in the waters. However, the ecological and health effects of these substances through drinking water at the relatively lower concentrations still need further notice in light of their possible biological magnifications. Therefore, the long-term source control in the water and adding advanced treatment process for drinking water supplies should be given special attention in this area.

REFERENCES

1. M. Nikonorow, H. Mazur, and H. Piekacz, "Effect of orally administered plasticizers and polyvinyl chloride stabilizers in the rat," Toxicology and Applied Pharmacology, vol. 26, no. 2, pp. 253–259, 1973.
2. M. Vitali, M. Guidotti, G. Macilenti, and C. Cremisini, "Phthalate esters in freshwaters as markers of contamination sources—a site study in Italy," Environment International, vol. 23, no. 3, pp. 337–347, 1997.
3. G. Latini, C. de Felice, G. Presta et al., "In utero exposure to di-(2-ethylhexyl) phthalate and duration of human pregnancy," Environmental Health Perspectives, vol. 111, no. 14, pp. 1783–1785, 2003.
4. C. E. Mackintosh, J. A. Maldonado, M. G. Ikonomou, and F. A. P. C. Gobas, "Sorption of phthalate esters and PCBs in a marine ecosystem," Environmental Science and Technology, vol. 40, no. 11, pp. 3481–3488, 2006.
5. M. Clara, G. Windhofer, W. Hartl et al., "Occurrence of phthalates in surface runoff, untreated and treated wastewater and fate during wastewater treatment," Chemosphere, vol. 78, no. 9, pp. 1078–1084, 2010.
6. M. M. Abdel Daiem, J. Rivera-Utrilla, R. Ocampo-Pérez, J. D. Méndez-Díaz, and M. Sánchez-Polo, "Environmental impact of phthalic acid esters and their removal

from water and sediments by different technologies—a review," Journal of Environmental Management, vol. 109, pp. 164–178, 2012.
7. L. Chen, Y. Zhao, L. Li, B. Chen, and Y. Zhang, "Exposure assessment of phthalates in non-occupational populations in China," Science of the Total Environment, vol. 427-428, pp. 60–69, 2012.
8. M. J. Teil, M. Blanchard, and M. Chevreuil, "Atmospheric fate of phthalate esters in an urban area (Paris-France)," Science of the Total Environment, vol. 354, no. 2-3, pp. 212–223, 2006.
9. P. Roslev, K. Vorkamp, J. Aarup, K. Frederiksen, and P. H. Nielsen, "Degradation of phthalate esters in an activated sludge wastewater treatment plant," Water Research, vol. 41, no. 5, pp. 969–976, 2007.
10. J. Gasperi, S. Garnaud, V. Rocher, and R. Moilleron, "Priority pollutants in wastewater and combined sewer overflow," Science of the Total Environment, vol. 407, no. 1, pp. 263–272, 2008.
11. F. Zeng, K. Cui, Z. Xie et al., "Occurrence of phthalate esters in water and sediment of urban lakes in a subtropical city, Guangzhou, South China," Environment International, vol. 34, no. 3, pp. 372–380, 2008.
12. F. Zeng, K. Cui, Z. Xie et al., "Phthalate esters (PAEs): emerging organic contaminants in agricultural soils in peri-urban areas around Guangzhou, China," Environmental Pollution, vol. 156, no. 2, pp. 425–434, 2008.
13. G. Wildbrett, "Diffusion of phthalic acid esters from PVC milk tubing," Environmental Health Perspectives, vol. 3, pp. 29–35, 1973.
14. H. Fromme, T. Küchler, T. Otto, K. Pilz, J. Müller, and A. Wenzel, "Occurrence of phthalates and bisphenol A and F in the environment," Water Research, vol. 36, no. 6, pp. 1429–1438, 2002.
15. A. D. Vethaak, J. Lahr, S. M. Schrap et al., "An integrated assessment of estrogenic contamination and biological effects in the aquatic environment of The Netherlands," Chemosphere, vol. 59, no. 4, pp. 511–524, 2005.
16. C. Dargnat, M. Blanchard, M. Chevreuil, and M. J. Teil, "Occurrence of phthalate esters in the Seine River estuary (France)," Hydrological Processes, vol. 23, no. 8, pp. 1192–1201, 2009.
17. T. Suzuki, K. Yaguchi, S. Suzuki, and T. Suga, "Monitoring of phthalic acid monoesters in river water by solid-phase extraction and GC-MS determination," Environmental Science and Technology, vol. 35, no. 18, pp. 3757–3763, 2001.
18. S. Y. Yuan, C. Liu, C. S. Liao, and B. V. Chang, "Occurrence and microbial degradation of phthalate esters in Taiwan river sediments," Chemosphere, vol. 49, no. 10, pp. 1295–1299, 2002.
19. B. Li, C. Qu, and J. Bi, "Identification of trace organic pollutants in drinking water and the associated human health risks in Jiangsu Province, China," Bulletin of Environmental Contamination and Toxicology, pp. 1–5, 2012.
20. F. Wang, X. Xia, and Y. Sha, "Distribution of phthalic acid esters in Wuhan section of the Yangtze River, China," Journal of Hazardous Materials, vol. 154, no. 1–3, pp. 317–324, 2008.
21. Y. J. Sha, X. H. Xia, and X. Q. Xiao, "Distribution characters of phthalic acid ester in the waters middle and lower reaches of the Yellow River," China Environmental Science, vol. 26, no. 1, pp. 120–124, 2006.

22. J. Lu, L.-B. Hao, C.-Z. Wang, W. Li, R.-J. Bai, and D. Yan, "Distribution characteristics of phthalic acid esters in middle and lower reaches of no. 2 Songhua River," Environmental Science & Technology, vol. 12, article 014, 2007.
23. X. J. Zhu and Y. Y. Qiu, "Measuring the phthalates of xiangjiang river using liquid-liquid extraction gas chromatography," Advanced Materials Research, vol. 301, pp. 752–755, 2011.
24. B. L. Yuan, X. Z. Li, and N. Graham, "Aqueous oxidation of dimethyl phthalate in a Fe(VI)-TiO2-UV reaction system," Water Research, vol. 42, no. 6-7, pp. 1413–1420, 2008.
25. Z. Yunrui, Z. Wanpeng, L. Fudong, W. Jianbing, and Y. Shaoxia, "Catalytic activity of Ru/Al2O3 for ozonation of dimethyl phthalate in aqueous solution," Chemosphere, vol. 66, no. 1, pp. 145–150, 2007.
26. H. N. Gavala, U. Yenal, and B. K. Ahring, "Thermal and enzymatic pretreatment of sludge containing phthalate esters prior to mesophilic anaerobic digestion," Biotechnology and Bioengineering, vol. 85, no. 5, pp. 561–567, 2004.
27. Y. Guo, Q. Wu, and K. Kannan, "Phthalate metabolites in urine from China, and implications for human exposures," Environment International, vol. 37, no. 5, pp. 893–898, 2011.

CHAPTER 4

Assessment of Domestic Wastewater Disposal in Some Selected Wards of Maiduguri Metropolis, Borno State, Nigeria

ABBA KAGU, HAUWA LAWAN BADAWI, AND JIMME M. ABBA

4.1 INTRODUCTION

4.1.1 BACKGROUND TO THE STUDY

Water is one of the basic necessities for human survival, but the uses of water leads to generation of its output which is wastewater. Karen (2008) pointed out that, of all the planets in the solar system, Earth is the only planet known as "the water planet". However, the larger part of its human population either suffers lack of adequate potable water supply or mismanages the little within their reach most especially in the urban areas. Ross (1972) asserts that, water uses generally range from domestic, industrial to agricultural purpose. In the urban areas for instance, the water

---

use are more of domestic and industrial activities, as such, large quantity of wastewater is generated with its serious problem of disposal on the environment and the health of people. According to Karen (2008) therefore, wastewater could simply be referred to as "water that contains waste from homes or industries". This confirms with the assertion of Brower and Ende (1990) that wastes from water could be attributed largely to sewage which consists of human waste, other organic wastes and detergents. From the views of these scholars, domestic wastewater could be referred to as all kinds of water generated as aftermath of water usage from bathing, washing and food preparation at home.

In Borno State for instance, most of the population is concentrated in Maiduguri and this is not unconnected with the fact that Maiduguri is the state capital, as most of the infrastructural facilities are located in its confinement. As a result of this and coupled with the lack of monitoring of urban development and poverty, gross environmental mismanagement set-in with serious consequences on human health and environmental quality. Wastewater generation and its improper disposal then becomes one of such environmental problems in Maiduguri, paving ways to distortion of its environment and posing health threats. The open disposal of wastewater provides convenient ground for breeding germs, disease vectors and an eye sore with offensive odour. This is affirmed by Pink (2006) who stated that, as a result of improper wastewater disposal in most cities in the developing countries, larger proportion of their population are exposed to varied forms of diseases that are claiming lives and distorting their scenery.

However, in the developed countries, wastewater generated is disposed off through centralized channel facilities where they are treated and recycled for effective water management. Whereas, in developing countries most of the urban settlements were either not effectively planned or not served with effective wastewater disposal facilities (Karen, 2008). From this assertion Maiduguri metropolis could be termed as one of such urban areas found in the developing countries. In this urban, effective wastewater management facilities are not available as such wastewater was either allowed to flow freely into open space or into poorly constructed drainage network.

4.1.2 STATEMENT OF THE PROBLEM

Maiduguri with the population 733,176, as at 2007, has the highest number of people in terms of population compared with the local government areas in the state (NPC, 2007).This dynamic population of Maiduguri is directly proportional to its demand for water supply for both domestic and industrial needs by the residents. Consequently, the consumption of water equally leads to wastewater generations and disposal which are most of the time done without environmental consciousness. Generally, wastewater is generated from various points ranging from homes, hospitals, markets to industries. In most of the wards in Maiduguri, it is common to see wastewater being disposed off indiscriminately in open spaces and uncontrolled evacuation gutters and soak- away. All these actions are not without their implication on the health of the residents and their environmental quality. This is because improper wastewater disposals serve as the breeding ground for disease vectors such as mosquitoes, flies and other organisms. The sight of this wastewater equally distorts the beauty scenery of the environment, block road or path and generates offensive smell or odour. Bulunkutu, Gwange and Hausari wards in Maiduguri for instance were among the worst hit wards where the indiscriminant wastewater disposal is encouraged due to lack of good drainage network. It is against this background that the study was necessitated particularly in these wards as there was limited study that captured assessment of domestic wastewater disposal particularly in the effected wards mentioned (Gwange, Hausari and Maisandari wards) so as to suggest effective ways of managing the wastewater in the area.

4.1.3 AIM AND OBJECTIVES

The aim of this study is to examine the disposal of domestic wastewater in some selected wards of Maiduguri. To do this, the study intends to achieve the following specific objectives:

1. Quantify the amount of domestic wastewater generated in Gwange, Hausari and Maisandari wards.
2. Study and describe the methods used in disposing wastewater.
3. Highlight the issues due to poor drainage in the town.

4.2 LITERATURE REVIEW

Wastewater is defined in so many ways by different authors. UNEP (2010) viewed Wastewater as any water that has been adversely affected in quality by anthropogenic influence. It comprises liquid waste discharged by domestic residences, commercial properties, industry and/or agriculture and can encompass a wide range of potential contaminants and concentrations. According to Larry (2002) wastewater is the water that has been used and is no longer needed for any other particular purpose.

The population of people in households or areas contributes to domestic wastewater generation and disposal in places with enough water supply (Kenneth, 1993). For example a household of ten persons will use more water than that of three or four persons, consequently, the generation of wastewater will be more in the household with many persons, wards and areas with high population densities also generate more wastewater than areas of less population. This was supported by Mollison (1998) that the disposal of wastewater especially in urban areas of both developed and developing countries is a problem that continues to grow with development of nations and growth of population. The demand and uses of modern facilities increase the rate of wastewater disposal. In areas and homes with more facilities and high demand for water, wastewater disposal tends to be high (http\domestic\disposal\wastewater\htm). Wastewater disposal can also be influenced by quantity or volume of water discharged by a dwelling or household. Hammer (1977) stated that the volume of wastewater from residential areas vary depending on the type of dwelling. Large flows come from family houses that have several bathrooms, automatic washing machine and other water using appliances. Availability of disposal systems is a factor that affects wastewater disposal.

Oluwande (1971) stated that "collection of wastewater is best achieved by a full sewage system". If there is no sewage disposal or transportation system, wastewater will be disposed off anyhow. This could be either by free throw into water bodies which are harmful to human and animals and equally pollute the environment. Some drainage networks are not properly maintained and such leads to their blockage either by sand, refuse or stones.

UNEP (2010) stated that, in terms of wastewater disposal, in Africa about 80% of water consumption of those connected to the sewer ends up discharged into the municipal sewer. For instance in semi-arid and drought prone Gaborone the return flow is 50-65% thus, if poor people are connected in great numbers, the resultant reduced sewer flows could upset the operation of the sewer system because of the little water being used to keep waste flowing. He added in Lagos, wastewater is discharged to the Lagos lagoon and even sullage (grey water) water is discharged in open drain, throughout some of the urban catchments. Again in the largest peri urban settlement of Nairobi known as Kibera, drainage is virtually non-existent and during the rains in April/May and December, the areas are hardly accessible due to storm water and sullage nuisance.

In ideal situation the sewage is channeled or piped out of cities for treatment. Bulk of sewage contains water as the main component while other constituents include organic wastes and chemicals. Sewage water pollution is one of the major problems in cities. This is because; sewage water is drained off into rivers without treatment. Careless disposal of sewage water leads to a chain of problems such as spreading of diseases, euthrophication, increase in BOD etc (www.buzzle.com).

Considering the magnitudes of the effect or impact of wastewater, over 8 million people on the planet, disposing of sewage waste is a major problem. In developing countries, many people still lack clean water and basic sanitation (hygienic toilet facilities) sewage disposal affects people's immediate environments and leads to water-related illnesses such as diarrhea that kills 3-4 million children each year (According to WHO, water-related diseases could kill 135 million people by 2010). Moreover, an estimated 700 million people have no access to proper toilet and 1,000 Indian children die of diarrhea sickness every day (The Economist, 2008). Some 90% of China's cities suffer from some degree of water pollution (Chinadaily.com, 2005) and nearly 500 million people lack access to safe drinking water (The New York Times, 2007). In addition to acute problems of water pollution in developing countries, industrialized countries continue to struggle with pollution problems as well. Most people in the developed world have flush toilets that drain sewage waste quickly and hygienically away from their homes (www.explainthatstuff.com).

Lastly, around half of all ocean pollution is caused by sewage and wastewater. Each year, the world generates 4 billion tons of industrial waste, much of which is dumped untreated into rivers, oceans and other waterways. (www.explainthatstuff.com). Improper handling of wastewater is the main reason behind the pollution of water. The sewage is drained off in large quantities into rivers. It slows down the process of dilution of the constituents present in water; which in turn, stagnates the river (www.buzzle.com). Water consumption and wastewater generation depletes outer resources and has a destructive impact on the environment. Recent attention has aimed at preserving water resources and preventing pollution through several routes. Restrictions on wastewater discharge into the environment, recycling, reuse and regeneration of wastewater streams are now common practices towards achieving these objectives.

Conclusively, the indiscriminate disposal of wastewater creates poor sanitary conditions and environmental pollution. Therefore, there is the need to alleviate most of the pressures being placed on sewage drainage, treatment and disposal systems by the increasing pollution in general using modern and proper drainage channels which will help households in no small measures, in disposing off their wastewater in such a way that will drastically reduce health hazards associated with domestic wastewater disposal in particular.

4.3 STUDY AREA AND METHODOLOGY

4.3.1 STUDY AREA

Maiduguri urban, with a land mass of 137.356 Sq km (NPC, 2010), is located between latitude N 11°46'18" to N 11°53'21" and longitude E 13°03'23" to E 13°14'19" (Google Earth, 2012). The area lies within the lake Chad Basin formation, which is an area formed as a result of down-warping during the Pleistocene period (Waziri, 2007).

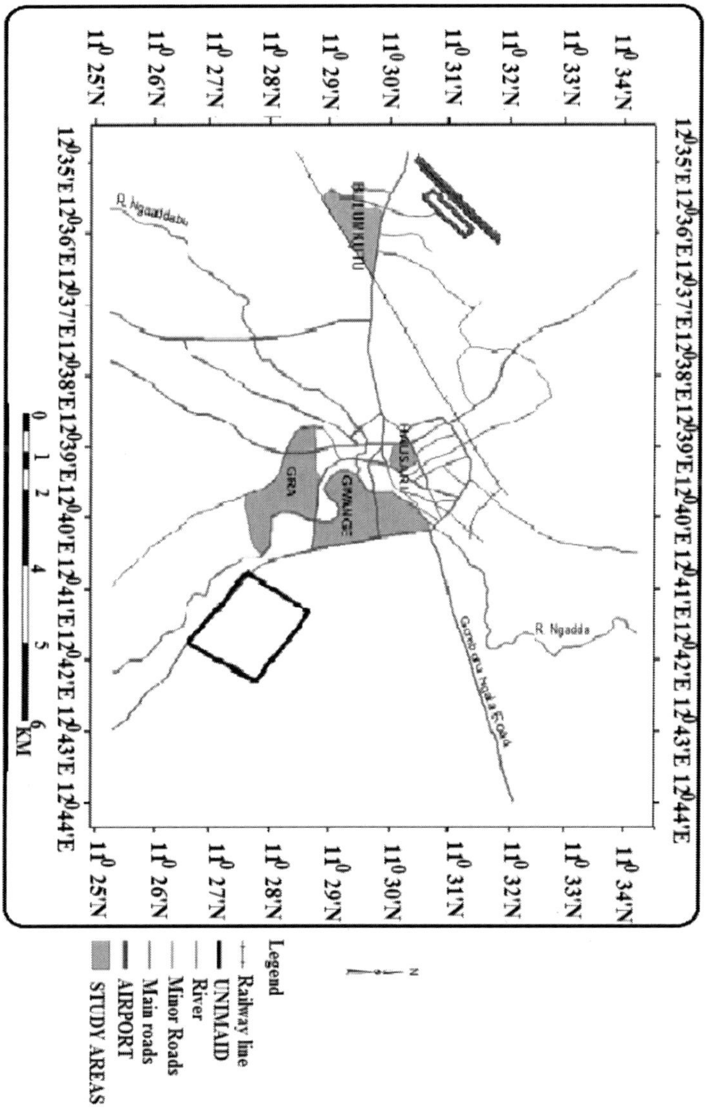

FIGURE 1: Maiduguri showing the study area Source: Digitized from Google earth image, 2012.

4.3.2 METHODOLOGY

The major source of data for this study was the primary source which was used in this research. The primary data were generated through a structured questionnaire distributed to the inhabitants in the Study area. This was administered based on households, designed to capture the objectives of the study. Purposive, stratified and accidental sampling techniques were employed at different stages for obtaining data in the study area. Based on the size of Maiduguri and the number of the wards, as well as for convenience and effective management of data. Three wards were selected using purposive sampling and 120 samples each were selected from each of these wards to give a total of 360. In the selection of the required sampled household therefore, each of these wards was stratified using its major streets. For instance, in Gwange wards there were three streets, therefore 40 households were accidentally selected from each of those streets. Hausari has four streets therefore, 30 households were accidentally selected to give a total of 120 respondents while in Maisandari, 55 households were accidentally selected in GRA and 65 household were equally accidentally selected in Bulumkutu respectively. The interviewed respondents were the household heads or any adult in the household. Figure 1 shows Maiduguri Township locating the sample wards.

The data obtained from the field was analyzed using simple descriptive and inferential statistics; the simple descriptive includes Charts (histogram), frequency and percentages to compare the data that have been obtained. The inferential statistics includes analysis of variance (ANOVA) which was used to show the mean response of the resident in relation to wastewater generation and disposal in the three wards, in other to determine the variation in the wastewater generated.

4.4 RESULTS AND DISCUSSION

4.4.1 DAILY QUANTITY OF WATER USED IN HOUSEHOLDS

The quantity of water use daily in a household differs from one household to another depending on the number of persons in that household and the use water is put to. Table 1 shows the responses from the residents.

TABLE 1: Daily quantity of water used in households in liters Source: Field work, 2011.

Volume in Litres/day	Maisandari		Gwange		Hausari	
	f	%	F	%	f	%
Bathing						
20-80	23	19.2	40	33.3	41	34.2
81-160	63	52.5	55	45.8	41	34.2
161-240	33	27.5	19	15.8	16	13.3
240&above	1	0.8	6	5.0	22	18.3
Washing						
20-80	114	95.0	101	84.2	95	79.2
81-160	19	15.8	5	4.2	13	10.8
161-240	-	-	1	0.8	8	6.7
240&above	-	-	-	-	4	3.3
Food preparation						
20-80	119	99.2	116	967	109	90.8
81-160	1	0.8	4	3.3	7	5.8
161-240	-	-	-	-	4	3.3
Others						
20-80	118	98.3	73	60	73	60.8
81-160	1	0.8	1	0.8	9	7.5
161-240	1	0.8	-	-	8	6.7
240&above	-	-	-	-	5	4.2

Table 1 shows the quantity of water use daily in the households. From the Table 1 therefore, most of the residents in the three wards use 81–160 litres daily per household for bathing, while 20–80 liters is being used for washing daily per household in the study area. From the table it is observed that the main uses of water which leads to high generation of wastewater in the three wards is bathing with 52.2%, 45.8% and 34.2% respectively. The availability of water and population of persons per household plays an important role in the volume of water to be used. If there are 5–10 persons in a family they will consume more water than the family of 4 and below. While 20–80 liters of water per day is being used for food preparation and other purposes with the percentages of 99.2%, 96.7% and 90.8% respectively.

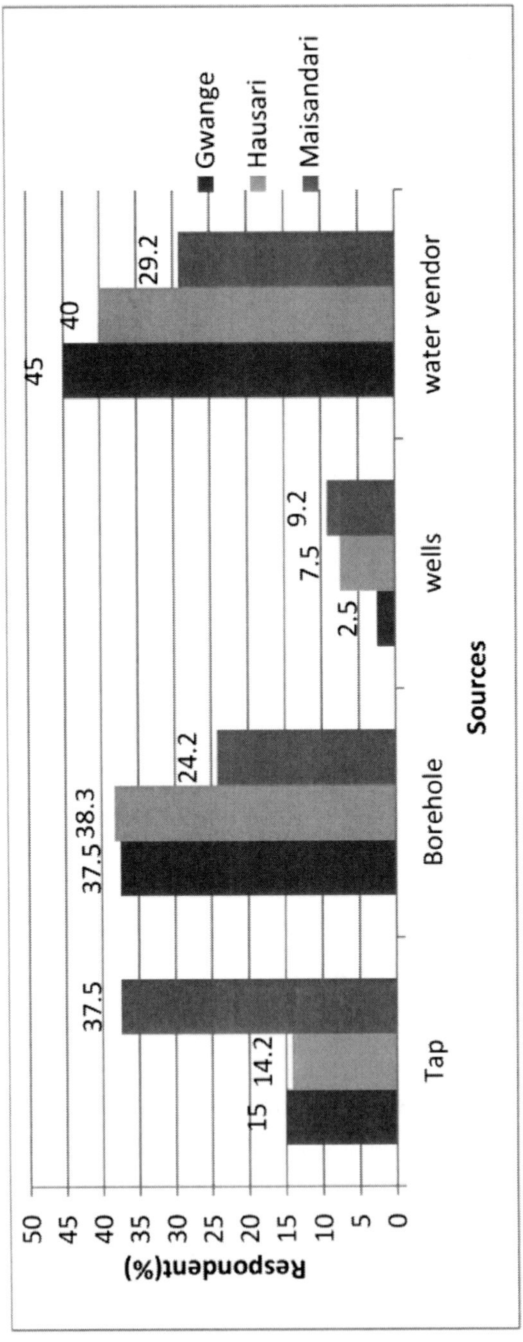

FIGURE 2: Sources of water supply Source: Field work, 2011.

4.4.2 RESULTS OF ANALYSIS OF VARIANCE (ANOVA)

Analysis of Variance was used to show the mean responses of the residents in relation to the wastewater generation in the three wards (Gwange, Hausari and Maisandari) with a view to determine the variation in the wastewater generated. The variables for the analysis include bathing, washing, food preparation and other purposes. From the results obtained, the mean responses of the residents with regard to the wastewater generated from bathing, washing, food preparation and others, As revealed, the mean responses shows clearly that wastewater generated from bathing in Maisandari has the highest value of 2.1583[a] at 0.05 level of significance which implies that most of the wastewater generated from bathing in the study area comes from Maisandari. This is because, the developed part of the Maisandari (GRA) has enough water supply and uses modern facilities like showers, and other water using appliances. This might have been the reason for the increase in the rate of wastewater generation and disposal. This is in consonance with Hammer (1977) who stated that "Wastewater disposal can be influenced by quantity or volume of water discharged by a household". The volumes of wastewater from residential areas vary depending on the type of dwelling. Large flows come from family houses that have several bathrooms, washing machine and other water using appliances.

Gwange with the value 2.1000[ab] at 0.05 significance level comes next to Maisandari (2.1583[a]). Statistically, there is no significant difference between the two. This may not be unconnected with the fact that Gwange is planned area and some of the residents either have a water storage facility or uses modern facilities. The large family size of most of the residents might have also contributes to wastewater generation and disposal in the area. Hausari with 1.9295[b] at 0.05 significance level is the least in terms of wastewater generated through bathing. This may be as a result that most of them do not use showers, having no taps and largely depend on vendors for their daily water supply; thus water being a scarce resource to them was most of the time being quantified for use.

Moreover wastewater generated from washing is high in Maisandari with the value 1.3417[a] at 0.05, level of significance. This is higher than the values of Gwange (1.1583[b]) and Hausari (1.0583[b]) at 0.05 level of significance (though statistically, there is no significant difference between the two). In terms of wastewater generated as a result food preparation, Maisandari still has the highest value with (1.1250[a]) at 0.05 significance level as compared with Gwange (1.0083[b]) and Hausari (1.0333[b]) at 0.05 significant levels respectively. This result therefore implies that there is no significant difference between Gwange and Hausari wards but the different exist between the two and Maisandari. Generally, wastewater generated from food preparation including washing of vegetables, meat, fish and other food products.

It equally revealed that, Maisandari 1.1250[a] at 0.05 significant level in terms of wastewater generated from others purposes. While Gwange with 0.5000[b] and Hausari with 1.1000[b] clearly indicated that Maisandari is the highest in terms of wastewater generation and disposal. This may not be unconnected with the fact that the availability of water in Maisandari ward encouraged high utilization of water for brushing of teeth, ablution and body cleaning after using toilet. This action contributes to large quantity of wastewater generation and disposal.

It is evident that wastewater generated in the study area was higher in Maisandari than Gwange and Hausari wards. This may likely be due to higher quantity of water supplies in that ward and the uses of modern facilities. This resultant effect is the increase in the rate of wastewater generation and disposal. This directly contributes significantly to health complications of the residents and the reduction of environmental quality.

4.4.3 SOURCES OF WATER SUPPLY

Wastewater accumulation could only be possible if there is water supply. Figure 2 therefore shows the sources of water supply in Gwange, Hausari and Maisandari wards.

Figure 2 shows the major sources of water supply in the three wards of the study area. From the Figure 2 therefore, water vendor with 45% is the major source of water in Gwange, while well water with 2.5% formed the lowest in terms of source of water supply. This implies that majority of the residents depend on water vendors to supply water for livelihood. Figure 2 shows that water vendors with the highest percent (40%) formed the major source of water supply for residents in Hausari ward, while those who source water from well (8%) formed the lowest. In Maisandari ward, tap water (37.5%) is the highest in terms of source and well accounted for 9%. From the result therefore, it is clear that residents in both Gwange and Hausari wards largely depend on water vendors for their source of portable water supply. This may likely be as a result that for over decades, most of the boreholes and taps have stopped functioning and people of the area resorted to other alternative sources. However, in Maisandari ward, the source of their portable water supply shows a clear distinction. The ward is divided into two based on the settlement planning and economic status. For instance, in the Government Reservation Area (GRA) most of the residents depend on tap water (37.5%) while in the poorly settled area such as Bulunkutu Tsallake, tap water does not exist, as such most of them depend on well and water vendors (38.4%) for portable water supply. This has largely affected the quantity of water the two areas used and the quantity they generate and dispose.

4.4.4 SOURCES OF WASTEWATER

Wastewater generation depends on the number of people in the family; the availability of the water and the uses to which the water is put to. Figure 3 therefore, presents the sources of wastewater generated in the study area.

From Figure 3, the main uses of water that leads to large quantity of wastewater generation and disposal in the study area are bathing with 49.2% in Gwange; washing clothes with 52.3% in Hausari; washing utensils with 15% in Maisandari and flushing toilets with 7.5% in Maisandari respectively. This confirmed the assertion of Ghoreishi and Haghighi (2003) who were of the view that domestic wastewater is generated by the type of dwellings and commercial facilities and activities taking place. These activities/facilities include washing, bathing and other related activities. In some parts of the study area for instance, it is common among the residents of the poorly settled areas to recycle wastewater generated from some of these activities for other purposes and disposed them only when the recycled water becomes too bad for any use.

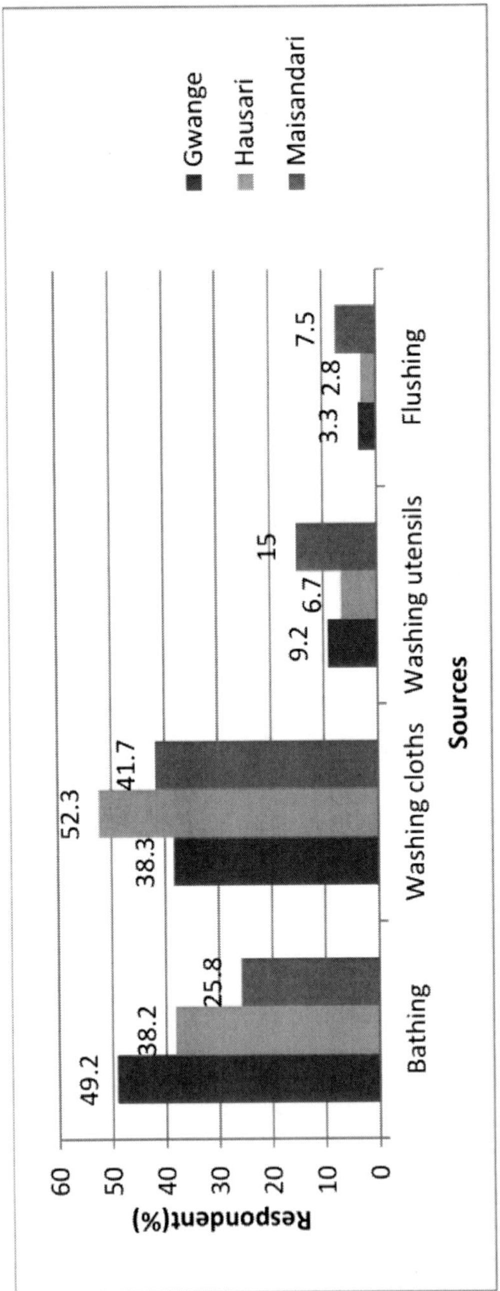

FIGURE 3: Sources of wastewater Source: Field work, 2011.

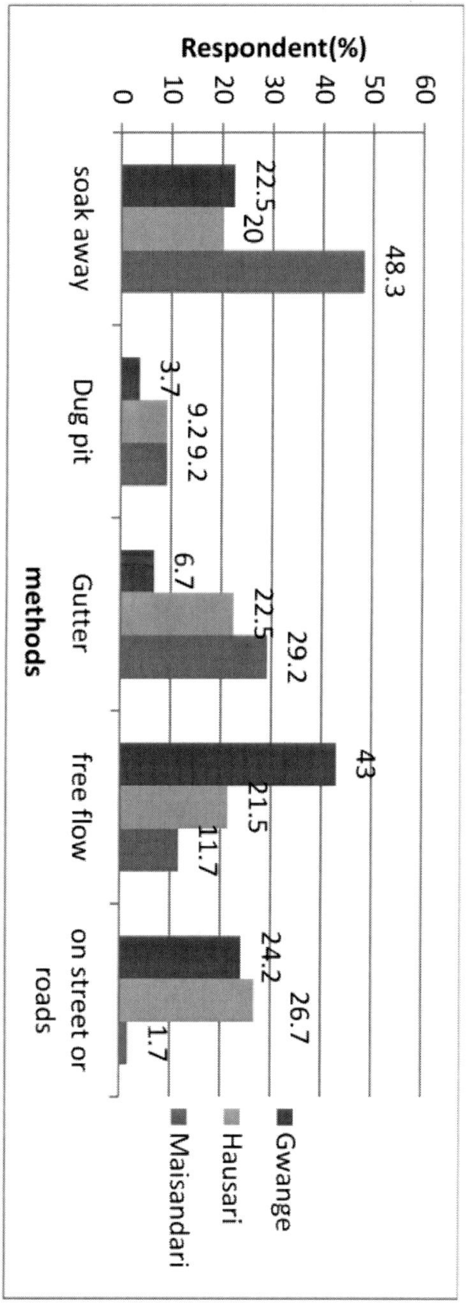

FIGURE 4: Methods of wastewater disposal (Source: Field work, 2011)

PLATE 1: A typical free flow of wastewater in Gwange ward

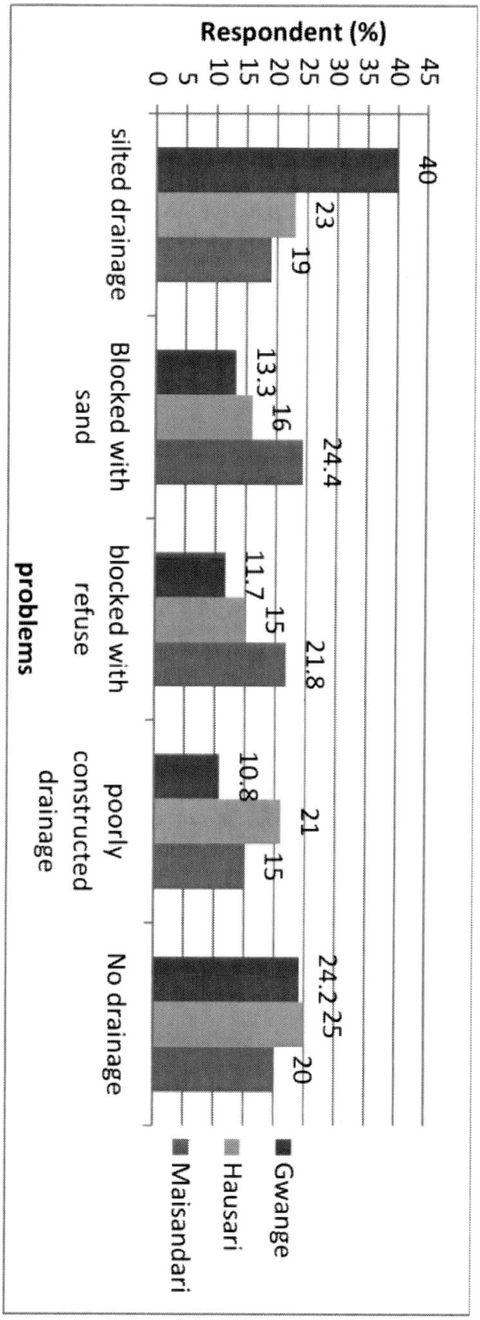

FIGURE 5: Drainage problems Source: Field work, 2011.

4.4.5 METHODS OF WASTEWATER DISPOSAL

Domestic wastewater disposal varies from place to place. Figure 4 presents the methods of wastewater disposal in Gwange, Hausari and Maisandari Wards in the Maiduguri.

From Figure 4, the most commonly used method of wastewater disposal in Gwange is through free flow with 43%, in Hausari Ward the method is on street or road with 26.7%. This is because there are no or poorly constructed drainage systems in the area. While Maisandari Ward most of the residents particularly those in the developed areas of Maisandari (GRA), dispose their wastewater through soak away with 48.3% while the least developed areas of Maisandari dispose theirs in Dug pit, Gutters, Free flow, and on street or road respectively. This is also due to the fact, there is no drainage or poorly constructed drainages, as the people, particularly in the least developed areas of Maisandari do not have effective channels through which wastewater generated in their houses are channeled to, whereas the developed areas of Maisandari have the efficient and effective channel through which water is emptied or disposed. Plate 1 shows a typical free flow of wastewater in Gwange ward of the study area.

4.4.6 DRAINAGE PROBLEMS

The frequent generation of wastewater leads to the problems of disposal; to overcome these problems, drainage of good capacity is required and the drainage most be, free from all sorts of hindrance to serve its purpose, but reverse is the case in most places. The drainages are poorly constructed or no drainage system at all. Figure 5 shows the responses on these problems in the study area.

Drainage problems faced by the residents of Gwange is caused by silted drainage (40%), there is presence of drainage but siltation is the major problem, in Hausari ward (25%), there are no drainage system, while Maisandari has blocked drainages 24.4% as shown in Plates 2 and 3. This is in accordance with Kodiya and Monguno (1997) who reported that the problem of drainage is experienced more in the urban centers due to the nature of settlements and interruptions with the natural channel resulting in the development of urban drainage system. Proper drainage management greatly helps to protect the physical environment. Indeed, an effective drainage management is an important link between man and his environment. More so the scenery of an environment depends on the nature of drainage existing in that environment and that is why any drainage problem is considered a serious environmental problem that has both spatial and temporal dimensions.

4.4.7 MAJOR FACTORS AFFECTING WASTEWATER DISPOSAL

Several factors interplay in wastewater disposal; these include population, inadequate drainages, and availability of water supply. In areas with

PLATE 2: A typical blocked drainage with sand in Hausari

enough water supplies having high population with less disposal facilities, wastewater is indiscriminately disposed. Figure 6 shows the responses on how population, drainage, lack of awareness and availability of water supply affects wastewater disposal.

Wastewater disposal is influenced by several factors of which some were examined in this study. Lack of drainage system and over population was observed to be the most. The aforementioned factors were seriously lamented of by the respondents in the study area as a source of concern in wastewater management. About 55.0%, 63.9% and 53.3% of the residents in the three wards reported that, lack of drainage is the major factor affecting wastewater disposal, while population accounted for 40.8%, 20.2% and 27.5% respectively.

PLATE 3: A typical Blocked Drainage with refuse in Maisandari

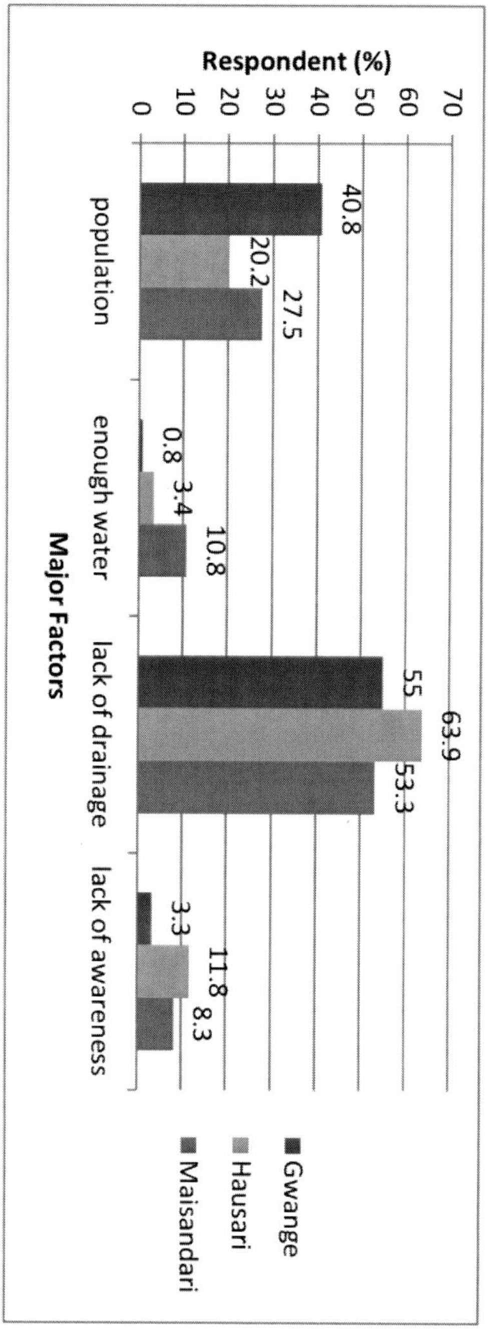

FIGURE 6: Major factors affecting wastewater disposal Source: Field work, 2011.

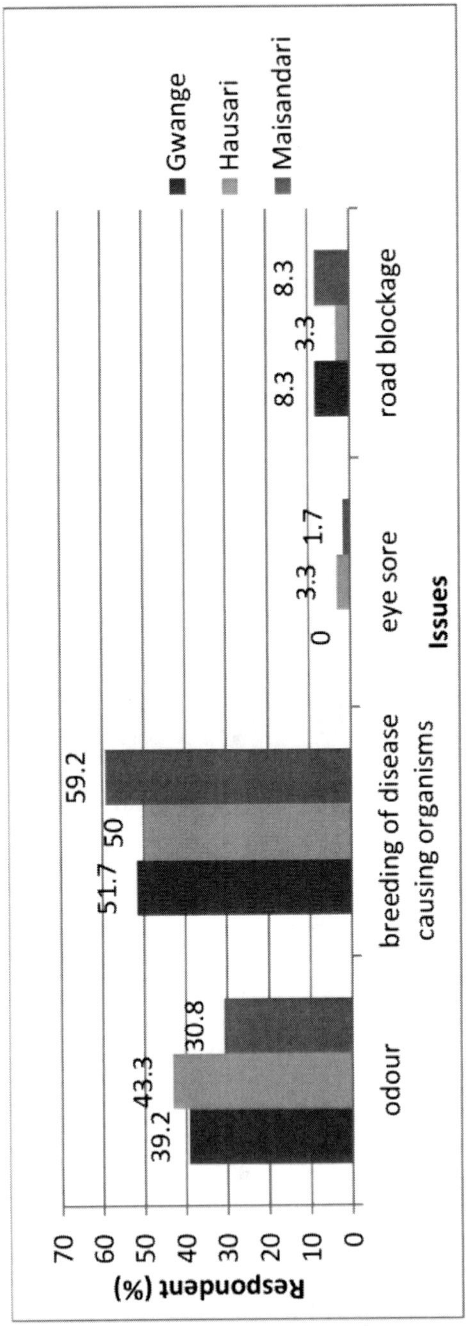

FIGURE 7: Issues due to poor drainage Source: Field work, 2011.

4.4.8 ISSUES DUE TO POOR DRAINAGE

The problem of blocked and poorly constructed drains has so many effects on the residents. These problems include pollution, breeding of disease causing organisms, eye sore and blockage of road. Wastewater that is stagnant in the drains generate into smelly odour and serves as breeding ground for diseases. Figure 7 shows the responses on the issue of poor drainage.

Figure 7 presents the effects of wastewater disposal in the study area. It can be seen that breeding ground for diseases and odour are the most common effects, the percentages of breeding ground for diseases are 51.7%, 50% and 59%, while pollution accounted for 39.2%, 43.3% and 30.8% respectively. In consonance with Figure 5 the various factors mentioned might have resulted into the prevalence in to the effect on both human health and the environment as shown in the Figure 5. This study confirmed the environmental studies by the UNEP (2010) which noted that, wastewater disposal is one of the main factors that encourages poor health and environmental pollution with serious consequences on human livelihood and environmental safety. Domestic wastewater usually contains certain amount of contaminants and vector-borne organisms such as disease causing bacteria, infection, viruses, parasitic organisms and other pathogens as well as toxic metals, household chemicals, and excess nutrients such as nitrates. These contaminants have negative effect on the environment, drinking water and cause health hazards to people of the study area (www.wastewaterdisposal.com).

4.5 CONCLUSION AND RECOMMENDATIONS

4.5.1 CONCLUSION

From the findings of the study, it is evident that improper wastewater management and disposal in Maiduguri constitutes a serious hazard on human as well as the environment. Therefore, wastewater generated should be managed and disposed properly in order to avoid those hazards on the people and create a friendly and healthy environment.

4.5.2 RECOMMENDATIONS

Based on the findings, the following recommendations are put forward:

- Proper maintenance of the existing disposal systems by the government and individuals or the community should be done. The ministry of health in collaboration with the Borno State Sanitation Board should embark on environmental education campaign so as to create public awareness on the health implications of the improper disposal of wastewater.
- Wastewater should be flushed and drained regularly by individuals or community through weekly environmental sanitation exercise under strict supervision of the State Environmental Sanitation Board and Ministry of Health so as to help to improve the quality of the environment.
- Government should provide adequate and efficient drainage networks in all the affected areas. The design methods should be those with cover-slaps to prevent future indiscriminate dumping of refuse that are creating blockage of free flow of liquid waste.
- Government should provide drainages in the urban areas which are channelized into a central outlet system where it can easily be treated before final disposal for other use.
- The State legislature should enact law forbidding indiscriminate dumping of waste. Defaulters should by same law be made to face severe penalties.

REFERENCES

1. Brower, J. E., Zar, J. H., & Ende, C. V. (1990). Field and Laboratory Method for General Ecology. W.C. Brown Publishers.
2. Buzzle. (2010). Sewage water pollution. Retrieved May 21, 2010, from www.buzzle.com
3. Chinadaily. (2005). Water Pollution in Cities. Retrieved from http://www.Chinadaily.Com.cn/English/doc/2005-06-07/content_449451.htm
4. Explainthatstuff. (2009). Water pollution: An introduction to cause, effects, types and solution. Retrieved May 21, 2010, from www.explainthatstuff.com
5. Ghoreishi, S. M., & Haghighi, R. (2003). Chemical Catalytic Reaction and Biological Oxidation for Treatment of Non-biodegradable Textile Effluent. Chemical Engineering Journal, 95, 163-169. http://dx.doi.org/10.1016/S1385-8947(03)00100-1
6. Hammer, J. M. (Ed.). (1977). Water quality and pollution Waste and water technology (pp. 143-168). New York: John Wiley and Sons.
7. Karen, A. (2008). Environmental science. Holt, Rinechart and Winston. A Harcount education company, 1080N mopac express way building 3, Austin, Texas.

8. Kodiya, H. M., & Monguno, A. K. (1997). Environmental Change and Natural Hazards in Borno. In M. M. Daura (Ed.), A Study of 1994 Maiduguri Flood. Maiduguri: Later Day Publishers.
9. Larry, J. A. (2002). Biology Treatment of Wastewater. West Sussex: Ellis horwood Ltd.
10. Mollison, K. D. (1998). Wastewater Treatment. Mac graw-hill Book.
11. NPC. (2007). Nation Population Commission: Population Census of the Federal Republic of Nigeria. NPC, Abuja.
12. Oluwande, A. P. (1971). Cheap Sewage Disposal in Developing Countries. Ibadan: Ibadan University Press.
13. Pink, D. H. (2006). Investing in Tomorrow's Liquid gold. Retrieved form http://finance.yahoo.com/columnist/article/trenddesk/3748
14. Ross, R. D. (1972). Air Pollution and Industry.
15. The Economist. (2008). Infrastrucure is india's biggest handicap. Retrieved December 11, 2008, from http://www.economist.com/special reports/displaystory
16. The New York Times. (2007). Pollution Reaches Deadly Extremes. Retrieved form http://www.nytimes.com/2007/08/26/world/asia/26china.html
17. UNEP. (2010). United Nations Environmental Programme. International Source book on environmentally sound technologies for wastewater and storm water management (UNEP).
18. Wastewater disposal. (2010). Effects of domestic wastewater disposal. Retrieved May 21, 2010, from www.wastewaterdisposal.com
19. Waziri, M. (2007). Trends in Population Dynamics and Implication for Contemporary Socio economic Development in the Chad Basin. Paper Presented at the Kanem Borno Millennium Conference, Maiduguri. In M. W. A. Kagu., & K. M. Abubakar (Eds.), Issue in the Geography of Borno State.
20. Wikipedia. (2010). Wastewater, the free Encyclopedia. Retrieved from http://en.Wikipedia.Org/wiki/wastewater

CHAPTER 5

Detection of Free-Living Amoebae Using Amoebal Enrichment in a Wastewater Treatment Plant of Gauteng Province, South Africa

P. MUCHESA, O. MWAMBA, T. G. BARNARD, AND C. BARTIE

5.1 INTRODUCTION

Free-living amoebae (FLA) are unicellular protozoa that exist in high numbers in aquatic environments where they play a useful role as predators of bacteria, algae, viruses, and fungi [1]. They have been isolated from process water systems such as cooling towers, hospital water networks, and drinking and wastewater water plants [2–4]. *Naegleria fowleri, Balamuthia mandrillaris*, and *Acanthamoeba* are some of the FLA species known to be pathogenic to humans [5–7]. *Acanthamoeba* species are the causative agent of amoebic keratitis (AK) and granulomatous amoebic encephalitis (GAE) while *Naegleria fowleri* and *Balamuthia mandrillaris* have been associated with amoebic meningoencephalitis (PAM) and GAE, respectively [6]. This study focused on detection of these pathogenic FLA, specificallythe *Acanthamoeba* sp., in a wastewater treatment plant using an optimized amoebal enrichment technique.

Most FLA have two developmental stages (some FLA also have a flagellate intermediate form): an active trophozoite stage and a dormant cyst stage. Trophozoites actively feed through phagocytosis and pinocytosis on microorganisms and small organic particles in the environment [8, 9]. The cyst stage occurs when environmental conditions are unfavourable, for example, in extremes of temperature, osmotic pressure, and pH, or when nutrient levels are depleted. FLA can survive in the cyst stage for extended periods of time, only to become active trophozoites when environmental conditions become favourable again [6, 10]. These amoebal cysts contain cellulose which forms a physical protective barrier making them resistant to a wide variety of water treatment regimes. Some studies have reported survival of amoebal cysts after clarification, rapid filtration, and ultrafiltration processes, as well as after biocide treatment. Biocides such as chlorine, chlorine dioxide, monochloramine, ozone, copper-silver nitrate, and ultraviolet light have shown limited success against a variety of amoebal cysts in water treatment systems [11–13]. This has huge implications in water treatment systems for drinking water and sewage treatment in South Africa which relies heavily on chlorine as biocidal for water treatment.

Free-living amoebae can also act as reservoirs of pathogenic bacteria such as methicillin-resistant *Staphylococcus aureus, Vibrio cholerae*, *Legionella* species including *Legionella pneumophila*, and environmental *Mycobacterium* species as reviewed by Goñi et al. [14]. These "amoebae resistant bacteria" (ARB) are able to infect and resist the digestive process of FLA, survive, multiply, and exit FLA enabling them to spread and colonize aquatic water systems [15–17]. The list of confirmed ARB currently stands at 102 species and continues to grow [18]. ARB use their amoebal hosts for nutrition and protection (when amoebae form cysts) during harsh environmental conditions such as in the presence of biocides like chlorine used in water treatment. Some genera, particularly, *L. pneumophila* and members of the *M. avium* complex, are believed to increase their own virulence during passage through their amoebal hosts [19, 20]. FLA, therefore, can act as proliferators and distributors of pathogenic bacteria in water systems other than being pathogenic themselves.

International research programs have consequently focused on the coexistence of FLA and ARB and the effects this relationship might have on traditional water quality testing techniques which look for the presence/absence of faecal indicators and protozoan parasites [15, 19]. Amoebal enrichment techniques have been used successfully, to selectively grow FLA and recover ARB from environmental samples [15, 21]. However, no studies to date in South Africa have applied amoebal enrichment techniques to selectively grow indigenous FLA in water systems, presenting a need to optimize this technique using local conditions.

Although the presence of FLA in natural environmental waters and manmade water systems has been well documented worldwide, few studies have reported on the occurrence of FLA in wastewater treatment plants [22–24]. Therefore, there is a need to obtain more information regarding the occurrence of FLA in wastewater treatment plants. This work included in this study is the first to determine the occurrence of FLA in a wastewater treatment plant in South Africa. The investigations in this study are divided into two parts: the first includes the optimization and establishment of amoebal enrichment techniques to isolate FLA under laboratory conditions using seeded samples, whereas the second includes the application of optimized conditions to isolate FLA potentially containing pathogenic ARB, at different stages of a wastewater treatment plant taking into consideration seasonal differences.

5.2 MATERIALS AND METHODS

5.2.1 OPTIMIZATION AND SEEDING EXPERIMENTS

This study was an exploratory study to investigate the possibility for the presence of amoebae. It was decided to use *A. castellanii* as it is easily identified morphologically (the basis of isolation in this study) compared to other FLA which require further methods like polymerase chain reaction (PCR) to confirm them. The most appropriate temperature, food source,and concentration method for growth of *Acanthamoeba castellanii* (ATCC 30010) type strain were determined as indicated below.

5.2.1.1 OPTIMIZATION OF LABORATORY CONDITIONS FOR A. CASTELLANII GROWTH

The type strain *Acanthamoeba castellanii* Neff (ATCC 30010) was obtained from the American Type Culture Collection (Rockville, MD, USA). The strain was reconstituted and grown in tissue culture flask (Nunc, USA) containing 5 mL plate count broth (PCB) (Merck, SA) according to the manufacturer's instructions. The reconstituted amoebae were incubated in triplicate experiments at three temperatures including room temperature, 32°C, and 37°C. The growth medium was replaced at weekly intervals to maintain a constant supply of fresh axenic amoebal trophozoites. In order to compare membrane filtration and centrifugation as methods for sample concentration, two split samples were prepared from 500 mL water samples which were seeded with *A. castellanii* (ATCC 30010). One portion was concentrated by membrane filtration through 0.45 μm pore size cellulose nitrate membranes (Millipore, SA) and the other by centrifugation at 1000 g for 20 minutes (Biovac Neofuge 15R, Vacutec, SA). Membrane filtration and incubation at 32°C were found to be optimal and thus used for further experiments. In order to determine the most appropriate food source for recovering amoebae, the concentrated samples were inoculated into living or heat-killed *E. coli*. The *Escherichia coli* (ATCC 25922) type strain was obtained from the American Type Culture Collection (Rockville, MD, USA). The type strain was reconstituted and maintained according to supplier's instructions before being inoculated onto nutrient agar (NA) and incubated at 37°C overnight. The stock plates were sealed and stored at 4–10°C until use. The type strain was maintained by weekly subculturing onto fresh NA plates. Heat-killed *E. coli* was prepared by placing a suspension of the type strain (*E. coli*) in a boiling water bath for 20 minutes immediately before use. Non-nutrient agar (NNA) plates were inoculated with 100 μL of living or heat-killed *E. coli* (HKEC) by spreading the suspension evenly over the surface. For quality control purposes, nutrient agar plates were also inoculated with HKEC and incubated overnight at 37°C.

One split sample (500 mL) was passed through a cellulose nitrate membrane (Millipore, SA) with a pore size of 0.45 μm. The membrane was cut into three pieces and each piece was placed upside down onto a NNA-E. coli plate. The plates were incubated aerobically at room temperature (22–25°C), 32°C, and 37°C. The other split sample was centrifuged at 1000 g for 20 minutes as recommended in the Health Protection Agency protocol [25]. The supernatant was aseptically removed by aspiration leaving approximately 2 mL covering the pellet. This was mixed thoroughly; 100 μl of this mixture was inoculated onto NNA-*E. coli* plates and incubated aerobically at 32°C, 37°C, and room temperature.

5.2.1.2 SEEDING EXPERIMENTS

The optimized food source and temperature were used for the growth of *A. castellanii* as mentioned above. Trophozoites were harvested by centrifugation at 1000 g for 20 minutes to obtain a pellet. The pellet was washed three timeswith 1 mL sterile Page's amoebal saline (PAS) and centrifuged at 1000 g for 20 minutes after each wash. The resulting pellet was resuspended in 1 mL sterile PAS. Nine sterile distilled water samples (500 mL each) were then seeded with 100 μL suspension of *A. castellanii.* After seeding, each water sample was divided into 10 equal portions of 50 mL to represent split samples (n = 90); each seeded sample was treated identically thereafter.

5.2.2 AMOEBAL ENRICHMENT

Amoebal enrichment technique used was adapted from previous studies [2, 21]. Briefly, the 90 from seeding experiments were concentrated by filtration using a 0.45 μm pore size cellulose nitrate membrane (Millipore, SA). The membrane was placed upside down onto a NNA-HKEC plate with a few drops of sterile PAS, incubated aerobically at 32°C, and checked daily under light or inverted microscope for the appearance of amoebal trophozoites and cysts. The density of amoebal growth on the plates was recorded as (the average in 10 fields) <10 per field (+), 10–100 per field (++), or >100 per field (+++). Plates with amoebal growth were purified by aseptically cutting small agar plugs, placing them upside down onto fresh NNA-HKEC plates, and incubating as before. Once purified, amoeba were removed from the agar by gentle scraping, resuspended in sterile PAS, and washed at least three times at 1000 g for 20 minutes to remove extracellular bacteria and debris. The concentrate was then resuspended in 1 mL sterile PAS, inoculated into a sterile 24-well flat-bottomed microtiter plate (Nunc, USA), and again incubated at 32°C. The plates were checked for the morphological appearance of trophozoites and/or cysts under an inverted microscope (Leica, Germany), equipped with a 40x objective, at regular intervals. Fifty microliters of the amoebae suspension was harvested from the microtiter plate, heat-fixed on microscope slides, and Giemsa-stained to screen for the presence of amoebal trophozoites and/or cysts.

5.2.3 ENVIRONMENTAL SAMPLES

5.2.3.1 SAMPLE COLLECTION

The wastewater treatment plant (Figure 1) consists of a screen/grit channel, primary sedimentation, thickeners for raw sludge, thickeners for waste activated sludge, and bioreactors incorporating the three stages, configuration, final clarification, and maturation ponds. A total of 172 samples were collected over 4 seasons: autumn (41), winter (43), spring (44), and summer (44) during May, July, September, and November of 2010 at a wastewater treatment plant in Gauteng, South Africa. The samples were collected over different days for a total in a month/season. Along the treatment plant, samples (500 mL each) were collected from influent (16), bioreactor feed (20), anaerobic zone of the bioreactor (16), anoxic zone of the bioreactor (16), the two aerators (32), bioreactor effluent (16), bioreactor final effluent (11), and the maturation ponds (45).

The concentrations of chlorine residual in the treated effluents were determined on-site using the Lovibond Comparator system 2000 (Cydna laboratory, SA). Sample bottles for final effluent and maturation ponds contained 0.1% sodium thiosulphate (3% solution) to neutralize residual chlorine. At each sampling point, the temperature and pH were recorded on-site, respectively, with a portable thermometer and pH meter. Samples were processed within 24 hours of collection.

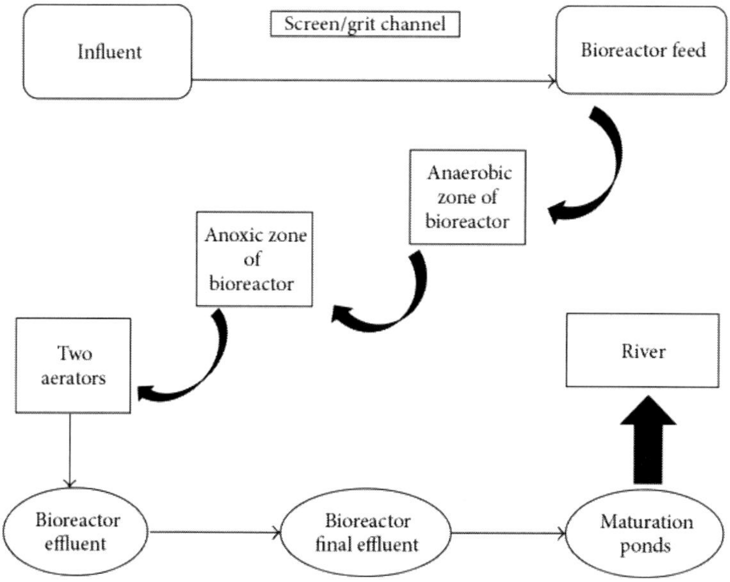

FIGURE 1: Schematic diagram of wastewater treatment plant indicating the sampling points of the present study.

5.2.3.2 SAMPLE PROCESSING

All samples were analyzed according to the methods established with seeded samples. Samples (500 mL each) were filtered through a 0.45 µm pore size cellulose nitrate membrane (Millipore, SA). The membrane filters were then inoculated onto NNA-HKEC plates and incubated at 32°C to allow for the growth of indigenous amoeba. When amoebal trophozoites and/or cysts were observed, they were subcultured by aseptically cutting small agar plugs, placing them upside down onto fresh NNA-HKEC plates. Subculturing was repeated 3 to 4 times to allow purification of amoebae isolates. Once purified, amoebae were resuspended in sterile PAS, inoculated into a sterile 24-well flat-bottomed microtiter plate (Nunc, USA), and again incubated at 32°C. The plates were checked daily for the morphological appearance of trophozoites and/or cysts under an inverted microscope (Leica, Germany), equipped with a 40x objective. A suspension of 50 μL from the microtiter plate containing the amoebae was heat-fixed on microscope slides and Giemsa-stained to screen for the presence of amoebal trophozoites and/or cysts potentially containing intracellular bacteria.

5.2.4 MOLECULAR METHODS

5.2.4.1 DNA EXTRACTION

A total of 30 environmental samples, 22 positive for *Acanthamoeba* sp. and 8 random samples negative for *Acanthamoeba* sp., were selected for molecular analysis. Amoebae DNA were extracted without pretreatment from environmental samples. Volumes of 700 μL of the sample were centrifuged for 2 minutes at 12000 ×g to concentrate cells. The supernatant was discarded and 700 μL of the lysis buffer was added to the pellet, mixed, and incubated for 10 minutes at 70°C. Volumes of 250 μL of 100% ethanol were added and incubated at 56°C for another 10 minutes. To prepare the spin column, 50 μL of celite was added, vortexed, and incubated at room temperature for 30 minutes with mixing every 30 s. The spin column was then placed in a clean 2 mL Eppendorf tube. A third of the solution (400 μL) was loaded into the spin column, centrifuged for 30 s at 12000 ×g before discarding the elute (this step was repeated until the column was fully loaded). Wash buffer (400 μl) was added, centrifuged for 30 s at 12000 ×g before discarding the elute (this step was repeated). Volumes of 400 μL of 70% (v/v) ethanol were then added to the column, centrifuged for 30 s at 12000 ×g before the elute was discarded (this step was also repeated). The column was dried by centrifuging at 12000 ×g for 2 minutes before being transferred to a clean 1.5 mL Eppendorf tube. TE (Tris and EDTA) buffer (100 μl) was then added to the column and incubated for approximately 2 minutes at 56°C. The column was discarded after the solution was centrifuged for 2 minutes 12000 ×g. The extracted DNA was then stored at −20°C and used for further applications. DNA was quantified using the Quanit-It HS assay kit (Invitrogen, SA) according to the manufacturer's instructions.

5.2.4.2 PCR ASSAYS

Polymerase chain reactions were performed in 50 μL reaction tubes containing specific primer sets as described elsewhere [2]. The primer set, Ami6F1 5'CCAGCTCCAATAGCGTATATT3' and Ami9R1 5'GTTGAGTCGAATTAAGCCGC3', was used to amplify the 18S rRNA gene. These primers yield a fragment of approximately 700 bp. The primers at a concentration of 10 μM each were transferred into a reaction tube containing 10 μL template DNA, 2 mM $MgCl_2$, 2.5 U of taq DNA polymerase (Life Technologies, SA), 100 μM each deoxynucleoside triphosphate and 8 μL ultrapure water (Fermentas, Canada), and 0.2 μL hotStar Taq Polymerase. A positive control was used in each experiment which comprised all the reagents mentioned above other than template DNA which was replaced with 10 μL of genomic DNA extracted from the reference strains of *Acanthamoeba* spp. A negative control was also used in each experiment which comprised all the reagents mentioned above other than template DNA which was replaced with 10 μL of PCR water. To detect *Acanthamoeba* sp., the reaction tubes were initially activated at 95°C for 15 minutes followed by 40 cycles of amplification using denaturation at 94°C for 45 s. Annealing was done at 57°C for 45 s and extension at 72°C for 1 minute followed by a final extension cycle at 72°C for 3 minutes. DNA was analyzed in a horizontal 1% (w/v) agarose slab gel (FP Agarose from Promega) with ethidium bromide (0.5 μg/mL) in a TAE (40 mM Tris acetate; 2 mM EDTA, pH 8.3) buffered system. 5 μL of 100 bp DNA marker (Fermentas O'GeneRuler DNA ladder, Canada) was loaded into the first well of the gel and into the remaining wells 10 μL each sample (including positive and negative controls) mixed with 3 μL of loading dye (Fermentas Orange × 6 Loading Dye, Canada).

5.2.5 STATISTICAL ANALYSIS

Statistical analysis was carried out to compare amoebae recovered using live or heat-killed *E. coli* at different temperatures (room temperature, 32°C, and 37°C) and concentration techniques (centrifugation or filtration). The Stata v11 statistical software was used and results were presented in a tabular format (STATA software, version 7.0; Stata Corporation, College Station, TX). Pearson chi-square test was used to test for association between categorical variables. The interpretation was performed at 95% confidence limit. All tests of significance and correlations were considered statistically significant at P values of <0.001.

TABLE 1: Densities of amoebae recovered using different food source, concentration methods and temperatures.

Condition	Amoebae recovered (no. of plates)		
	Low (+)	High (++)	Very High (+++)
Live *E. coli*	0	29	21
HK *E. coli*	1	29	20
Filtration	2	25	23
Centrifugation	31	19	0
RT	15	24	11
32°C	0	28	22
37°C	0	32	18

RT = room temperature.

5.3 RESULTS

5.3.1 SEEDED SAMPLES

Acanthamoeba was identified by both the polygonal shaped walls in the cyst form and the finger-like acanthapodia in the trophozoite form in all seeded samples (Figure 2). However, there were differences in the densities of amoebae recovered when using different concentration methods and incubation temperatures (Table 1). High densities of amoebae were observed when filtration was used as a concentration method, with only 2 plates classified as low, 25 as high, and 23 as very high. When centrifugation was used low densities of amoebae were observed with as many as 31 plates classified as low, 19 as high, and none as very high. Greater amoebae recovery densities were observed in samples incubated at 32°C and 37°C compared to those at room temperature ($P < 0.001$). There was, however, no significant difference between 32°C and 37°C with respect to amoebae densities, despite the fact that amoebae at 37°C encysted rapidly after 3 days compared to amoebae at 32°C which took one week to form cysts. No significant difference between live or heat-killed *E. coli* with respect to amoebae recovered was also observed. Consequently, we decided to use HK-*E. coli* for the following reasons: (i) to ensure that we do not observe actual "food" bacteria in the amoeba while setting up the method and (ii) also due to the high level of contamination in the wastewater samples, we felt that using HK-*E. coli* would be more appropriate to give a more reliable indication of intracellular bacteria in the samples.

5.3.2 ENVIRONMENTAL SAMPLES

5.3.2.1 PHYSICOCHEMICAL PARAMETERS

The mean water temperature of samples taken at the wastewater treatment plant was 18.3°C, 12.5°C, 20.6°C, and 25.3°C in autumn, winter, spring and summer, respectively (Table 2). The water temperature was significantly different ($P < 0.0001$) amongst the seasons when samples were collected, ranging from as low as 6.6°C in winter to as high as 27.7°C in summer. In contrast, pH was not seasonally dependent as the mean pH for autumn, winter, spring and summer was 7.19, 7.28, 7.20, and 7.18, respectively. However, among the different sampling points the pH varied from 6.33 to 8.13 (Table 2).

5.3.2.2 CONCENTRATION OF RESIDUAL CHLORINE IN THE ENVIRONMENTAL SAMPLES

Table 3 illustrates free chlorine residual concentrations at the different sampling points during the study period. Residual chlorine concentration ranged between 0.01 and 1.10 mg/L throughout the sampling period, with the final effluent having the highest concentration of 1.10 mg/L. The ranges for pond 1, pond 2, and river were below 0.1 mg/L. The mean chlorine residual concentration decreased from final effluent (0.37 mg/L) to maturation pond 3 (0.03 mg/L) where the treated water enters the river.

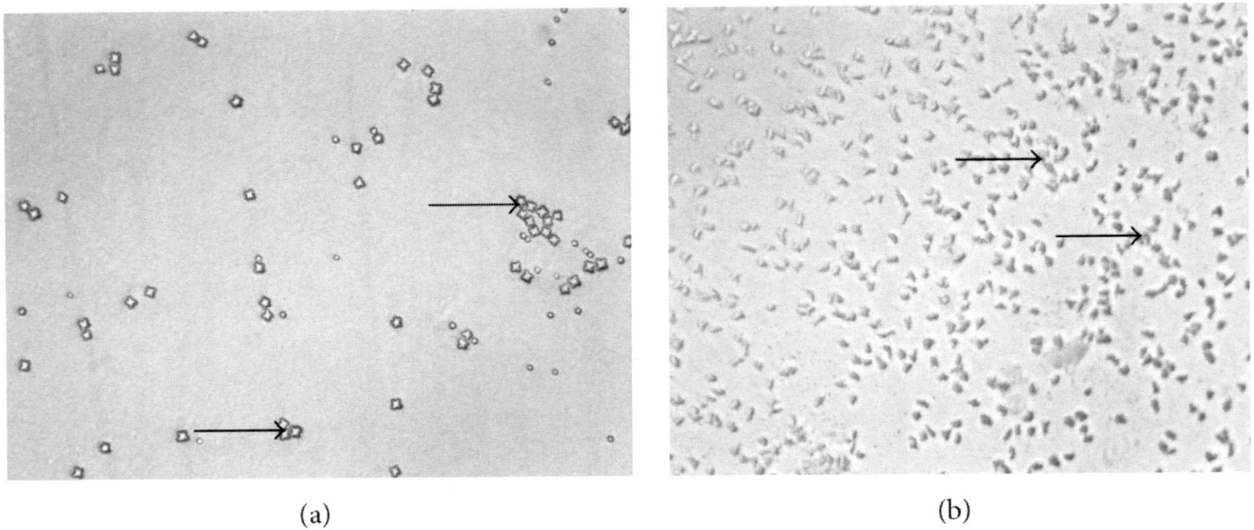

FIGURE 2: (a) Typical star shaped *Acanthamoeba* cysts (arrow) and (b) *Acanthamoeba* trophozoites (arrow) observed on HK-E. coli-NNA plates from all samples seeded with *A. castellanii* type strain, microscope at ×10 magnification (Leica, Germany).

TABLE 2: Water temperature and pH during sampling for different seasons.

Sampling season	Sample number	Water temperature (°C)		Water pH	
		Range	Mean	Range	Mean
Autumn	41	12.6–22.9	18.3[a]	6.33–7.90	7.19
Winter	43	6.6–18.2	12.5[b]	6.65–8.13	7.28
Spring	44	12.7–23.6	20.6[c]	6.67–7.76	7.20
Summer	44	21.8–27.7	25.3[d]	6.77–7.70	7.18
Probability (P)			<0.001[e]		NS

[a, b, c, and d]: mean values in the same column not sharing the same superscript are statistically significantly different $P < 0.0001$. [e]: values of $P < 0.001$. NS: not statistically significant difference.

TABLE 3: Concentrations of free residual chlorine at different sampling points.

Sample source	Free chlorine residual (mg/L)	
	Ranges	Means
Effluent	0.05–1.10	0.37
Maturation pond 1	0.02–0.07	0.04
Maturation pond 2	0.01–0.05	0.03
Maturation pond 3	0.01–0.05	0.03

5.3.2.3 ISOLATION OF FREE-LIVING AMOEBAE

Free-living amoebae identified on Giemsa stain were surrounded by numerous extracellular bacteria (Figure 3). *Acanthamoeba* sp. was identified by both the polygonal shaped walls in the cyst form and the finger-like acanthapodia in the trophozoite form in all the wastewater samples (Figure 4). A total of 150 (87.2%) samples were positive for free-living amoebae. Twenty-two (12.8%) samples were identified as *Acanthamoeba* sp. with the highest number recorded in autumn. Although FLA were isolated in 43 (100%) of the winter and spring samples, none of the samples had

Acanthamoeba sp. which were only isolated in autumn, 21 (51.2%), and summer, 1 (2.2%) (Table 4).

TABLE 4: Isolation of amoebae at different seasons in a wastewater treatment plant.

Season	Sample number	*Acanthamoeba* sp.	Other FLA
Autumn	41	21 (51.2%)	20 (48.8%)
Winter	43	—	43 (100%)
Spring	44	—	44 (100%)
Summer	44	1 (2.2%)	43 (97.8%)

FLA were isolated in all sampled stages of the wastewater treatment plant using the amoebal enrichment technique. From Figure 5, a total of 16 (9.3%) samples collected were positive for FLA from the influent, 20 (11.6%) from the bioreactor feed, 16 (9.3%) from the anaerobic zone, 16 (9.3%) from the anoxic zone, 32 (18.6%) from the aerators, 16 (9.3%) from the bioreactor effluent, 11 (6.4%) from the bioreactor final effluent, and 45 (26.2%) from the maturation pond. No *Acanthamoeba* sp. was isolated in the bioreactor feed (Figure 5). Intracellular small bacterial-like organisms were observed in amoebae isolated in 30 (17.4%) of the environmental samples. According to the samples that were positive for intracellular bacteria, 24 (80%) were from autumn with the rest of samples spreading evenly among the other seasons.

5.3.2.4 DETECTION OF FLA BY PCR

All samples detected morphologically as *Acanthamoeba* sp. were also positive using PCR. Only one of the eight samples not detected by morphology as *Acanthamoeba* sp. was positive using PCR. The primer set, Ami6F1 and Ami9R1, amplified a 700 bp (approximate) fragment for *Acanthamoeba* DNA. The positive control (*A. castellanii*) was amplified while the negative control (distilled water) was not amplified (Figure 6).

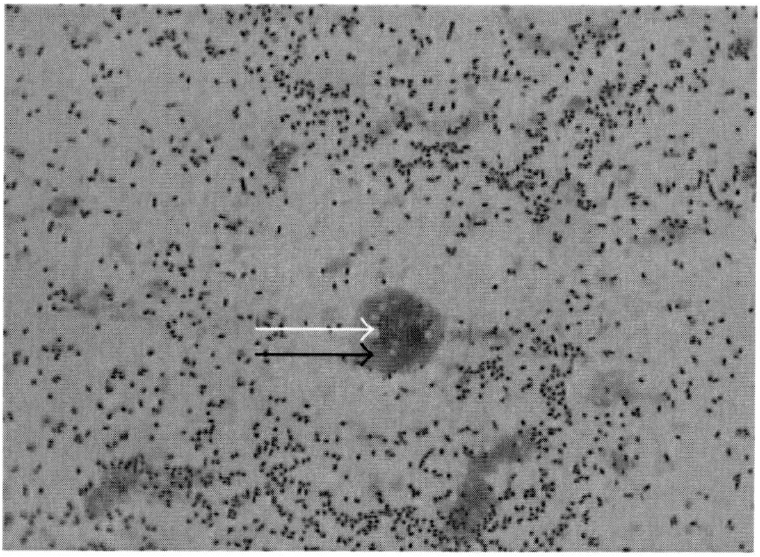

FIGURE 3: Amoebal trophozoite (black arrow) on Giemsa stain with round vacuoles (white arrow), ×100 (Olympus, Japan).

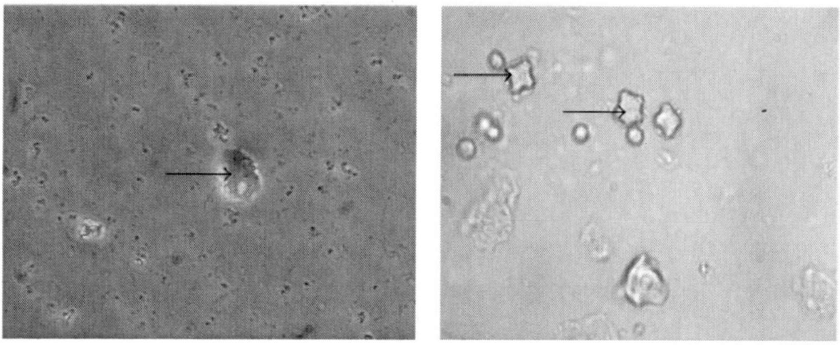

(a) (b)

Figure 4: (a) Typical *Acanthamoeba* trophozoites (arrow) and (b) Typical star shaped Acanthamoeba cysts (arrow) observed on HK-E. coli-NNA plates from environmental samples, phase contrast ×40 (Leica, Germany).

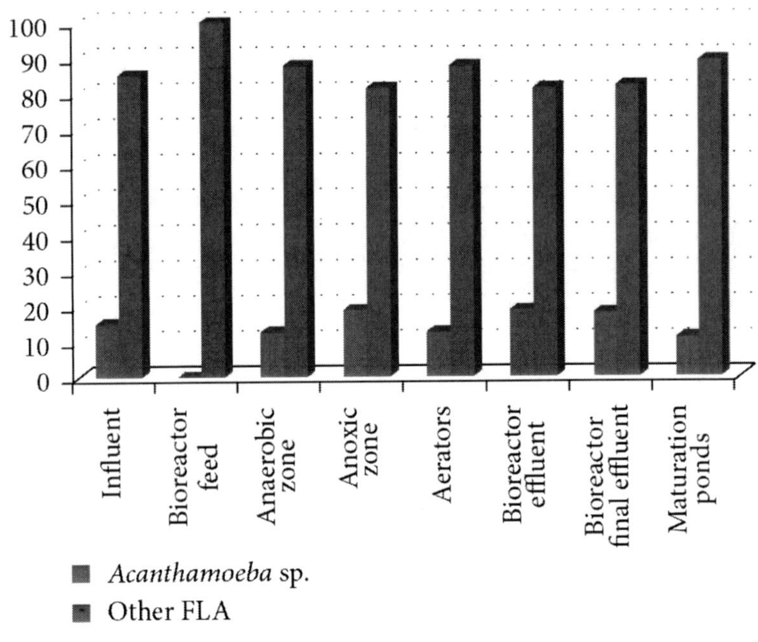

FIGURE 5: Percent of *Acanthamoeba* sp. and FLA isolated at different sampling points of a wastewater treatment plant in South Africa.

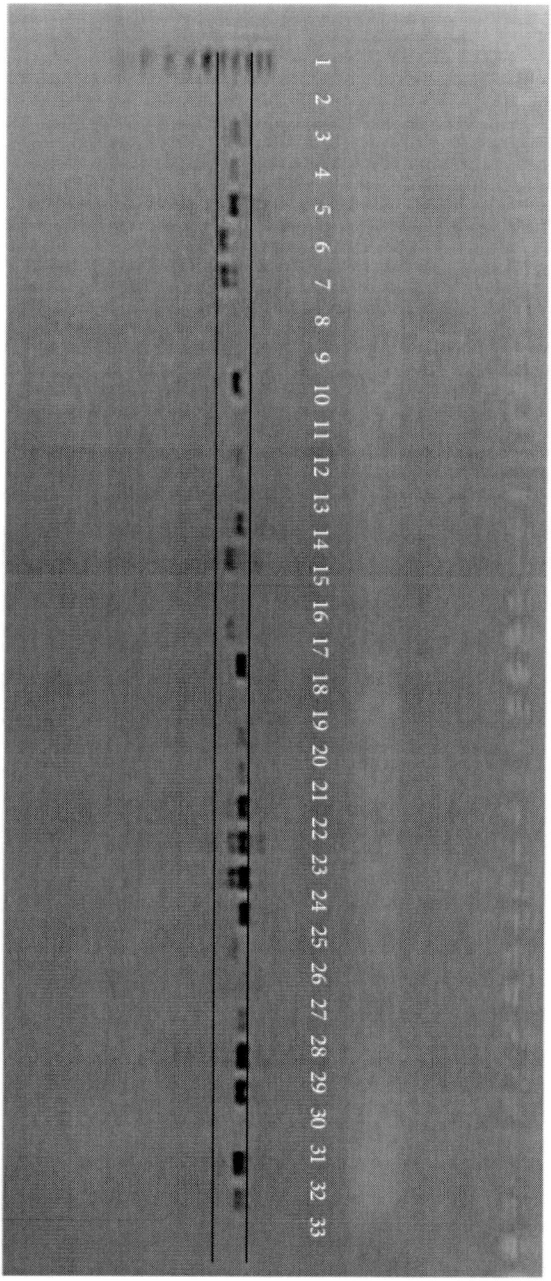

FIGURE 6: Gel picture showing PCR amplification of *Acanthamoeba* using primers Ami6F1 and Ami9R1, 1: 100 bp ladder (Fermentas), 2: negative control, and 3: positive control.

5.4 DISCUSSION

5.4.1 OPTIMIZATION OF CONDITIONS FOR AMOEBAL ENRICHMENT

The amoebal enrichment technique has been used to isolate FLA from environment using different conditions for concentration, temperature, and food source. Well-known concentration methods, membrane filtration and centrifugation, have been used to isolate and concentrate FLA in environmental water samples. In the present study, the efficiency of these concentration methods in recovering amoebae in seeded water samples was compared. Results showed that samples that were filtered significantly recovered more amoebae as compared to samples that used centrifugation as a concentration method. Our findings are similar to those reported by the Health Protection Agency [25] in which membrane filtration was found to be more efficient in recovering amoebae when compared to centrifugation. A study by [26] has also shown that amoebae can be recovered more easily when samples are processed by filtration rather than centrifugation. However, the concentration method used also depends to some extent on sample volumes. Studies that have used centrifugation have concentrated relatively small sample volumes of 50 mL compared to filtration which has been used to concentrate relatively high sample volumes of 500 mL and 1000 mL [23, 27]. Volumes of 500 mL used for concentration in our study might explain why filtration was more efficient in recovering amoebae compared to centrifugation. The amoebae yield by centrifugation could therefore be improved by dividing the sample into smaller sample portions. There is no consensus in studies required to obtain a high recovery of amoebae on the speed and the time used when centrifugation is used. A report by the Health Protection Agency [25] recommends that water samples be centrifuged at 750 g for 20 minutes. However, studies by [28, 29] have used 120 g for 10 minutes and 1200 g for 15 minutes, respectively, to concentrate amoebae from environmental samples.

Temperature is another important factor that influences the growth of amoebae. For example, a study by Khan [30] has showed that nine strains of *Acanthamoeba* sp. grew at temperatures ranging from 10 to 37°C, with the pathogenic varieties surviving at higher temperatures (>37°C). This study indicates a relatively wide temperature range tolerated by amoebae. Environmental *Acanthamoeba* isolates of a study in Slovakia have also been shown to grow at 23°C, 32°C, and 37°C [31]. These findings agree with our study in which growth of amoebae was established at room temperature (22°C–25°C), 32°C, and 37°C. However we managed to recover more amoebae at 32°C and 37°C compared to room temperature which had a low recovery of amoebae. Although *A. castellanii* formed cysts after 3 days of incubation at 37°C, its ability to grow at this temperature showed its pathogenic potential for humans. With these findings, this study showed that amoebae can be grown at 32°C and/or 37°C giving important indications of the pathogenic potential of these organisms.

In general, FLA and *Acanthamoeba* sp. can be readily cultivated on nonnutrient agarcontaining a lawn of killed or living Gram negative bacteria. In the present study, heat-killed (HK) *E. coli* and living *E. coli* were compared as food sources for amoebal growth. There was no significant difference in the amoebaerecovered when alive or HK-E. coli were used as food sources. In contrast, a study done by Pickup et al. [32] has shown growth rates of *A. castellanii* to be higher on living cells of *E. coli*, *P. aeruginosa, K. aerogenes*, and *S. aureus* compared with those on the heat-killed bacterial cells. The effectiveness of live and heat-killed bacterial suspensions on the growth of *Acanthamoeba* has also been shown in another study by Selvam et al. [33]. In their study, live *P. aeruginosa*, *E. coli*, and *Bacillus* sp. yielded a higher total number of *Acanthamoeba* compared to heat-killed *P. aeruginosa*, *E. coli*, and *Bacillus* species.

5.4.2 FLA IN WASTEWATER

The occurrence of FLA such as *Acanthamoeba* and *Naegleria* species has been previously reported in manmade sources like cooling towers, swimming pools, hospital water networks, and drinking water plants as well as in natural sources like rivers and lakes. However, few studies have reported on the occurrence of FLA in wastewater treatment plants and particularly in sewage water. Bose et al. [22] did a characterization study of potentially pathogenic FLA in sewage samples which resulted in the isolation of a pathogenic strain of *A. castellanii* and a nonpathogenic strain of *A. astronyxis*. Another study done by Ramirez et al. [23] isolated thirteen species of FLA (pathogenic and nonpathogenic) that included three species of *Acanthamoeba*: *A. castellanii*, *A. culbertsoni,* and *A. polyphaga*, from an activated sludge plant. A more recent study by García et al. [24] characterized potentially pathogenic *Acanthamoeba*, *Hartmannella*, and *Naegleria* from sewage effluents of Spanish wastewater treatment plants despite disinfection with chlorine. In the present study, FLA and *Acanthamoeba* sp. were isolated at different stages of a wastewater treatment plant (including sewage effluents), showing consistence with previous studies. Using the culture-based method, amoebal enrichment, this study focused on *Acanthamoeba* sp. which was further confirmed by PCR. It is well known that PCR methods are more rapid and capable of detecting even nonculturable cells and allow genus discrimination of the isolates [34]. The sensitivity of the molecular analysis is supported in our study by one positive sample for *Acanthamoeba* sp. out of the eight analyzed which could not be detected by amoebal enrichment. However, amoebal enrichment makes the organism available for further classification and allows testing of infectivity in human macrophages and testing antibiotic susceptibility [15].

All samples collected in the present study were positive for FLA in all the four seasons. However, *Acanthamoeba* sp. was only isolated during the autumn and summer in our study compared to a study by Ettinger et al. [35] which isolated *Acanthamoeba* sp. in spring and summer. This shows that seasonal changes may affect the prevalence of FLA in environmental water as the temperature fluctuates. Isolation of amoebae from chlorinated samples in our study shows that some amoebal cells can survive the wastewater treatment process even after chlorination, resulting in the discharge of amoebae to receiving water bodies such as rivers. The survival of amoebae to chlorination is because of amoebal cyst walls containing cellulose that forms a physical barrier against chlorine [18]. The chlorine residual concentration in this study ranged between 0.01 and 1.10 mg/L in the final effluent which fell well outside the limit of free chlorine residual applicable in South Africa. The residual chlorine range of discharged wastewater into a water resource such as a river should vary from 0.3 mg/L to 0.6 mg/L [36]. However, a study done by Storey et al. [11] showed that amoebal cysts can survive chlorine concentrations as high as 100 mg/L for 10 minutes. Therefore, the currently applied effluent concentrations of free chlorine residual may not result in the inactivation of amoebal cysts. In addition, Thomas et al. [12] also demonstrated that FLA, including *Acanthamoeba* sp., *Hartmannella* sp., and *Vahlkampfia* sp., can resist treatment with ozone, chlorine dioxide, monochloramine, copper-silver, and chlorine.

The resistance of FLA to biocidal treatments such as chlorine has major implications for disease transmission as some FLA species can potential infections of the central nervous system, skin and eye. In addition to their role as pathogens, FLA are known to serve as natural hosts and vectors of various pathogenic intracellular bacteria [15]. In the present study, typical amoebal trophozoites and cysts containing live bacteria were observed in 30 (17.4%) of the environmental samples suggesting the intracellular existence of these amoeba resistant bacteria. The number of samples positive for intracellular bacteria in this study might also be underestimated because of the presence of other organisms in the environmental samples which may have skewed our results. However, what is important is that even if the number of positive samples was underestimated, 17.4% were still a relatively high number of samples to contain potential pathogenic intracellular bacteria. This in turn reduces the microbiological quality of the receiving water body (river in this case) as pathogenic microorganisms are released from the water treatment plant. This also increases the risk to the health of communities living in the vicinity of the river that uses the water for multiple purposes which include drinking, agricultural, and recreational purposes [37].

5.5 CONCLUSION

In this investigation, the amoebal enrichment technique was successfully optimized with seeded samples using filtration as a concentrating method, HK-*E. coli* as a food source for amoebae, and 32°C and/or 37°C as the incubation temperature. Using these optimized amoebal enrichment conditions, FLA (with some harboring potential pathogenic bacteria) were detected at all stages of the wastewater treatment plant. *Acanthamoeba* sp. was only detected in summer and autumn, showing that their prevalence is temperature dependent. The presence of amoebae in 87.2% of the environmental samples in this study shows that the current wastewater treatment process is not adequate for the removal and disinfection of amoebae using chlorine. Future studies should focus on the detection of FLA in wastewater effluents using both culture and molecular analysis to identify potentially pathogenic FLA entering receiving water bodies where humans could be exposed.

REFERENCES

1. S. Rodriguez-Zaragoza, "Ecology of free-living amoebae," Critical Reviews in Microbiology, vol. 20, no. 3, pp. 225–241, 1994.
2. V. Thomas, K. Herrera-Rimann, D. S. Blanc, and G. Greub, "Biodiversity of amoebae and amoeba-resisting bacteria in a hospital water network," Applied and Environmental Microbiology, vol. 72, no. 4, pp. 2428–2438, 2006.
3. D. Corsaro, V. Feroldi, G. Saucedo, F. Ribas, J.-F. Loret, and G. Greub, "Novel Chlamydiales strains isolated from a water treatment plant," Environmental Microbiology, vol. 11, no. 1, pp. 188–200, 2009.
4. A. Garcia, P. Goñi, J. Cieloszyk et al., "Identification of free-living amoebae and amoeba-associated bacteria from reservoirs and water treatment plants by molecular techniques," Environmental Science and Technology, vol. 47, no. 7, pp. 3132–3140, 2013.
5. F. L. Schuster and G. S. Visvesvara, "Free-living amoebae as opportunistic and non-opportunistic pathogens of humans and animals," International Journal for Parasitology, vol. 34, no. 9, pp. 1001–1027, 2004.
6. G. S. Visvesvara, H. Moura, and F. L. Schuster, "Pathogenic and opportunistic free-living amoebae: Acanthamoeba spp., Balamuthia mandrillaris, Naegleria fowleri, and Sappinia diploidea," FEMS Immunology and Medical Microbiology, vol. 50, no. 1, pp. 1–26, 2007.
7. B. da Rocha-Azevedo, H. Tanowitz, and F. Marciano-cabral, "Diagnosis of infections caused by pathogenic free living amoebae," Interdisciplinary Perspectives on Infectious Diseases, vol. 2009, Article ID 251406, p. 10, 2009.
8. N. A. Khan, "Acanthamoeba: biology and increasing importance in human health," FEMS Microbiology Reviews, vol. 30, no. 4, pp. 564–595, 2006.
9. F. A. Yousuf, R. Siddiqui, and N. A. Khan, "Acanthamoeba castellanii of the T4 genotype is a potential environmental host for Enterobacter aerogenes and Aeromonas hydrophila," Parasites and Vectors, vol. 6, no. 1, article 169, 2013.
10. F. Marciano-Cabral and G. Cabral, "Acanthamoeba spp. as agents of disease in humans," Clinical Microbiology Reviews, vol. 16, no. 2, pp. 273–307, 2003.
11. M. V. Storey, J. Winiecka-Krusnell, N. J. Ashbolt, and T.-A. Stenström, "The efficacy of heat and chlorine treatment against thermotolerant Acanthamoebae and Legionellae," Scandinavian Journal of Infectious Diseases, vol. 36, no. 9, pp. 656–662, 2004.

12. V. Thomas, T. Bouchez, V. Nicolas, S. Robert, J. F. Loret, and Y. Lévi, "Amoebae in domestic water systems: resistance to disinfection treatments and implication in Legionella persistence," Journal of Applied Microbiology, vol. 97, no. 5, pp. 950–963, 2004.
13. J.-F. Loret, M. Jousset, S. Robert et al., "Elimination of free-living amoebae by drinking water treatment processes," European Journal of Water Quality, vol. 39, no. 1, pp. 37–50, 2008.
14. P. Goñi, M. T. Fernández, and E. Rubio, "Identifying endosymbiont bacteria associated with free-living amoebae," Environmental Microbiology, vol. 16, no. 2, pp. 339–349, 2014.
15. G. Greub and D. Raoult, "Microorganisms resistant to free living amoebae," Clinical Microbiology Reviews, vol. 17, no. 2, pp. 413–433, 2004.
16. H. Abd, A. Saeed, A. Weintraub, G. B. Nair, and G. Sandström, "Vibrio cholerae O1 strains are facultative intracellular bacteria, able to survive and multiply symbiotically inside the aquatic free-living amoeba Acanthamoeba castellanii," FEMS Microbiology Ecology, vol. 60, no. 1, pp. 33–39, 2007.
17. R. Lone, K. Syed, R. Abdul, S. A. Sheikh, and F. Shah, "Unusual case of methicillin resistant Staphylococcus aureus and Acanthamoeba keratitis in a non-contact lens wearer from Kashmir, India," BMJ Case Reports, 2009.
18. V. Thomas, G. McDonnell, S. P. Denyer, and J.-Y. Maillard, "Free-living amoebae and their intracellular pathogenic microorganisms: risks for water quality," FEMS Microbiology Reviews, vol. 34, no. 3, pp. 231–259, 2010.
19. F. L. Schuster, "Cultivation of pathogenic and opportunistic free-living amebas," Clinical Microbiology Reviews, vol. 15, no. 3, pp. 342–354, 2002.
20. J.-F. Loret and G. Greub, "Free-living amoebae: biological by-passes in water treatment," International Journal of Hygiene and Environmental Health, vol. 213, no. 3, pp. 167–175, 2010.
21. F. Lamoth and G. Greub, "Amoebal pathogens as emerging causal agents of pneumonia," FEMS Microbiology Reviews, vol. 34, no. 3, pp. 260–280, 2010.
22. K. Bose, D. K. Ghosh, K. N. Ghosh, A. Bhattacharya, and S. R. Das, "Characterization of potentially pathogenic free-living amoebae in sewage samples of Calcutta, India," Brazilian Journal of Medical and Biological Research, vol. 23, no. 12, pp. 1271–1278, 1990.
23. E. Ramirez, A. Warren, F. Rivera et al., "An investigation of the pathogenic and non-pathogenic free-living amoebae in an activated-sludge plant," Water, Air, and Soil Pollution, vol. 69, no. 1-2, pp. 135–139, 1993.
24. A. García, P. Goñi, A. Clavel, S. Lobez, M. T. Fernandez, and M. P. Ormad, "Potentially pathogenic free-living amoebae (FLA) isolated in Spanish wastewater treatment plants," Environmental Microbiology Reports, vol. 3, no. 5, pp. 622–626, 2011.
25. Health Protection Agency, "Operating Procedure: isolation and identification of Acanthamoeba species," Reference no W 17 i2.2, Standards Unit, Evaluations and Standards Laboratory on behalf of the Water Working Group and the Environmental Surveillance Unit, CDSC, 2003.
26. J. Winiecka-Krusnell and E. Linder, "Acanthamoeba keratitis: increased sensitivity of the detection of parasites by modified cultivation procedure," Scandinavian Journal of Infectious Diseases, vol. 30, no. 6, pp. 639–641, 1998.

27. I. Pagnier, D. Raoult, and B. La Scola, "Isolation and identification of amoeba-resisting bacteria from water in human environment by using an Acanthamoeba polyphaga co-culture procedure," Environmental Microbiology, vol. 10, no. 5, pp. 1135–1144, 2008.
28. D. T. John and M. J. Howard, "Techniques for isolating thermotolerant and pathogenic freeliving amebae," Folia Parasitologica, vol. 43, no. 4, pp. 267–271, 1996.
29. E. Ramirez, E. Robles, P. Bonilla et al., "Occurrence of pathogenic free-living amoebae and bacterial indicators in a constructed wetland treating domestic wastewater from a single household," Engineering in Life Sciences, vol. 5, no. 3, pp. 253–258, 2005.
30. N. A. Khan, "Pathogenicity, morphology, and differentiation of Acanthamoeba," Current Microbiology, vol. 43, no. 6, pp. 391–395, 2001.
31. V. Nagyová, A. Nagy, Š. Janeček, and J. Timko, "Morphological, physiological, molecular and phylogenetic characterization of new environmental isolates of Acanthamoeba spp. from the region of Bratislava, Slovakia," Biologia, vol. 65, no. 1, pp. 81–91, 2010.
32. Z. L. Pickup, R. Pickup, and J. D. Parry, "Growth of Acanthamoeba castellanii and Hartmannella vermiformis on live, heat-killed and DTAF-stained bacterial prey," FEMS Microbiology Ecology, vol. 61, no. 2, pp. 264–272, 2007.
33. K. P. Selvam, C. S. Shobana, K. Ravikumari, P. Manikandan, R. Rajendran, and K. Rajaduraipand, "Evaluation of various axenic and monoxenic media on the cultivation of Acanthamoeba," International Journal of Biotechnology and Biochemistry, vol. 6, no. 1, pp. 1075–1082, 2011.
34. D. Rivière, F. M. Szczebara, J.-M. Berjeaud, J. Frère, and Y. Héchard, "Development of a real-time PCR assay for quantification of Acanthamoeba trophozoites and cysts," Journal of Microbiological Methods, vol. 64, no. 1, pp. 78–83, 2006.
35. M. R. Ettinger, S. R. Webb, S. A. Harris, S. P. McIninch, G. C. Garman, and B. L. Brown, "Distribution of free-living amoebae in James River, Virginia, USA," Parasitology Research, vol. 89, no. 1, pp. 6–15, 2003.
36. C. L. Obi, J. O. Igumbor, M. N. B. Momba, and A. Samie, "Interplay of factors involving chlorine dose, turbidity flow capacity and pH on microbial quality of drinking water in small water treatment plants," Water SA, vol. 34, no. 5, pp. 565–572, 2008.
37. S. Toze, "Reuse of effluent water-benefits and risks, new directions for a diverse planet," in Proceedings of the 4th International Crop Science Congress, Brisbane, Australia, September-October 2004.

CHAPTER 6

Water and Wastewater Management and Biomass to Energy Conversion in a Meat Processing Plant in Brazil: A Case Study

HUMBERTO J. JOSÉ, REGINA F. P. M. MOREIRA, DANIELLE B. LUIZ, ELAINE VIRMOND, AZIZA K. GENENA, SILVIA L. F. ANDERSEN, RENNIO F. DE SENA, AND HORST F. R. SCHRÖDER

6.1 INTRODUCTION

A commitment to sustainability and an understanding of the concepts of "cleaner production" are current requirements for achieving environmentally-friendly industrial practices. Such concepts promote the minimization of fresh water consumption, a reduction in wastewater production and the recycling of wastes. Hence, in a world where water scarcity and climate change are a reality, actions to protect fresh water resources and enhance renewable energy capacity are mandatory for any type and size

José HJ, Moreira RFPM, Luiz DB, Virmond E, Genena AK, Andersen SLF, de Sena RF, and Schröder HFR (2013). Water and Wastewater Management and Biomass to Energy Conversion in a Meat Processing Plant in Brazil – A Case Study. From Food Industry, *Dr. Innocenzo Muzzalupo (Ed.), ISBN: 978-953-51-0911-2, InTech, DOI: 10.5772/53163.*

of industry. With reference to solid wastes, social and environmental responsibility goes beyond the obligations determined by law and relies on substantial technical research to establish a strict environmental management policy.

Meat processing plants worldwide use approximately 62 Mm^3 per year of water. Only a small amount of this quantity becomes a component of the final product. The remaining part becomes wastewater with high biological and chemical oxygen demands, high fat content and high concentrations of dry residue, sedimentary and total suspended matter as well as nitrogen and chloride compounds (Sroka et al., 2004). Of the components usually found in these effluents, blood can be considered as the most problematic due to its capacity to inhibit floc formation during physicochemical wastewater treatment and its high biochemical (BOD_5, biochemical oxygen demand during decomposition over a 5-day period) and chemical oxygen demand (COD). In fact, even with correct handling during meat processing, this activity generates 2.0 and 0.5 liters of blood as effluent for each bovine animal and pig, respectively (Tritt & Schuchardt, 1992). The treatment of both the solid wastes and the wastewater from the meat processing industry represents one of the greatest concerns associated with the agro-industrial sector globally, mainly due to the restrictions that international trade regulations have imposed over their use and the related environmental issues.

In order to meet this challenge, one of the largest meat processing companies in Brazilian initiated a series of investments in scientific research to improve its environmental performance. Biomass as an energy source, air pollution control, and water and wastewater management were the main issues addressed in research projects carried out from 2003 to 2010.

The Brazilian agro-industrial sector consumes large amounts of fresh water and produces large amounts of residues and by-products, which can potentially be used as energy sources. The Brazilian legislation itself admits the need for water management in industrial plants to implement cleaner production techniques, which include the conscious uses of water. There are several legal documents that promote the recognition of water as public property and a finite resource with economic value. These legal norms and legislation are gathered in a single official document called "Set of legal regulations: water resources" (Brazil, 2011) and promote:

(1) the rationalization of water use and its conservation, reconditioning and sustainable management; (2) investment in pollution control, reuse, protection and conservation as well as the use of clean technologies to protect water resources; (3) the practice of water reuse to reduce discharges of pollutants into receiving waters, conserving water resources for public supply and other uses which demand high quality water; (4) the practice of water reuse to reduce the costs associated with pollution, contributing to the protection of the environment and public health; and (5) the creation of guidelines to regulate and encourage the practice of direct reuse of non-potable water. Official Brazilian reports highlight that the costs of water treatment have been raised by the contamination of water resources and water shortages (aspects of quality and quantity) in certain regions of the country. Consequently, they emphasize that high quality water should not be used in activities that tolerate water of lower quality (Brazil, 2011).

Regarding the solid waste materials generated in agro-industries, these are commonly generated during the processing of crops, but are also produced by all sectors of the food industry including everything from meat production to confectionery, such as peelings and scraps from fruit and vegetables, food that does not meet quality control standards, pulp and fiber from sugar and starch extraction, sludge from physicochemical and biological wastewater treatment and filter sludge. The co-digestion of energy crops and a variety of residual biomasses may be a good integrated solution for energy recovery from such waste materials, particularly with wastes that are unsuitable for direct disposal on land, as proposed by Schievano et al. (2009). These authors evaluated the suitability and the costs associated with many substitutes for energy crops in biogas production such as: swine manure, municipal solid waste, olive oil sludge, glycerine from biodiesel production and other agro-industrial by-products and residues. They concluded that farms could implant biogas plants to treat their own biomass generated and other urban and agro-industrial organic wastes, providing power for the neighborhood and improvements in the agrarian economy.

The use of the biosolids originating from the physicochemical treatment of meat processing wastewater can reduce the costs associated with its disposal (which has been prohibited in many locations by strict regulations) as it can directly and significantly reduce the mass and volume of such wastes, allowing energy recovery and generally lower toxic gas emissions when compared to fossil fuels. As long as emissions are below the specified legislative limits, changing energy policies lend support to the use of this type of biomass as a fuel source, as part of a move towards achieving low carbon economies.

The EIA Annual Energy Outlook 2011 reported that the global marketed energy consumption is expected to rise by nearly 50 percent from 2009 through 2035 (US EIA, 2011). Unless the world energy matrix is altered, fossil fuels will account for 90% of this increase.

The requirement to reduce carbon dioxide emissions has sparked interest in the use of many types of biomass as alternative energy sources. Since biomass is produced by the photosynthetic reduction of carbon dioxide, its utilization as biofuel can essentially be carbon neutral with respect to the build-up of atmospheric greenhouse gases, increasing both the demand for the characterization of alternative fuels and encouraging the proliferation of scientific papers concerned with this subject (Demirbas, 2004, 2005; de Sena et al., 2008, 2009; Floriani et al., 2010; Obernberger et al., 2006; Virmond et al., 2010, 2011 2012a, 2012b; Werther et al., 2000).

Brazil is currently implementing advanced programs aimed at the use of biomass energy, and several experimental and commercial projects are being implemented, such as those presented by Lora and Andrade (2009), to provide important information in order to overcome the technical and commercial barriers which inhibit the extensive implementation of bioenergy. The solid wastes produced by the meat industry have been applied mostly to the production of animal feed, which include the slaughter wastes and the wastewater treatment solids as main ingredients (Johns, 1995; Tritt and Schuchardt, 1992). However, diseases such as BSE (Bovine Spongiform Encephalopathy) have led to restrictions over the use of these wastes for feed production.

The first actions taken by the case study meat processing company, between the years 2003 and 2004, as shown by de Sena et al. (2008), were related to the in-depth investigation of the physicochemical treatment carried out at the wastewater plant with regard to its solids removal, mainly to achieve an increase in the chlorine-free biomass obtainment with a view to its utilization as a biomass fuel for steam generation. The data obtained indicate that the raw wastewater has a high organic load comprised basically of blood and organic materials that cause the red color, the greater part of the turbidity, the high concentration of total solids, oils and greases, the BOD_5 and the COD. The combustion of these wastes, especially the sludge from the wastewater treatment plants, might be a nobler utilization for economic reasons, however, many parameters related to the combustion must be monitored due to the formation of pollutants such as polychlorinated dibenzodioxins (PCDD), polychlorinated dibenzofurans (PCDF), volatile organic compounds (VOCs), NOx, and SO_2. The authors showed that the physicochemical treatment carried out at the meat processing wastewater plant provides around 20% of sludge, an organic solid residue, using the chlorine-free coagulant ferric sulfate (instead of aluminum or ferric chloride). In order to avoid discharge and subsequent environmental problems, the authors performed a preliminary combustion test with a mixture of biosolids and sawdust in a mass ratio of 4:1. The results suggested that the use of the biosolids as an alternative energy source would offer a favorable solution, reducing disposal and processing costs, as well as avoiding environmental and health problems for staff and the community close to these processing plants, thus establishing a cheaper and cleaner energy source for the meat industry segment (de Sena et al., 2008).

Another point related to sustainability in the meat processing industry is the associated farm residues, like pig and chicken manure. A reasonable solution for these wastes is their anaerobic digestion to produce biogas and/or fertilizers. Many farms in Brazil are implementing biodigestors in order to obtain biogas to produce electrical or thermal energy. Boersma et al. (1981), for example, studied the energy recovery from biogas produced from pig waste and verified an economy of 86%, showing a very good potential for this kind of solution. Also, carbon credits could be sold when the biodigestion process together with the energy recovery are applied to the farms.

Biogas is composed basically of methane, carbon dioxide, hydrogen sulfide, and other components in lower concentration. The gas production and the proportion of each compound are dependent on the biodigestor parameters and the chemical composition of the substrates (Lucas Jr., 1994).

A typical composition of a biogas is 55-65% of CH_4, 35-45% of CO_2 and a low concentration of H_2S. The presence of H_2S can cause corrosion problems when using the biogas as a fuel and also, when it is emitted to the atmosphere, its greenhouse potential is 21 times higher than that of CO_2. A high concentration of methane is desirable, as its presence increases the calorific value of the gas, making it more attractive for energy production.

In this context, this chapter was designed to highlight complementary research projects that have been carried out between 2005 and 2010 to implement actions to reduce the fresh water consumption, promoting water recycling and reuse, and to further investigate the application of biomass residues as energy sources and gaseous emissions in combustion processes.

6.2 CASE STUDY: MEAT PROCESSING PLANT

The industrial plant which formed the basis of this case study is located in the west of Santa Catarina State (southern Brazil), where water pollution and overexploitation, the uneven distribution of rainfall over the seasons and long periods of drought, especially in summer, have become a significant problem. The activities of the meat processing plant of this case study include the slaughtering and processing of poultry and swine, while the poultry hatchery plant includes all activities involved in poultry growth: breeding, hatching, rearing, food production and waste handling.

The meat processing plant has its own drinking water treatment plant (DWTP) and wastewater treatment plant (WWTP). The major water resource of this unit is from a river called Rio do Peixe. Its DWTP produces around 8,600 $m^3\ d^{-1}$ of drinking water, and the WWTP treats around 7,900 $m^3\ d^{-1}$ of wastewater. As described by de Sena et al. (2008), after the flotation process with a continuous capacity of 350 $m^3\ h^{-1}$, the treated effluent undergoes a biological treatment, while the biosolids are transported by pumps to a three-phase centrifugal system, which separates oil, water and solid parts (biomass). Afterwards, the biomass is dried in an industrial rotating granulator drier with an operating capacity of 400 kg h^{-1} (model Bruthus, Albrecht, Brazil), where the moisture content was reduced from approximately from 80 wt% to 10-20 wt% in order to make the burning process feasible.

Figure 1 shows the wastewater treatment process of the case study meat processing plant.

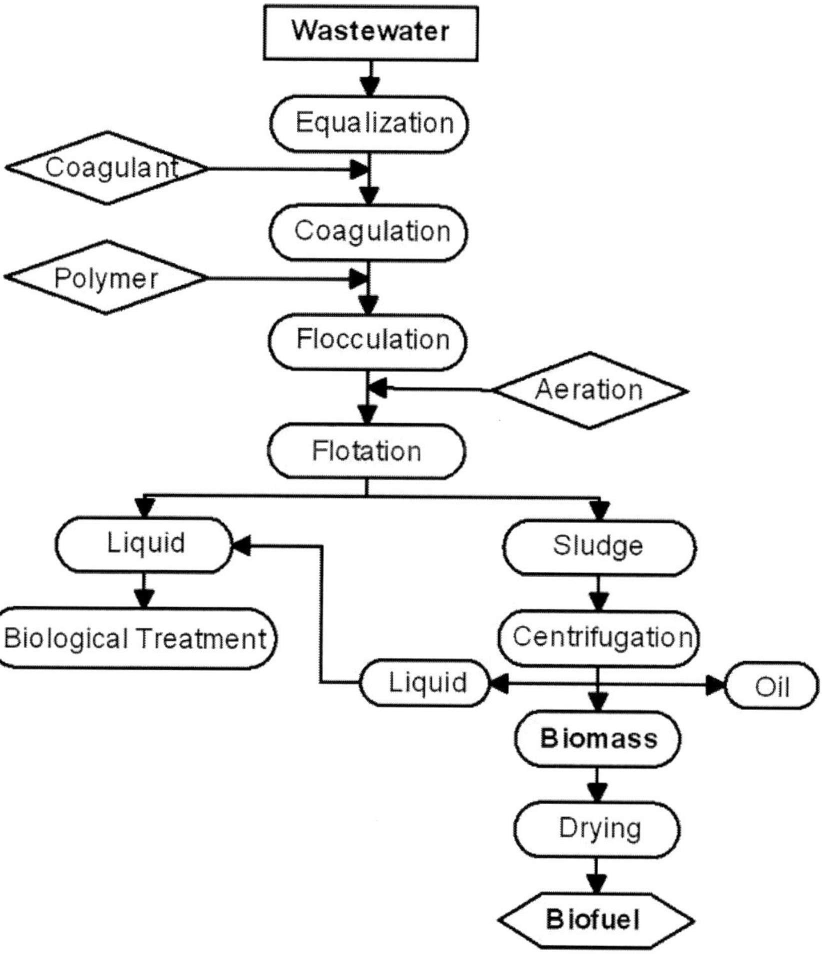

FIGURE 1: Processes involved in the wastewater treatment plant and obtainment of biofuel (de Sena et al., 2008)

The wastewater treatment plants of meat processing units in Brazil usually undergo the same type of treatment process, where a flotation system is the most commonly used solid-liquid separation step, due to the natural characteristics of these effluents, which possess high oil and grease contents. To increase the flotation performance the use of coagulants and coagulation aids are mandatory. Dissolved air flotation (DAF) has become an attractive separation process because of its well-known higher efficiency in terms of organic matter abatement, although the increase in costs associated with the production of micro-bubbles and system maintenance must be considered. Flotation processes are preferred in relation to sedimentation considering their faster solids separation, the lower moisture content of the sludge produced and the smaller area requirements. Coagulants themselves are very efficient for floc formation during the coagulation process, but the use of coagulation aids (e.g., anionic polyacrylamide polymers) after the rapid mixing of the coagulant-wastewater to disperse the coagulant, have been shown to increase of floc size and to provide higher floc stability and high solid separation rates. Since the coagulation and flocculation (addition of coagulation aids during gentle dispersion) are successive steps applied to neutralize the suspended particles and achieve strong flocs, the addition of these reagents must be carefully and precisely controlled to enhance solid-liquid separation. If some variables related to the process efficiency are not properly controlled, such as pH, reagent concentrations, mixing speed and contact time, the whole process will be unsuccessful. During the increase in floc size air bubbles of different diameters are incorporated into the flocs and this is responsible for the flotation phenomenon. Flotation efficiencies may vary from 60 to 95% of organic matter removal, according to the technology applied.

When the flotation of the solids is complete, the froth on the surface is separated from the water and skimmed off. It is collected in chambers and is pumped to a three-phase centrifuge where another polymer, a cationic polyacrylamide, is added to improve the oil-water-sludge separation. The water undergoes biological treatment and the oil is collected and sold as a raw material for the soap and detergent manufacturing industries. The remaining solid fraction is the sludge which was formerly used as an ingredient for animal food and feeds, especially the pet segment. However, due to the above-mentioned restrictions regarding its use in feeds, there are currently two other available options for the correct discharge of this so-called waste: combustion/incineration or land disposal. The combustion of the sludge for steam generation was the option chosen in this case study due to both economic and environmental aspects, since the use of an existing waste as part of the fuel content will decrease the fuel costs for internal energy supply, and the amount of sludge added to the fuel used (wood chips) could be properly controlled with regard to the gaseous emissions. On the other hand, land disposal might bring extra costs associated with transportation and long-term storage. All of the results obtained, as well as their pros and cons, are discussed in detail in the following sections.

The importance of the Brazilian poultry industry can be verified by its strong presence in the rural regions, mainly in the southern and south-eastern states. In many cities, poultry production is the main economic activity. The poultry hatchery unit of this case study, as in the case of the meat processing unit, also has its own WWTP. The wastewater originated from the processes of this unit is characterized by a high organic content, with the presence of nutrients such as nitrogen and phosphorus, as well as persistent organic compounds such as the residues of sanitizing products (e.g. pesticides) and veterinary drugs (Genena, 2009). The treatment system for the poultry hatchery wastewater comprises a screening stage (primary treatment), followed by equalization and finally biological treatment (secondary treatment: stabilization ponds). The treated wastewater is then discharged into a river (surface water).

6.3 WATER AND WASTEWATER MANAGEMENT

The water and wastewater management (W2M) proposed for the pilot plant aimed to minimize the water consumption and evaluate the possibilities for water and wastewater reuse in the food industry. The W2M, described in a previous publication (Luiz et al., 2012a), proposed strategies for water management in slaughterhouses considering the restrictions imposed by Brazilian legislation and hygiene concerns particular to the food industry. The objective was to present alternatives for the minimization of water consumption and wastewater production.

The proposed W2M is a practical model of industrial water management, which consists of seven stages:

1. Collection and analysis of documents;
2. Measurement of water consumption and wastewater production (water balance);
3. Verification of the points of greatest water consumption;
4. Minimization of water consumption with emphasis on the points of greatest water consumption;
5. Evaluation of the potential for water reuse and recycling without reconditioning;
6. Evaluation of the potential for water reuse and recycling after reconditioning; and
7. Maintenance of water management.

The points identified as being associated with major water consumption were: (1) pre-cooling of giblets, (2) washing of poultry carcass before pre-chilling, (3) transportation of giblets, poultry necks and feet, and (4) washing of swine carcass after buckling. The potential for reducing the fresh water consumption in-line with the current Brazilian legislation in these four process steps was approximately 806 m3 d-1 (Luiz et al., 2012a).

After the minimization of water use, the most important action is the evaluation of direct recycling and reuse of wastewater without reconditioning or treatment (direct reuse). The direct reuse could be "in processes without direct contact with food products, that is, in non-potable uses (e.g., as cooling water, for flushing toilets or as irrigation around the plant), thus saving fresh potable water" (Luiz et al., 2012a). Hence, according to the water balance carried out, "the wastewater with the possibility for direct or indirect recycling or reuse was evaluated physically, chemically and microbiologically to verify if and where it could be recycled and reused" (Luiz et al., 2012a). The four types of wastewaters which offered the possibility of reuse originated from: (1) the defrosting of refrigerating and freezing chambers, (2) the purging of condensers, (3) the cooling of smoke fumigator chimneys, and (4) the sealing and cooling of vacuum pumps. These residues had similar water quality parameters; hence they could be mixed before reuse, totaling approximately 1,383 m^3 d^{-1} of wastewater. Depending on the final use, this mixed wastewater could be reused without major treatment or following simple filtration; thus, this approach can be considered as direct wastewater reuse.

The theoretical reduction in water consumption, after applying the principles of water minimization and wastewater reuse, was 25.6%, representing a financial saving of around $434,000 per year (Table 1). However, new regulations need to be elaborated together with national environmental, sanitary and water supply agencies, processing industries and research institutions aiming at the legalization and promotion of water reuse in the food industry.

TABLE 1: Water and financial savings

Condition	Water flow (m^3 day^{-1})	Water saving (%)	Annual Costs[1] ($)
Production in 2007	8616.0	-	1,539,353
Theoretical production after water minimization	7810.0	9.4	1,366,000
Theoretical production after wastewater reuse	7216.8	16.0	1,256,996
Theoretical production after water minimization and wastewater reuse	6410.0	25.4	1,104,731

[1]*Considering costs in 2007: $0.10 and $0.42 per m^3 to treat water (DWTP) and wastewater (WWTP), respectively, in the case study meat processing plant. Data reproduced from Luiz et al., 2012a.*

6.3.1 TERTIARY AND ADVANCED TREATMENT FOR INDIRECT WASTEWATER REUSE

Additionally, tertiary treatments are a good alternative to produce high quality indirect reuse water, reducing the percent of fresh water consumption. Tertiary treatments can be applied to recondition secondary effluents (i.e., after secondary activated-sludge treatment), further increasing the possibilities for indirect wastewater reuse inside or outside the building. For example, an industrial wastewater treatment plant can produce high quality tertiary wastewater to be used as reuse water in its processes that do not involve contact with the food product, without the risk of adverse effects in terms of the product quality and human health. Alternatively, it can provide this high quality reuse water for another industrial activity, which does not require fresh potable water for all of its processes.

To improve the quality of the wastewater to be reused, it is necessary to disinfect it and to decrease or eliminate the concentration of biologically persistent organic compounds. The inefficient removal of these organic compounds from the wastewater before reuse or discharge into natural watercourse is promoting their accumulation in fresh water bodies and causing environmental and human health problems and, especially, harming the aquatic animals (Esplugas et al, 2007; Liu et al., 2009; Luiz et al., 2009, 2010, 2011; Oller et al., 2011).

Biologically resistant pollutants or persistent organic pollutants (POPs) are compounds which are not eliminated through the metabolic activity of living organisms (mainly bacteria and fungi) in natural waters and soils (Oller et al., 2011). Thus, conventional primary (removal of suspended compounds) and secondary (such as activated sludge) wastewater treatments are inefficient in removing these pollutants (Luiz et al., 2009). These compounds are present in municipal wastewaters primarily as pharmaceuticals and personal care products (PPCPs) (Esplugas et al., 2007), and also in industrial wastewaters which contain a large number of synthetic and toxic compounds, mainly polar and non-polar hazardous compounds, pharmaceuticals, phenols, pesticides, endocrine disruptor compounds (EDCs), and non-biodegradable and toxic chlorinated solvents (Esplugas et al., 2007; Liu et al., 2009; Luiz et al., 2009, 2010, 2011, 2012a; Petrović et al., 2003).

Due to the large variety of recalcitrant organic contaminants, the tertiary treatment applied to produce water for reuse must be exceptionally efficient. The advanced oxidation processes (AOPs) are an excellent alternative. During AOPs, highly reactive oxidizing radicals are formed, mainly hydroxyl radicals (•OH) (Koning et al., 2008). These radicals are non-selective, promoting the oxidation of all organic and inorganic contaminants and, in the presence of a sufficient amount of oxidant and optimized reaction conditions, complete mineralization can be reached, the final products being CO_2, H_2O and inorganic anions. Thus, AOPs are applied to totally or partially remove recalcitrant organic compounds, increasing the biodegradability of wastewater (Rizzo, 2011).

6.3.1.1 PROPOSED TERTIARY TREATMENTS OF SLAUGHTERHOUSE SECONDARY WASTEWATER

In previous studies carried out by our group we evaluated the different options of tertiary treatments to produce high quality reuse water to be used in processes without contact with food products, that is, non-potable uses (e.g. as cooling water, boiler feed water, toilet flushing water or for irrigation around the plant) (Cornel et al., 2011; Luiz et al., 2009, 2011, 2012a). However, since Brazilian legislation only allows the use of fresh potable water in the food industry, our research, using real wastewater and aiming to obtain reclaimed water with drinking water quality which adhered to Brazilian legislation, was carried out in bench-scale and pilot-scale.

The tertiary treatments evaluated for the slaughterhouse secondary wastewater included: UV, H_2O_2, O_3, and AOPs (H_2O_2/UV, O_3/UV; O_3/H_2O_2/UV; TiO_2/UV; and H_2O_2/TiO_2/UV). Additionally, for the best combinations, the kinetics of the photo-induced degradation of color, UV_{254}, total organic carbon (TOC) and/or total coliforms were evaluated (Luiz et al., 2009, 2010, 2011, 2012a,b).

Two main problems were encountered during this research. The first issue was the variation in the quality of the target slaughterhouse wastewater over time, which affected the treatment efficiency and the determination of the best treatment (Luiz et al., 2011). The second issue was the high concentration of nitrate and nitrite: 45.9(±17.7) mg NO_3^--N L^{-1} and 3.74-3.77 mg NO_2^--N L^{-1}, respectively (Luiz et al., 2012). The Brazilian drinking water standard (Brazilian Ministry of Health Administrative Ruling 518/2004) allows 10 mg NO_3^--N L^{-1} and 1 mg NO_2^--N L^{-1}, respectively.

In order to remove the recalcitrant organic compounds and the nitrate/nitrite, to reduce the color and turbidity, and to disinfect the secondary wastewater in a single treatment, micro-filtration followed by an AOP employing H_2O_2/TiO_2/UV was identified as the best combination evaluated (Figure 2).

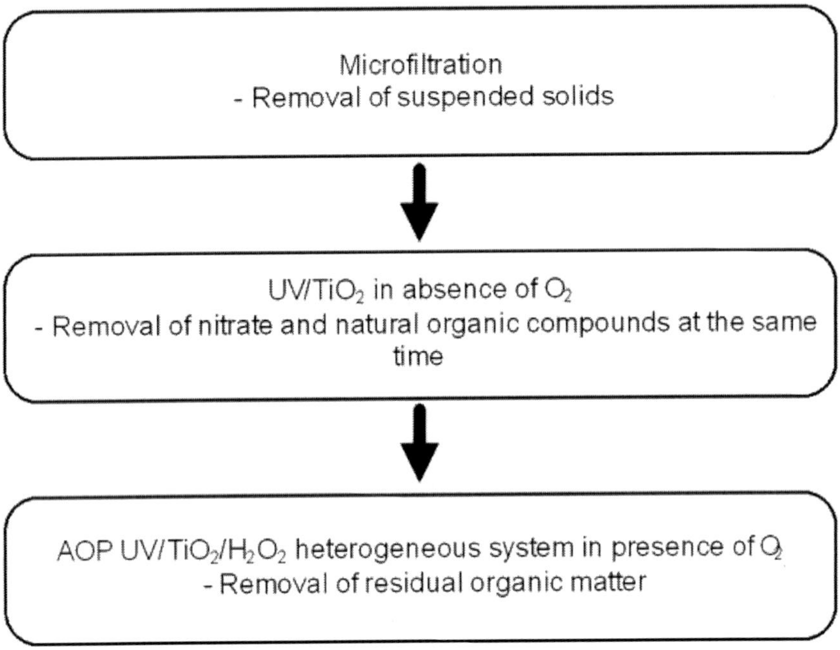

FIGURE 2: Process proposed for the treatment of secondary wastewaters with high concentration of nitrate/nitrite and recalcitrant organic compounds.

The photocatalytic removal of nitrate/nitrite is more effective in the absence of dissolved oxygen, because if O_2 is present this oxidant agent will be a better final electron acceptor than nitrate or nitrite. The catalyst is activated by the absorption of high energy photons, promoting the excitation of the electrons from the valence band (VB) to the conduction band (CB), and consequently an electron (e-) and a positive hole (h+) are formed in the CB and VB, respectively (Luiz et al., 2012b). The electron reduces the oxidizing agent adsorbed on the catalyst, and the hole oxidizes the organic compound or H_2O. In the latter case, the oxidation of H_2O produces •OH radicals, which will also oxidize organic matter (Ahmed et al., 2010).

Therefore, during the photocatalytic removal of nitrate/nitrite by UV/TiO_2, nitrate and nitrite ions will be the final electron acceptor and they will be reduced to N_2 gas. The natural organic compounds or residual, biologically persistent, organic pollutants (in the case of industrial wastewaters) will be the hole scavengers (electron donors, reducing agent). In cases where the natural concentration of organic compounds in the aquatic medium is not sufficient to promote the reduction of nitrate/nitrite to below the desired concentration, a carbon source should be added. Formic acid is a good alternative since its residue can be completely decomposed into the harmless compounds CO_2 and H_2O (Luiz et al., 2012b; Rengaraj and Li, 2007; Sá et al., 2009; Wehbe et al., 2009; Zhang et al, 2005). Finally, the heterogeneous AOP system UV/TiO_2/H_2O_2 was applied in the presence of O_2 to remove residual organic matter and achieve the required standard of drinking water quality (Luiz et al., 2012b).

The proposed treatment was also successful in removing recalcitrant organic compounds present in the secondary treated slaughterhouse wastewater, which include antibiotics, pharmaceuticals and personal care products which are commonly found in industrial, but predominantly in sanitary and domestic, wastewater (Luiz et al., 2009). One such compound found was the macrolide antibiotic erythromycin A and its removal and degradation products resulting from direct ozone attack and hydroxyl radical attack (AOPs O_3/UV, O_3/H_2O_2 and UV/H_2O_2) were evaluated. However, the research indicated that the degradation of organic micropollutants, such as erythromycin, in the AOP may be faster than under ozone treatment, because the hydroxyl radical attack (AOP treatments) is not selective and is usually diffusion-controlled. On the other hand, the direct attack of ozone is selective and is typically targeted toward functional groups with a lone valence electron pair where the electrophilic addition of ozone occurs (unsaturated compounds with carbon-carbon double or triple bonds - π bonds, aromatic rings, amines and sulfides) (Luiz et al., 2010, 2012a).

6.3.1.2 POULTRY HATCHERY WASTEWATER TREATMENT

Industrial wastewater is generally comprised of various effluent streams generated at different points in a particular process. Its physicochemical characteristics can present considerable variation over time due to, for instance, changes in operating procedures and cleaning activities. Therefore, the complexity and variation of its composition are typical attributes of industrial wastewater (Genena, 2009).

The poultry hatchery wastewater of this case study was collected and passed through the stages of screening and equalization. The wastewater variability was investigated over a period of 48 h and its quality was evaluated by chemical oxygen demand (COD) analysis, which is an overall pollution indicator and represents the amount of organic matter present in the sample. The COD values ranged from 218±2 to 997±5 mg O_2 L^{-1}, which confirms the high variability in the nature of the poultry hatchery wastewater (Genena, 2009).

Wastewater in the poultry hatcheries originates mainly from the washing of equipment and utensils. Therefore, a series of diverse compounds may be present, such as veterinary drugs administered to the animals through feed and excreted by them in urine, and sanitizer agents and pesticides used in the cleaning and disinfection of the work environment (Genena, 2009). These compounds, which are persistent compounds, are very harmful to the environment, presenting high toxicity and bioaccumulation (Almeida et al., 2004). They are complex and often difficult to degrade in the biological treatment systems commonly present in industrial wastewater treatment plants.

The food industry is constantly seeking ways to improve the quality of its wastewater through changes in treatment systems. The growing concern regarding emerging and persistent compounds has resulted in researchers focusing their attention on alternative methods of wastewater treatment to minimize or avoid the discharge of these pollutants into water resources, since the biological treatment processes typically used by the food industry are not able to destroy these types of compounds (Genena et al., 2011). The application of oxidative elimination methods, e.g., direct oxidation with ozone or hydrogen peroxide and AOP have been highlighted as strong alternatives for the treatment of wastewater containing compounds which do not degrade easily (de Sena et al., 2009; Genena, 2009, Genena et al., 2011; Luiz et al., 2009, 2010; Tambosi et al., 2009).

The proposal for the use of physicochemical processes for the treatment of the poultry hatchery wastewater of this case study was based on the value of 4.6 for the COD/BOD_5 ratio (low biodegradability) and the presence of persistent compounds. Therefore, the application of different AOPs (H_2O_2/Fe_2^+ – Fenton, $H_2O_2/Fe_2^+/UV$ – photo-Fenton and H_2O_2/UV) for the poultry hatchery wastewater treatment was investigated. The wastewater treatment process by photo-Fenton reaction was found to be the most appropriate, resulting in better organic matter removal efficiency (approximately 91.9% of COD and 66.3% of TOC). Additionally, the COD/BOD_5 ratio obtained for the treated wastewater indicates that all physicochemical treatments applied improved the biodegradability, i.e., there was an increase in the amount of material susceptible to degradation by biological processes, reaching a value of 1.5 in the photo-Fenton process. Thus, the biological process can be considered as a post-treatment stage, which would reduce the total costs of the wastewater treatment process (Genena, 2009).

An important consideration in the degradation processes is the potential for the generation of toxic intermediates or compounds which are even more toxic than their parent molecule, and thus it is necessary to monitor the process using toxicity assays (Bila et al., 2005). The Daphnia magna acute toxicity evaluation showed that all treatments promoted a significant reduction in the wastewater toxicity effects, and a 94% reduction was reached in photo-Fenton process (Genena, 2009).

Photo-Fenton and Fenton processes result in the formation of a sludge, which is usually deposited in landfills. Thus, better alternatives are being proposed and among them is sludge combustion for power generation. However, in this case study the amount of sludge obtained was insufficient for the determination of its calorific power (Genena, 2009).

The poultry hatchery wastewater was submitted to analysis by liquid chromatography coupled to mass spectrometry (LC/MS) with the objective of investigating the presence of persistent compounds. The presence of imazalil (pesticide) was confirmed among the investigated compounds. Imazalil is an organochloride compound used as a fungicide in the industry for sanitization (Genena, 2009, Genena et al., 2011). Organochlorine pesticides are typical persistent organic pollutants and are the subject of worldwide concern due to their persistence, bioaccumulation and potential negative impacts on humans and animals (Guan et al., 2009; Zhang et al., 2007). The biological treatment of wastewater containing micropollutants, like pesticides, is often very complicated or even impossible, because many pesticides are highly toxic to wastewater biocoenosis (Genena et al., 2011).

The treatment of ultrapure water to remove imazalil has been investigated applying the photo-Fenton (AOP) and ozonation processes. Tert-butanol (t-BuOH) was used in the ozonation process as an •OH scavenger to ensure that the study was focused only on the direct attack of imazalil by molecular ozone. For both processes the detection and identification of by-products were carried out, applying sophisticated analytical techniques such as LC/MS and LC/MS^n (liquid chromatography coupled to mass or multiple tandem mass spectrometry). The toxicity induced by these by-products was also investigated. For each process of oxidative treatment, four degradation products not yet known were detected and their structures were elucidated. The toxicity analysis (*Daphnia magna* assays) revealed a decrease in toxicity over time for both treatments, indicating that the by-products were not more toxic than their parent molecules (Genena, 2009, Genena et al., 2011).

6.4 BIOMASS-TO-ENERGY ACTIONS

The biosolids originating from the wastewater treatment system of the meat processing plant, sawdust and their mixture in a mass ratio of 1:9 (w/w) were characterized as fuels. The correlations between the fuel properties, the operating parameters for the combustion and the gaseous emissions were then investigated in order to evaluate the feasibility of applying this organic residue as a substitute fuel for thermal energy generation.

TABLE 2: Biomass properties

	Units	SD[1]	LFP1	LFPSD[1]:9[2]
Proximate analysis[3]				
Ash	(wt%, db)	0.43	12.30	1.75
Moisture	(wt%, raw)	19.97	15.00	50.23
Volatiles	(wt%, daf)	79.78	85.29	83.08
Fixed carbon	(wt%, daf)	20.22	9.58	17.01
Ultimate analysis[3]				
Carbon	(wt%, daf)	55.30	58.04	51.06
Hydrogen	(wt%, daf)	7.14	9.01	6.64
Nitrogen	(wt%, daf)	0.21	9.24	1.36
Sulfur	(wt%, daf)	< 0.01	0.34	0.03*
Oxygen	(wt%, daf)	37.34	22.68	40.94
Chlorine	(wt%, daf)	< 0.01	0.18	< 0.01
Fluorine	(wt%, daf)	< 0.20	< 0.20	n.d.
Phosphorus	(wt%, daf)	0.01	1.03	n.d.
Lower Heating Value[3]				
LHV	(MJ kg^{-1}, daf)	16.62	25.77	20.31
LHV	(MJ kg^{-1}, raw)	16.55	22.60	19.76

*[1]Data reproduced from Virmond et al. (2011); [2]Data reproduced from Floriani et al. (2010); [3]Maximum experimental uncertainties equal to 0.30%; db is on a Dry Basis; daf is on a Dry and Ash Free basis; *Value previously presented by Floriani et al. (2010) corrected; n.d. is Not Determined; LHV is Lower Heating Value*

6.4.1 BIOMASS PROPERTIES

The fuel properties often form the basis for the selection of the most appropriate technology for the biomass-to-energy conversion process. Depending on these properties, a biomass fuel may not be suitable for specific conversion options, partially for technical and sometimes for environmental reasons. The characteristics of the biomass are influenced by its origin and also by the entire processing system preceding any conversion step. Biomass presents a wide variation in its physical and chemical properties. Many publications have investigated the effects of the biomass properties on thermal conversion processes (Demirbas, 2004; Jenkins et al., 1998; Obernberger et al., 2006; van Paasen et al., 2006; Werther et al., 2000; Werther, 2007). The use of biomass as a fuel in combustion processes is frequently desirable in the agro-industry sector because the residues, such as wastewater sludge, usually present high calorific value. However, burning biomass containing different mineral matter compositions may create various problems which can affect the boiler operation or make the firing of the biomass in conventional combustion systems unprofitable.

Wood and wood-based materials are extensively used as fuel for thermal energy generation particularly in the Brazilian food industry, which requires large amounts of steam.

In order to evaluate the potential for the utilization of the biosolids originating from the case study plant for co-combustion with sawdust, a sample of the biosolids obtained from the physicochemical treatment (LFP) was chemically and physically characterized, and its composition was compared to that of sawdust (SD), taken as a reference fuel. Additionally, a sample of a mixture of LFP and SD in a mass ratio of 1:9 w/w (LFPSD1:9) was also characterized and the results compared to SD and LFP properties.

The methodology applied for the biomass characterization and the results obtained for LFP, SD and LFPSD1:9 were reported by Floriani et al. (2010) and Virmond et al. (2008, 2011), and are summarized in Table 2.

Carbon (C), hydrogen (H) and oxygen (O) are the main components of solid biofuels. Carbon and hydrogen contribute positively to the HHV (higher heating value). The content of hydrogen also influences the LHV (lower heating value) due to the formation of water. The content of greases was also measured in the LFP composition (34.39 wt%, raw) and it contributes considerably to the high energy content of the LFP (LHV of 25.77 MJ kg^{-1}, daf). The presence of chlorine in the biomass (0.18 wt%) occurs due to the utilization of chlorine-based products for hygiene purposes at the plant and is incorporated into the wastewater as well as into the remaining biosolids (LFP). The nitrogen content of the fuel mixture LFPSD1:9 (1.36 wt%), even though much lower than the concentration found in LFP, can still cause problems in terms of NOx emission during its combustion. The variability of components present in the biomass is mainly due to the chemical compounds used as ingredients during meat processing operations, especially salts and additives. The sulfur content in LFP is mainly due to the conversion of sulfur-containing proteins, but some may remain from the precipitation agent used in the wastewater treatment (ferric sulfate).

In previous publications (Floriani et al., 2010; Virmond et al., 2008, 2011), the authors have addressed the effects of the LFP ash composition on the fouling and slagging tendency in the combustion systems, showing that the occurrence of this problem can be reduced when burning a mixture of the biosolids with wood residues such as SD compared to LFP alone.

As shown in Table 3, the ash melting temperatures of LFP are much lower than the values estimated for LFPSD1:9 through mass balance analysis considering a homogeneous mixture, and its utilization is recommended in low blending proportions.

TABLE 3: Biomass ash properties

	Units	SD[1]	LFP[1]	LFPSD1:9[2]
Ash composition[3]				
Fe_2O_3	(wt%, db)	4.44	32.40	9.34
CaO	(wt%, db)	31.27	17.40	22.77
MgO	(wt%, db)	11.64	1.30	4.59
Na_2O	(wt%, db)	1.67	1.70	1.14
K_2O	(wt%, db)	10.44	1.70	8.77
SiO_2	(wt%, db)	15.69	4.90	17.84
Al_2O_3	(wt%, db)	12.30	1.70	8.22
TiO_2	(wt%, db)	3.94	0.00	3.07
P_2O_5	(wt%, db)	2.74	36.30	8.50
MnO	(wt%, db)	2.06	n.d.	n.d.
SO_4	(wt%, db)	3.02	n.d.	n.d.
Ash melting temperatures				
Deformation temperature	(°C)	>1150*	750	1335
Softening temperature	(°C)	>1170*	990	1359
Hemispherical temperature	(°C)	>1190*	1010	1361
Fluid temperature	(°C)	>1230*	1040	1364

[1]Data reproduced from Virmond et al. (2011); [2]Data reproduced from Floriani et al. (2010); [3]Maximum experimental uncertainties equal to 0.30%; db is on a dry basis; n.d. is not determined; [4]Data reproduced from Llorente & García (2005) for eucalyptus sample

It was observed that the main element found in the sludge ash was phosphorus, followed by iron. This is considered a problem because P forms compounds with lower melting temperature, which may have influenced the results presented in Table 3. As expected, the mixture of biomasses maintained a relatively high ash melting temperature, which is a desirable aspect when considering the combustion of solid fuels. Additionally, the design of the equipment and the definition of the operating conditions are extremely important to control, or even avoid, the occurrence of such problems.

Besides the determination of the biomass properties for biofuel applications, the organic and inorganic contents of the sludge generated from the WWTP were characterized, since it is necessary to assure that sludge containing high pollutant loads is not applied as fertilizer, in order to avoid contamination of agricultural soil and cultivated plants, i.e., to avoid the transfer of contaminants into the food chain. Inorganic and organic pollutants not removed during physicochemical wastewater treatment processes are either bio-chemically degraded or adsorbed by the sludge. The characterization of the sludge was reported by de Sena et al. (2009), where the trace metal, PAH, PCB and PCDD/PCDF concentrations in the WWTP sludge were determined. Trace metals might end up in the effluent from the meat processing plant through sources like equipment, sanitizers and cleaning agents, as well as equipment and pumps used in the wastewater treatment plant itself. Also, some metals such as arsenic, copper and zinc are occasionally added to animal feed as mineral food supplements and/or as growth promoters (US EPA, 2004). The group of PAHs, generated undesirably mostly during manifold incomplete incineration processes, includes numerous compounds with three or more condensed aromatic rings. PCBs synthesized for specific applications as non-inflammable insulators, hydrolic liquids and plasticizers are pollutants which today are ubiquitously found in the environment, although they were phased out from production worldwide at the end of the 1970s. PCDDs/PCDFs are not intentionally produced by humans but they are released into the atmosphere as sub-products of incineration and combustion processes, both domestic and industrial, when carbon, hydrogen, chlorine and oxygen together with copper as a catalyst are present. The incineration of municipal or clinical wastes, iron ore sinter plants and non-ferrous metal industries (Quass et al., 2000) as well as the chemical synthesis of chlorophenols and electrolysis of sodium chloride are sources of PCDDs/PCDFs. These compounds accumulate in sludge due to their extreme lipophilicity, and it is very difficult to assess the various sources of these compounds and their pathways into the environment and into the food chain (Klöpffer, 1996). Tables 4, 5 and 6 show the results of the biomass characterization, including 3 (three) types of sludge (BS) collected at different points in the WWTP. BSFlot refers to the sludge remaining after the flotation process, BSCent to the sludge remaining after the three-phase centrifugation, and BSBiol to the sludge collected from the activated sludge bioreactor (not primarily intended for combustion purposes).

TABLE 4: Results for the determination of trace metals and nutrients in the sludge samples

	Concentration (mg kg^{-1}, db)			
	Limit[1]	BSF_{lot}	BS_{Cent}	BS_{Biol}
Trace metal				
Hg	5	< 0.50	< 0.50	< 0.50
Cd	5	< 0.50	< 0.50	0.64
Cr	800	6.7	28.4	26.7
Cu*	800	16.2	29.8	182.1
Ni	200	1.9	9.9	22.0
Pb	500	1.3	3.4	6.1
Zn*	2000	88.2	183.8	1090.3
As	75	0.57	< 0.50	< 0.50
Mo*	75	0.50	1.7	4.4
Co*	5	< 0.50	< 0.50	4.1
Micronutrient				
K	-	427	599	6903
Fe	-	9360	25600	20900
Al	-	1750	498	3420
P	-	6350	15900	28400
Secondary nutrients				
Ca	-	1520	5080	18600
Mg	-	148	259	7185
S	-	3140	6630	9810

*db is Dry Basis; [1]Upper limit of pollution for the disposal of sewage sludge in the environment (EU, 2000); * Trace elements, biologically essential in small quantities. Reproduced from de Sena et al. (2009)*

TABLE 5: Results for the determination of PAHs and PCBs in the sludge samples

Compounds	Concentration (μg kg^{-1}, db)				
	TEF[1]	Limit[2]	BSF_{lot}	BS_{Cent}	BS_{Biol}
Polycyclic Aromatic Hydrocarbons (PAHs)					
Naphthalene (Nap)	-		110.0	< 40.0	< 30.0
Acenaphthylene (Acy)	-		84.0	< 30.0	< 30.0
Acenaphtene (Ace)	-		< 7.0	< 7.0	< 6.0
Fluorene (Flu)	0.001		20.0	< 7.0	< 6.0
Phenanthrene (Phe)	0.001		84.0	< 8.0	< 7.0
Anthracene (Ant)	-		< 2.0	8.0	8.0
Fluoranthene (Fla)	0.001		320.0	510.0	< 10.0
Pyrene (Pyr)	0.001		97.0	72.0	11.0
Chrysene (Cry)[3]	0.01		92.0	< 1.0	5.0
$\sum PAH_{LMW}$	-		816.0	616.0	113.0
Benzo[a]anthracene (BaA)[3]	0.1		24.0	< 1.0	4.0
Benzo[b]fluoranthene (BbF)[3]	0.1		38.0	25.0	< 3.0
Benzo[k]fluoranthene (BkF)[3]	0.1		11.0	< 1.0	< 1.0
Benzo[a]pyrene (BaP)[3,4]	1.0		2.0	< 1.0	5.0
Dibenzo[a,h]anthracene (DbA)[3,4]	5.0		< 4.0	< 4.0	< 4.0
Benzo[g,h,i]perylene (BgP)	0.1		12.0	< 6.0	< 6.0
Indeno[1,2,3-cd]pyrene (InD)[3]	0.1		< 10.0	< 10.0	< 10.0
$\sum PAH_{HMW}$	-		101.0	48.0	33.0
$\sum PAH$		≤ 6000	917.0	664.0	146.0
$\sum TEF_{PAH}$	-		11.0	3.0	6.0
Polychlorinated Biphenyls (PCB)					
PCB non-ortho					
PCB 77, 81, 126*, 169	< 0.1		< LOQ	< LOQ	< LOQ
PCB mono-ortho					
PCB 105, 114, 118, 123, 156, 167, 189	< 0.005		< LOQ	< LOQ	< LOQ
$\sum PCB$	-	≤ 800	< LOQ	< LOQ	< LOQ

*db is Dry Basis; [1]TEF for dioxin and dioxin-like compounds (Nisbet et al., 1992; van den Berg et al., 2006); [2]Limit for solid disposal onto soil (EU, 2001); [3]Carcinogenic isomers; [4]Isomers of PAH with TEF comparable to the most toxic polychlorinated dibenzo-dioxins and -furans; *Isomer of PCB with highest TEF; LOQ is Limit of Quantification. Reproduced from de Sena et al. (2009)*

TABLE 6: TEF and concentrations of PCDDs/PCDFs in the sludge samples

		Concentration (ng kg^{-1}, db)		
	TEF[1]	BS_{Flot}	BS_{Cent}	BS_{Biol}
Polychlorinated dibenzo-p-dioxins (PCDD)				
2,3,7,8 – TCDD[2]	1.0	0.1	0.4	0.4
1,2,3,7,8 - PeCDD	1.0	0.6	1.3	2.2
1,2,3,4,7,8 - HxCDD	0.1	0.3	1.5	1.7
1,2,3,7,8,9 - HxCDD	0.1	0.2	1.9	3.3
1,2,3,6,7,8 - HxCDD	0.1	0.7	1.3	1.8
1,2,3,4,6,7,8 - HpCDD	0.01	0.8	4.2	4.6
OCDD	0.0003	2.1	21.1	23.1
∑ PCDD	-	4.8	31.7	37.1
Polychlorinated dibenzofurans (PCDF)				
2,3,7,8 – TCDF	0.1	0.2	0.3	1.3
2,3,4,7,8 – PeCDF	0.3	0.5	2.1	2.1
1,2,3,7,8 – PeCDF	0.03	0.7	1.5	3.8
1,2,3,4,7,8 – HxCDF	0.1	0.6	1.2	2.0
1,2,3,6,7,8 – HxCDF	0.1	0.6	1.6	2.3
1,2,3,7,8,9 – HxCDF	0.1	1.7	2.9	3.1
2,3,4,6,7,8 - HxCDF	0.1	0.7	1.9	3.8
1,2,3,4,6,7,8 - HpCDF	0.01	2.6	4.2	5.3
1,2,3,4,7,8,9 – HpCDF	0.01	1.4	3.1	3.6
OCDF	0.0003	2.1	4.4	10.2
∑ PCDF	-	11.1	23.2	37.3
PCDD:PCDF Ratio	-	0.43	1.37	0.98
∑ TEF PCDD/PCDF	100*	1.4	3.8	5.4

*db is Dry Basis; [1]TEF for dioxins and dioxin-like compounds (van den Berg et al., 2006; WHO, 2005); [2]Isomer with highest acute toxicity; *TEF (ng kg^{-1}) limit for solid disposal onto soil (EU, 2001). Reproduced from de Sena et al. (2009)*

A study by de Sena et al. (2009) verified low pollution loads for the sludge (BS) originating from the WWTP, with respect to the most relevant inorganic and organic priority pollutants as monitored by the US EPA, at the case study meat processing plant located in the south of Brazil.

Although other pollutants such as veterinary drugs, pesticides and surfactants were not investigated in this first analytical approach, they are of high concern. However, this study was a preliminary report for future monitoring of other food processing segments located in different regions of Brazil.

6.4.2 BIOMASS COMBUSTION

Co-combustion of agro-industrial residues in thermal power plants is not necessarily a low cost alternative for the thermal treatment of wastes. There is the possibility of interaction between the components and the main fuels in such a way that either the operating behavior of the conversion system is improved or the emissions are reduced (Werther, 2007). The emission of pollutants generated during combustion is strongly related to the biomass properties. Pollutant formation mechanisms and many other parameters related to the combustion process must be monitored due to the formation of highly problematic compounds such as NOx, SO_2, benzene, toluene, ethyl-benzene, and (o-,m-,p-)xylenes (BTEX), polycyclic aromatic hydrocarbons (PAH), polychlorinated dibenzo-p-dioxins and polychlorinated dibenzofurans (PCDD/PCDF) (Chagger et al., 1998; Kumar et al., 2002; Mckay, 2002; Stanmore, 2004; Watanabe et al., 2004), and have to be controlled in order to comply with the stringent limits set by recent environmental legislation. Chlorine-associated, high-temperature corrosion and the potential corrosion problems associated with burning biomass fuels have been previously discussed (Nielsen et al., 2000). Fuel nitrogen can be a problem in terms of NOx emissions. The conversion of nitrogen in systems fired by solid fuels (mainly coal, but also biomass) has been reviewed in detail, as well as the combustion characteristics of different biomass fuels, the potential applications of renewable energy sources as the prime energy sources in various countries, and the problems associated with biomass combustion in boiler systems (Demirbas, 2004; Werther, 2000).

The pollutant emissions due to incomplete biomass combustion can be effectively controlled by an optimized combustion process, i.e., enhanced mixing, sufficient residence time at high temperatures (>850 °C), and low total excess air (Demirbas, 2005), as well as the appropriate choice of the combustion device.

Given the higher energy value of the biosolids (LHV equal to 22.60 MJ kg^{-1}) compared to sawdust (LHV equal to 16.55 MJ kg^{-1}), as shown in Table 2, the substitution of 10 wt% of sawdust with this residue can increase the thermal energy production by approximately 4% compared to the combustion of sawdust alone, leading to a sawdust saving of 1950 ton per year, besides the economic benefits related to reduced landfill disposal. However, the gaseous emissions have to be monitored so as not to infringe current legislations.

The evaluation of the feasibility of the utilization of the biosolids originating from the WWTP studied was performed through combustion testing of the biosolids as the sole fuel in a pilot-scale cyclone combustor (model Drako, Albrecht, Brazil) with a burning capacity of 100 kg h^{-1}, as described by Virmond et al. (2011), and in a furnace equipped with a reciprocating-grate coupled to a boiler at the meat processing plant with the capacity to process 12000 kg h^{-1}, as reported by Virmond (2007).

Grate firing is one of the main technologies that are currently used for biomass combustion aiming at heat and power production. Grate-fired boilers can fire a wide range of fuels with varying moisture content and show great potential in biomass combustion (Goërner, 2003). The plant-scale furnace was operated at a fuel feed rate of 2604 kg h^{-1} (moisture content of approximately 50 wt%) at 900 °C, with 59% excess air and without gas recirculation. The combustion test lasted for approximately 2 h after the system had reached steady state.

In the pilot-scale plant the combustion of SD was carried out at a fuel feed rate of 32 kg h^{-1} (moisture content of 9.16 wt%), with gas recirculation of 20% at an average temperature of 642 °C in a cyclone combustor (Drako, Albrecht, Brazil), which has been described by Virmond et al. (2011). The average temperature at the front stage of the combustor was 1052 °C and at the outlet it was 1024 °C, and an air-to-fuel (A/F) ratio of 9.08 (theoretical A/F ratio of 7.23 based on the fuel composition) was used. For the LFP combustion test, the conditions applied to the cyclone combustor were: fuel feed rate of 43 kg h^{-1} (moisture content of 11.44 wt%), with gas recirculation of 20% at an average temperature of 800 °C. The average temperature at the front stage of the combustor was 868 °C, at the outlet of the combustor it was 1080 °C, and the A/F ratio was 11.82 (theoretical A/F ratio of 7.76 based on the fuel composition). CO, O_2, SO_2, CxHy (measured as CH_4), NO, and NO_2 emissions were measured using a Greenline MK2 (Eurotron) analyzer and BTEX emission measurements were based on the adsorption and desorption of gases which were analyzed by gas chromatography. Emissions of BTEX were expressed as Total Organic Carbon (TOC). The detailed methodology for emissions sampling and analysis during the combustion tests, as well as the complete set of results, have been previously reported by Floriani et al. (2010) and Virmond et al. (2007, 2008) for LFPSD1:9 combustion, and by Virmond et al. (2011) for SD and LFP. Thus, only the main points are highlighted herein. The gaseous emissions observed in the combustion tests and the respective regulation limits were corrected to the reference oxygen content (O_2ref) of 7% and are given in Table 7.

TABLE 7: Gaseous emissions from the combustion tests at O_2ref = 7%

	CO (mg Nm^{-3})	CO_2(%)	CxHy (mg Nm^{-3})	NOx[1](mg Nm^{-3})	SO_2 (mg Nm^{-3})	TOC (mg Nm^{-3})
SD biomass[2]	93.58 ± 21.97	10.32 ± 0.03	0.00 ± 0.00	241.58 ± 26.31	0.00 ± 0.00	1.27 ± 0.57
LFP biomass[2]	63.33 ± 10.30	10.33 ± 0.05	0.00 ± 0.00	1727.43 ± 229.93	363.54 ± 90.80	1.23 ± 0.12
LFPSD1:9 biomass[3]	734.83 ± 12.39	10.39 ± 0.02	554.44 ± 7.91	497.94 ± 19.04	128.69 ± 4.31	1.72 ± 0.83
CONAMA[4]	124.88	n.a.	n.a.	560.00	280.00	n.a.
US EPA[5]	196.16	n.a.	n.a.	796.02	57.09	n.a.
17.BlmSchV (24 h; <50 MW)[6]	140.00	n.a.	n.a.	373.33	186.67	9.33
17.BlmSchV (24 h)[7]	70.00	n.a.	n.a.	280.00	70.00	14.00

[1]NOx expressed in terms of NO_2; TOC is Total Organic Carbon; [2]Data from Virmond et al. (2011); [3]Data from Floriani et al. (2010) and Virmond et al. (2007, 2008); [4]CONAMA 316/02, thermal treatment of wastes (CONAMA, 2002); n.a. is Not Applicable; [5]US EPA, solid waste incineration (US EPA, 2000); [6]17.BlmSchV 24 h, <50 MW, co-combustion of wastes (17.BlmSchV, 2003); [7]17.BlmSchV 24 h, direct combustion of wastes (17.BlmSchV, 2003)

Since no data on the combustion of meat processing or slaughterhouse wastes had been previously published, no comparison was possible. However, the results for the contaminants reported were compared with the emission limits established by national and international environmental agencies, such as the Brazilian guidelines issued by The National Council of the Environment (CONAMA, 2002) for gaseous emissions in the thermal treatment of wastes; the American guidelines issued by the US Environmental Protection Agency (US EPA, 2002) for emissions from commercial and industrial solid waste incineration units; and the German Guidelines 17.BlmSchV (17.BlmSchV, 2003) for emissions from biomass combustion and from biomass co-combustion. Concerning the combustion tests performed in the pilot-scale cyclone combustor with SD and LFP as fuels, the emissions of CO, CO_2, CxHy and TOC were well controlled and their concentrations remained below the regulation limits considered for both biomasses, except CO in the SD combustion test in relation to the German guidelines, which refer to waste incineration and are stricter than the other regulations.

The effect of the biomass composition on gaseous emissions was clearly observed, especially considering the N and S fuel contents in LFP, which led to concentrations of these pollutants being higher than the established limits.

The use of the biosolids originating from the meat processing plant investigated in this study as a fuel in the pilot cyclone combustor was shown to be feasible; however, further research is required concerning the control of SO_2 and NOx emissions to avoid exceeding the very strict emission limits as well as the occurrence of fouling and slagging.

In relation to the combustion test performed with the mixture of SD and LFP (LFPSD1:9), the high levels of CxHy and CO emitted indicate incomplete combustion. This can be attributed to the high moisture content of the biomass (50.23 wt%, as shown in Table 2), the lower combustion temperature (approximately 900 °C) compared to the other two tests performed in the pilot-scale combustor (approximately 1000 °C) and the absence of gas recirculation. Additionally, the control of the operating conditions of the large-scale plant is more difficult to achieve, due to the restricted testing time or minimal variation from the normal operation with the SD. Firing a biomass with low moisture content and flue gas recirculation could provide better oxidation conditions. Due to the lower nitrogen concentration found in LFPS1:9 compared to LFP, as well as to the low operating combustion temperature, NOx emissions remained below the limits established by CONAMA and US EPA.

The co-combustion of LFP and SD with lower-N fuel content reduced the NOx in the gaseous emissions compared to the burning of LFP alone. In fact, this option is the most feasible in Brazil considering the relatively high NOx emissions related to both the fuel nitrogen and to the fact that wood and wood-based materials are extensively used as fuel for thermal energy generation in the Brazilian food industry. Chlorine, PAHs and PCBs are among the elements or compounds that must be studied in greater depth as they are precursors to the formation of dioxins and partly of furans.

Industrial solid wastes must be disposed of safely, and co-firing them with SD has been shown in other studies by the authors, which are currently underway, to be profitable using the same pilot-scale cyclone combustor and different biomasses. The advantages are both a reduction in the consumption of primary fuels and the recovering of energy from wastes inside the plant, which would normally be disposed of in landfills, potentially causing environmental problems.

6.4.3 BIOGAS PRODUCTION AND COMBUSTION

During this study, only preliminary tests were performed, such as the determination of the chemical composition of the biogas and the exhaust gases when the biogas was burned. Measurements were performed at several partner pig farms close to the case study plant that rear pigs from 45 to 110 days of life. At these farms there were relatively small horizontal-type biodigestors (approximately 450 m^3) in which the input consisted only in pig wastes without previous treatment. The hydraulic retention time was approximately 30 days and the entrance flow 8 $m^3\ d^{-1}$.

Table 8 shows the average chemical composition and calorific values of the biogas. To analyze the composition of biogas, a biogas analysis kit was used (Alfakit, Brazil) which is based on colorimetric methods.

TABLE 8: Average chemical composition of biogas

	Average chemical composition	Reference values[1]
CH_4 (v/v%)	45-65	57
CO_2 (v/v%)	35-55	34
H_2S (v/v%)	>1020	-
CH_4/CO_2	0.82-1.86	1.7
HHV (kJ Nm^{-3})	17996-25995	-
LHV(kJ Nm^{-3})	16200-23400	-

[1]*Data reproduced from Silva et al. (2005); HHV is Higher Heating Value; LHV is Lower Heating Value*

At the time this study was completed the farms were burning only biogas to avoid harmful emissions and to provide a better disposal/reuse options for the waste. At all farms there were simple flares to burn the gas, and thus the gaseous emissions were evaluated. CO, SO_2, NO, NO_2, O_2, and CxHy were measured using a Greenline MK2 (Eurotron) analyzer and the sampling point was located at the top of the chimney. Results are not presented in this document because each burner presented different burning efficiencies and they were still in the adjustment phase. However, by the end of the project, all gaseous emissions were below the limits imposed by Brazilian Legislation. The SO_2 requires greater caution due to the presence of H_2S and therefore a pre-treatment has to be considered before the process can be considered adequate. H_2S can easily react with iron oxides and hydroxides, requiring the presence of water, and this can thus be considered a good method to remove the H_2S from biogas (Zichari, 2003).

6.5 CONCLUSIONS

The water balance analysis carried out considering all processing steps at the case study plant indicated that the minimization of the fresh water consumption at the four major consumption points could account for some 806 $m^3 d^{-1}$. Concerning wastewater reuse, the four streams identified as having a real possibility for reuse totaled approximately 1383 $m^3 d^{-1}$. These need simple reconditioning treatment before application to processes without direct contact with food products, that is, in non-potable uses (e.g. cooling water, toilet flushing water or irrigation around the plant), thus saving fresh potable water. The theoretical fresh water reduction after water minimization and wastewater reuse was 25.4% with a financial saving of around $434,622.00 per year.

Additionally, to reduce even further the percentage of fresh water consumption, indirect wastewater reuse could be carried out after reconditioning by applying tertiary treatments, such as advanced oxidation processes (AOPs), to treat the secondary effluent (after secondary activated sludge treatment). The tertiary-treated water effluent could then be used in other processes without contact with food products. The tertiary treatment model proposed was a heterogeneous AOP system ($UV/TiO_2/H_2O_2$) which can be applied to urban, rural or industrial effluents where the factors inhibiting their reuse as water of potable quality are the presence of suspended solids (even at low concentration), dissolved organic matter, recalcitrant micro-pollutants (trace compounds) and high concentrations of nitrate and nitrite. However, laboratory tests should be carried out with real wastewater to evaluate the efficiency of each process step.

For the treatment of the effluent from the case study poultry hatchery, a chemical or physicochemical process would be the best option due to the low biodegradability of the effluent ($COD/BOD_5 = 4.6$) and the presence of persistent compounds, which are not removed by biological processes. All treatments evaluated, particularly the photo-Fenton reaction, resulted in an increased biodegradability of the effluent, in other words, an increase in the portion of the material susceptible to degradation by biological processes. Thus, a biological process should be added as a final step in the effluent treatment, as a post-treatment mainly to remove the previously-formed more biodegradable compounds (with lower molar mass) and the nutrients that are not eliminated by the physicochemical process, e.g., nitrate. Also, a comparison of the pros and cons (especially costs and efficiency) of the two treatments (a) photo-Fenton + simple biological treatment (such as stabilization ponds) and (b) $UV/TiO_2/H_2O_2$ could be carried out in terms of their effectiveness in the treatment of slaughterhouse secondary effluent. Hence, the most important consideration to be evaluated is which treatment can ensure that the standards and limits set by legislation are achieved, thus avoiding undesirable impact on the environment (such as the discharge of persistent organic compounds into rivers) and providing economic benefits.

Regarding the solid wastes, the substitution of 10 wt% of the sawdust with the biosolids originating from the physicochemical wastewater treatment can increase the thermal energy production by approximately 4% compared to the combustion of sawdust alone, leading to an economy of 1950 tons per year of sawdust besides providing savings in relation to landfill disposal. Additionally, co-combustion is the most feasible option for energy recovery from this waste in Brazil, making it possible to control the burning process, to avoid the occurrence the fouling and slagging and to meet the emission limits established in the relevant legislation. Considering that wood and wood-based materials are extensively used as fuels for thermal energy generation in the Brazilian food industry, the mixture of such a small mass fraction of this solid waste with sawdust should not require significant changes to the current operating conditions.

Industrial solid wastes must be disposed of safely, and co-firing them with sawdust was shown to be profitable using a pilot-scale cyclone combustor in studies currently underway in our research group. The biogas produced from pig wastes has great potential to become another important bioenergy option for the Brazilian agroindustrial sector. Additionally, anaerobic digestion has other environmental benefits besides the production of a renewable energy carrier, which include the possibility of nutrient recycling and reduction of waste volumes. Nevertheless, studies are needed to investigate the effects of variations in the input to a biodigestor, how the waste composition influences the overall stability of the process and the product quality, and options for the biogas application.

The comprehensive technical-scientific analyses of the actions concerning water, wastewater and solid waste management carried out in the case study meat processing plant indicated that environmentally, financially and socially sustainable practices can be successfully implemented in any type and size of food processing plant.

REFERENCES

1. 17th Ordinance on the Implementation of the Federal Emission Control Act – Ordinance for Combustion and Co-combustion of Wastes of August 14th (Siebzehnte Verordnung zur Durchführung des Bundes-Immissionsschutzgesetzes. Verordnung über die Verbrennung und die Mitverbrennung von Abfällen) 17.BlmSchV, Bundesanzeiger, Bonn/Germany, 2003 (in German).
2. Ahmed, S., Rasul, M.G.; Martens, W.N., Brown, R., & Hashib, M.A. (2010). Heterogeneous photocatalytic degradation of phenols in wastewater: A review on current status and developments. Desalination, Vol. 261, No. 1-2, (Oct 2010), pp. 3-18, ISSN 0011-9164.
3. Almeida, E., Assalin, M.R., Rosa, M.A., & Duran, N. (2004). Tratamento de efluentes industriais por processos oxidativos na presença de ozônio. Química Nova, Vol. 27, No. 5, (Sept/Oct 2004), pp. 818–824, ISSN 0100-4042.
4. Bila, D.M., Montalvão, F., Silva, A.C., &Dezotti, M. (2005). Ozonation of a landfill leachate: evaluation of toxicity removal and biodegradability improvement. Journal of Hazardous Materials, Vol. 117, No. 2-3, (Jan 2005), pp. 235–242, ISSN 0304-3894.
5. Boersma, L., Gasper, E., Oldfield, J.E. & Cheeke, P.R. (1981) Methods for the recovery of nutrients and energy from swine manure. 1. Biogas. Netherlands Journal of Agricultural Science, Vol. 29, pp. 3–14, ISSN. 0028-2928.
6. Brazil, 2011. Set of legal rules: water resources, Ministry of Environment, Department of Water Resources and Urban Environment. 7th. ed., Brasília, DF, Brazil, 640 p. (in Portuguese). Retrieved from: http://www.cnrh.gov.br/sitio/index.php?option=com_docman&task=doc_download&gid=822.
7. Chagger, H. K., Kendall, A., McDonald, A., Pourkashanian, M., & Williams, A. (1998). Formation of dioxins and other semi-volatile organic compounds in biomass combustion. Applied Energy, Vol. 60, No. 2, (Jun 1998), pp. 101-114, ISSN 0306-2619.
8. CONAMA - National Council of the Environment (29 Oct 2002). Resolution 316, In: Ministry of the Environment (MMA), Brasília, DF, Brazil (in Portuguese), Date of access: 11 Jan 2009, Available from: <http://www.mma.gov.br/port/conama/legiabre.cfm?codlegi=338>.
9. Cornel, P., Meda, A., & Bieker, S. (2011). Wastewater as a Source of Energy, Nutrients, and Service Water, In: Treatise on Water Science, Peter Wilderer (ed.), Vol. 4, pp. 337-375, Oxford: Academic Press, ISBN: 9780444531933.
10. de Sena, R.F., Claudino, A., Moretti, K., Bonfanti, I.C.P., Moreira, R.F.P.M., & José, H.J. (2008). Biofuel application of biomass obtained from a meat industry wastewater plant through the flotation process - A case study. Resources, Conservation and Recycling Vol. 52, No. 3 (Jan 2008), pp. 557-569, ISSN 0921-3449.
11. de Sena, R.F., Tambosi, J.L., Genena, A.K., Moreira, R.F.P.M., Schröder, H.Fr., & José, H.J. (2009). Treatment of meat industry wastewater using dissolved air flotation and advanced oxidation processes monitored by GC–MS and LC–MS. Chemical Engineering Journal, Vol. 152, No. 1 (Oct 2009), pp. 151–157, ISSN 1385-8947.

12. de Sena, R.F., Tambosi, J.L., Floriani, S.L., Virmond, E., Schröder, H.Fr., Moreira, R.F.P.M., & José, H.J. (2009). Determination of inorganic and organic priority pollutants in biosolids from meat processing industry. Waste Management, Vol. 29, No. 9 (Sep 2009), pp. 2574-2581, ISSN 0956-053X.
13. Demirbas, A. (2004). Combustion characteristics of different biomass fuels. Progress in Energy and Combustion Science, Vol. 30, No. 2 (2004), pp. 219-230, ISSN 0360-1285.
14. Demirbas, A. (2005). Potential applications of renewable energy sources, biomass combustion problems in boiler power systems and combustion related environmental issues. Progress in Energy and Combustion Science, Vol. 31, No. 2 (2005), pp. 171-192, ISSN 0360-1285.
15. Esplugas, S., Bila, D.M., Krause, L.G., & Dezotti, M. (2007). Ozonation and advanced oxidation technologies to remove endocrine disrupting chemicals (EDCs) and pharmaceuticals and personal care products (PPCPs) in water effluents. Journal of Hazardous Materials, Vol. 149, No. 3 (19 Nov 2007), pp. 631-642, ISSN 0304-3894.
16. EU – European Union Commission. (2000). Working document on sludge, 3rd Draft. ENV.E.3/LM. Brussels, Belgium.
17. EU – European Union Commission. (2001). European Commission Directive, Joint Research Centre/Institute for Environment and Sustainability/Soil and Waste Unit, 2001. Organic contaminants in sewage sludge for agricultural use. Brussels, Belgium.
18. Floriani, S.L., Virmond, E., Althoff, C.A., Luiz, D.B., José, H.J., & Moreira, R.F.P.M. (2010). Potential of industrial solid wastes as an energy source and gaseous emissions evaluation in a pilot scale burner (ES2008-54355). Journal of Energy Resources Technology, Vol. 132, No. 1 (Mar 2010), pp. 011003-011010, ISSN 0195-0738.
19. Genena, A.K. (2009). Treatment of agroindustrial wastewater containing persistent compounds by coagulation-flocculation, Fenton, photo-Fenton, photo-peroxidation and ozonation (in Portuguese). Doctoral Thesis, Federal University of Santa Catarina, Florianópolis, Brazil. Date of access: 01 Aug 2012, Available from: http://www2.enq.ufsc.br/teses/d069.pdf (in Portuguese).
20. Genena, A.K., Luiz, D.B., Gebhardt, W., Moreira, R.F.P.M., José, H.J. & Schröder, H.Fr. (2011). Imazalil degradation upon applying ozone—transformation products, kinetics, and toxicity of treated aqueous solutions. Ozone: Science & Engineering, Vol. 33, No. 4 (Jul-Aug 2011), pp. 308-328, ISSN 0191-9512.
21. Goërner, K. (2003). Waste incineration: European state-of-the-art and new development. IFRF Combustion Journal, Vol. 32, Article No. 200303 (Jul 2003), 32 p., ISSN 1562-479X.
22. Guan, Y.-F., Wang, J.-Z., Ni, H.-G., & Zeng, E.Y. (2009). Organochlorine pesticides and polychlorinated biphenyls in riverine runoff of the Pearl River Delta, China: Assessment of mass loading, input source and environmental fate. Environmental Pollution, Vol. 157, No. 2 (Feb 2009), pp. 618–624, ISSN 0269-7491.
23. Jenkins, B.M., Baxter, L.L., Miles Jr, T.R. & Miles, T.R. (1998). Combustion properties of biomass. Fuel Processing Technology, Vol. 54, No. 1-3 (Mar 1998), pp. 17-46, ISSN 0378-3820.

24. Johns, M.R. (1995). Developments in wastewater treatment in the meat processing industry: a review. Bioresource Technology, Vol. 54, No. 3 (1995), pp. 203-216, ISSN 0960-8524.
25. Klöpffer, W. (1996). Environmental hazard assessment of chemicals and products. Part V. Anthropogenic chemicals in sewage sludge. Chemosphere, Vol. 33, No. 6, (Sep 1996), pp. 1067-1081, ISSN 0045-6535.
26. Koning, J., Bixio, D., Karabelas, A., Salgot, M., & Schäfer, A. (2008). Characterisation and assessment of water treatment technologies for reuse. Desalination, Vol.218, No. 1-3 (5 Jan 2008), pp. 92-104, ISSN 0011-9164.
27. Kumar, A., Purohit, P., Rana, S., & Kandpal, T.C. (2002). An approach to the estimation of the value of agricultural residues used as biofuels. Biomass and Bioenergy, Vol. 22, No. 3 (Mar 2002), pp. 195-203, ISSN 0961-9534.
28. Liu, Z.H., Kanjo, Y., & Mizutani, S. (2009). Removal mechanisms for endocrine disrupting compounds (EDCs) in wastewater treatment - physical means, biodegradation, and chemical advanced oxidation: A review. Science of the Total Environment, Vol. 407, No. 2 (1 Jan 2009), pp. 731-748, ISSN 0048-9697.
29. Llorente, M.J.F., & García, J.E.C. (2005). Comparing methods for predicting the sintering of biomass ash in combustion. Fuel, Vol. 84, No. 14-15 (Oct 2005), pp. 1893-1900, ISSN 0016-2361.
30. Lora, E.S., & Andrade, R.V. (2009). Biomass as energy source in Brazil. Renewable and Sustainable Energy Reviews, Vol. 13, No. 4 (May 2009), pp. 777-788, ISSN 1364-0321.
31. Lucas Júnior, J. (1994). Considerations on the use of swine manure as substrate for three anaerobic biodigestion systems in Portuguese). Doctoral Thesis, - Faculdade de Ciências Agrárias e Veterinárias, Universidade Estadual Paulista, Jaboticabal, 137 p.
32. Luiz, D.B., Genena, A.K., José, H.J., Moreira, R.F.P.M., & Schröder, H.Fr. (2009). Tertiary treatment of slaughterhouse effluent: degradation kinetics applying UV radiation or H2O2/UV. Water Science and Technology, Vol. 60, No. 7, pp. 1869-1874, ISSN 0273-1223.
33. Luiz, D.B., Genena, A.K., Virmond, E., José, H.J., Moreira, R.F.P.M., Gebhardt, W., & Schröder, H.Fr. (2010). Identification of degradation products of erythromycin A arising from ozone and AOP treatment. Water Environment Research, Vol. 82, pp. 797-805, ISSN 1061-4303.
34. Luiz, D.B., Silva, G.S., Vaz, E.A.C., José, H.J., & Moreira, R.F.P.M. (2011). Evaluation of hybrid treatments to produce high quality reuse water. Water Science and Technology, Vol. 63, No. 9, pp. 2046-2051, ISSN 0273-1223.
35. Luiz, D.B., José, H.J., & Moreira, R.F.P.M. (2012a). A discussion paper on challenges and proposals for advanced treatments for potabilization of wastewater in the food industry. In: Valdez, B., Zlatev, R., Schorr, M. (Org.). Scientific, Health and Social Aspects of the Food Industry. : InTech, p. 3-24, ISBN 978-953-307-916-5.
36. Luiz, D.B., Andersen, S.L.F., Berger, C., José, H.J., & Moreira, R.F.P.M. (2012b). Photocatalytic Reduction of Nitrate Ions in Water over Metal-Modified TIO2. Journal of Photochemistry & Photobiology, A: Chemistry. In press. ISSN: 1010-6030.

37. Mckay, G. (2002). Dioxin characterization, formation and minimization during municipal solid waste (MSW) incineration: review. Chemical Engineering Journal, Vol. 86, No. 3 (28 Apr 2002), pp. 343-368, ISSN 1385-8947.
38. Nielsen, H. P, Frandsen, F. J, Dam-Johansen, K., & Baxter, L.L. (2000). The implications of chlorine-associated corrosion on the operation of biomass-fired boilers. Progress in Energy and Combustion Science, Vol. 26, No. 3 (Jun 2000), pp. 283-298, ISSN 0360-1285.
39. Nisbet, I.C.T., & La Goy, P.K. (1992). Toxic equivalency factors (TEFs) for polycyclic aromatic hydrocarbons (PAHs). Regulatory Toxicology and Pharmacology, Vol. 16, No. 3 (Dec 1992), pp. 290–300, ISSN 0273-2300.
40. Obernberger, I., Brunner, T., & Bärnthaler, G. (2006). Chemical properties of solid biofuels - significance and impact. Biomass and Bioenergy, Vol. 30, No. 11 (Nov 2006), pp. 973-982, ISSN 0961-9534.
41. Oller, I., Malato, S., & Sánchez-Pérez, J.A. (2011). Combination of advanced oxidation processes and biological treatments for wastewater decontamination - A review. Science of the Total Environment, Vol. 409, No. 20 (15 Sep 2011), pp. 4141-4166, ISSN 0048-9697.
42. Petrović, M., Gonzalez, S., & Barceló, D. (2003). Analysis and removal of emerging contaminants in wastewater and drinking water. TrAC Trends in Analytical Chemistry, Vol. 22, No. 10 (Nov 2003), pp. 685-696, ISSN 0165-9936.
43. Rengaraj, S., & Li, X.Z. (2007). Enhanced photocatalytic reduction reaction over Bi3+–TiO2 nanoparticles in presence of formic acid as a hole scavenger. Chemosphere, Vol. 66, No. 5 (Jan 2007), pp. 930-938, ISSN 0045-6535.
44. Quaß, U., Fermann, M.W., & Bröker, G. (2000). Steps towards a European dioxin emission inventory. Chemosphere, Vol. 40, No. 9-11 (May-Jun 2000), pp. 1125–1129, ISSN 0045-6535.
45. Rizzo, L. (2011). Bioassays as a tool for evaluating advanced oxidation processes in water and wastewater treatment. Water Research, Vol. 45, No. 15 (1 Oct 2011), pp. 4311-43401, ISSN 0043-1354.
46. Sá, J., Agüera, C.A., Gross, S., & Anderson, J.A. (2009). Photocatalytic nitrate reduction over metal modified TiO2. Applied Catalysis B: Environmental, Vol. 85, No. 34- (Jan 2009), pp. 192-200, ISSN 0926-3373.
47. Silva, F.M., Lucas Junior, J.,Benicasa, M., & Oliveira, E. (2005). Performance of a water heating system using biogas (in Portuguese).. Engenharia Agrícola, Vol. 25, No. 3, pp. 608-614, ISSN 1808-4389.
48. Sroka, A., Kaminski, W., & Bohdziewicz, J. (2004). Biological treatment of meat industry wastewater. Desalination, Vol. 162, No. 10, pp. 85-91, ISSN 0011-9164.
49. Stanmore, B.R. (2004). The formation of dioxins in combustion systems - a review. Combustion and Flame, Vol. 136, No. 3 (Feb 2004), pp. 398-427, ISSN 0010-2180.
50. Schievano, A., D'Imporzano, G., & Adani, F. (2009). Substituting energy crops with organic wastes and agro-industrial residues for biogas production. Journal of Environmental Management, Vol. 90, No. 8 (Jun 2009), pp. 2537-2541, ISSN 0301-4797.
51. Tambosi, J.L., De Sena, R.F., Gebhardt, W., Moreira, R.F.P.M., José, H.J., & Schröder, H.Fr. (2009). Physicochemical and advanced oxidation processes – A comparison of elimination results of antibiotic compounds following an MBR treat-

ment. Ozone: Science & Engineering, Vol. 31, No. 6 (2009), pp. 428–435, ISSN 0191-9512.

52. Tritt, W.P., & Schuchardt, F. (1992). Materials flow and possibilities of treating liquid and solid wastes from slaughterhouses in Germany. A review. Bioresource Technology, Vol. 41, No. 3 (1992), pp. 235-245, ISSN 0960-8524.
53. US EIA – United States Energy Information Administration. (2011). Annual Energy Outlook 2011 with projections to 2035. Date of access: 12 Jun 2011, Available from: <http://www.eia.gov/forecasts/aeo/>.
54. US EPA - United States Environmental Protection Agency. (2004). Technical development document for the final effluent limitations guidelines and standards for the meat and poultry products point source category. EPA-821-R-04-011, Washington, DC.
55. US EPA - United States Environmental Protection Agency. (Dec 2000). Code of Federal Regulation. Title 40, Part 60. In: US Environmental Protection Agency. Date of access: 16 Jul 2009, Available from:<http://www.epa.gov/epacfr40/chapt-I.info>.
56. van den Berg, M., Birnbaum, L.S., Denison, M., De Vito, M., Farland, W., Feeley, M., Fiedler, H., Hakansson, H., Hanberg, A., Haws, L., Rose, M., Safe, S., Schrenk, D., Tohyama, C., Trischer, A., Tuomisto, J., Tysklind, M, Walker, N., & Peterson, R.E. (2006). The 2005 World Health Organization reevaluation of human and mammalian Toxic Equivalent Factors for dioxins and dioxin-like compounds. Toxicological Sciences, Vol. 93, No. 2 (Oct 2006), pp. 223-241, ISSN 1096-6080.
57. Van Paasen, S.V.B, Cieplik, M.K., & Phokawat, N.P. (2006). Gasification of non-woody biomass - economic and technical perspectives of chlorine and sulphur removal from product gas (non-confidential version), In: ECN-C-06-032, 2006. Date of access: 15 Apr 2010, Available from: <http://www.ecn.nl/docs/library/report/2006/e06032.pdf>.
58. Virmond, E. (2007). Aproveitamento do lodo de tratamento primário de efluentes de um frigorífico como fonte alternativa de energia. Master's Thesis, Federal University of Santa Catarina, Florianópolis, Brazil. Date of access: 01 Aug 2012, Available from: <http://www2.enq.ufsc.br/teses/m176.pdf > (in Portuguese).
59. Virmond, E., Floriani, S.L., Moreira, R.F.P.M., José, H.J., 2008. Co-combustion of wood chips with the biomass obtained from the physico-chemical treatment of slaughterhouse wastewater for steam generation - A case study. Proceedings of the 23rd International Conference on Solid Waste Technology and Management, ISSN 1091-8043, Philadelphia, Pennsylvania, USA, March 30- April 2, 2008, pp. 1177-1188.
60. Virmond, E., Schacker, R.L., Albrecht, W., Althoff, C.A., Souza, M., Moreira, R.F.P.M., & José, H.J. (2010). Combustion of apple juice wastes in a cyclone combustor for thermal energy generation (ES2009-90152). Journal of Energy Resources Technology, Vol. 132, No. 4 (Dec 2010), pp. 041401-041409, ISSN 0195-0738.
61. Virmond, E., Schacker, R.L., Albrecht, W., Althoff, C.A., Souza, M., Moreira, R.F.P.M., & José, H.J. (2011). Organic solid waste originating from the meat processing industry as an alternative energy source. Energy, Vol. 36, No. 6 (Jun 2011), pp. 3897-3906, ISSN 0360-5442.

62. Virmond, E., et al. (2012a). Characterisation of agroindustrial solid residues as biofuels and potential application in thermochemical processes. Waste Management (2012), http://dx.doi.org/10.1016/j.wasman.2012.05.014>, article in press.
63. Virmond, E., et al. (2012b). Valorization of agroindustrial solid residues and residues from biofuel production chains by thermochemical conversion: a review, citing Brazil as a case study. Brazilian Journal of Chemical Engineering, Accepted for publication, ISSN 0104- 6632.
64. Watanabe, N., Yamamoto, O., Sakai, M., &Fukuyama, J. (2004). Combustible and incombustible speciation of Cl and S in various components of municipal solid waste. Waste Management, Vol. 24, No. 6 (2004), pp. 623-632, ISSN 0956-053X.
65. Wehbe, N., Jaafar, M., Guillard, C., Herrmann, J.-M., Miachon, S., Puzenat, E., & Guilhaume, N. (2009). Comparative study of photocatalytic and nonphotocatalytic reduction of nitrates in water. Applied Catalysis A: General, Vol.368, No. 1-2 (31 Oct 2009), pp. 1-8, ISSN 0926-860X.
66. Werther, J., Saenger, M., Hartge, E-U., Ogada, T., & Siagi, Z. (2000). Combustion of agricultural residues. Progress in Energy and Combustion Science , Vol. 26, No. 1 (Feb 2000), pp. 1-27, ISSN 0360-1285.
67. Werther, J. (2007). Gaseous emissions from waste combustion. Journal of Hazardous Materials, Vol. 144, No. 3 (18 Jun 2007), pp. 604-613, ISSN 0304-3894.
68. WHO - World Health Organization. (2005). Project for the re-evaluation of human and mammalian toxic equivalency factors (TEFs) of dioxins and dioxin-like compounds. In: International Program on Chemical Safety (IPCS), WHO Headquarters. Geneva, Switzerland.
69. Zhang, F., Jin, R., Chen, J., Shao, C., Gao, W., Li, L., & Guan, N. (2005). High photocatalytic activity and selectivity for nitrogen in nitrate reduction on Ag/TiO2 catalyst with fine silver clusters. Journal of Catalysis, Vol. 232, No. 2 (10 Jun 2005), pp. 424-431, ISSN 0021-9517.
70. Zhang, G., Li, J., Cheng, H.R., Li, X.D., Xu, W.H., & Jones, K.C. (2007). Distribution of organochlorine pesticides in the northern South China Sea: implications for land outflow and air–sea exchange. Environmental Science and Technology, Vol. 41, No 11 (4 May 2007), pp. 3884–3890, ISSN 0013-936X.
71. Zichari, S.M. (2003). Removal of hydrogen sulfide from biogas using cow-manure compost. Master's Thesis, Faculty of the Graduate School of Cornell University, USA. Available from: http://www.green-trust.org/Al%20Rutan/MS-Thesis-Steve-Zicari.pdf

PART II

DEVELOPED NATIONS

CHAPTER 7

Intra- and Inter-Pandemic Variations of Antiviral, Antibiotics and Decongestants in Wastewater Treatment Plants and Receiving Rivers

ANDREW C. SINGER, JOSEF D. JÄRHULT, ROMAN GRABIC, GHAZANFAR A. KHAN, RICHARD H. LINDBERG, GANNA FEDOROVA, JERKER FICK, MICHAEL J. BOWES, BJÖRN OLSEN, AND HANNA SÖDERSTRÖM

7.1 INTRODUCTION

Pandemics are unique public health emergencies that can result in a large sudden increase in the use of a restricted set of pharmaceuticals within a short time period. In the case of an influenza pandemic, antiviral use will greatly exceed inter-pandemic use in most countries by several orders of magnitude, as few countries maintain significant inter-pandemic usage–Japan being a notable exception [1]. Depending on the severity of the pandemic, antibiotics have the potential to significantly exceed inter-pandemic usage for the treatment of secondary bacterial respiratory infections [2]. Decongestant usage is also predicted to increase with an increase in upper- and lower-respiratory tract infections [3].

TABLE 1: Drug dosage, pharmacokinetics and limit of quantification (LOQ) of study analytes.

ATC Code	Drug	Class of pharmaceutical	ADQ[a] (g)	% excreted[b]	LOQ (ng/L)
J01FA10	Azithromycin	Macrolide antibiotic	0.4	85%	1
J01DD01	Cefotaxime	Third-generation cephalosporin antibiotic	4	85%	10
J01MA02	Ciprofloxacin	Fluoroquinolone antibiotic	0.8	100%	5
J01FA09	Clarithromycin	Macrolide antibiotic	0.5	55%	4
J01AA02	Doxycycline	Tetracycline antibiotic	0.1	80%	1
J01 FA01	Erythromycin	Macrolide antibiotic	1	100%	4
J01MA06	Norfloxacin	Fluoroquinolone antibiotic	0.8	90%	1
J01MA01	Ofloxacin	Fluoroquinolone antibiotic	0.4	98%	4
J01AA06	Oxytetracycline	Tetracycline antibiotic	1	35%	4
J01 EC01	Sulfamethoxazole	Sulfonamide antibiotic	0.8	100%	2
J01EA01	Trimethoprim	Dihydrofolate reductase inhibitor	0.4	100%	1
R01AA08	Naphazoline	Vasoconstrictor decongestant	0.4	90%	1
R01AA05	Oxymetazoline	Vasoconstrictor decongestant	0.4	35%	7
R01AA07	Xylometazoline	Vasoconstrictor decongestant	0.8	90%	11
J05AH02	Oseltamivir carboxylate	Neuraminidase inhibitor antiviral	0.2	80%	2

[a]ADQ =is a measure of prescribing volume based upon prescribing behavior in England, which is similar, and often the same as the international standard defined daily dose (DDD).[b] [4,5].

Antibiotics, antivirals and decongestants are typically excreted as a large percentage of the parent dose in their bioactive form (mean: 82±22% for all drugs in this study (Table 1)) [4], [5]. The large load and high concentration of bioactive pharmaceuticals entering the wastewater and receiving rivers from widespread human consumption and excretion during a pandemic can potentially disrupt (micro)organisms through non-target effects [6]–[12] and cause the failure of wastewater treatment plants (WWTPs) to treat effluent to the required standard [13], [14], hasten the generation of antiviral resistance in wildfowl and other influenza-susceptible organisms [15]–[18], and accelerate the generation and spread of (novel) antibiotic resistance in the environment [2], [19], [20].

In this study, we measured eleven antibiotics, one antiviral and three decongestants (see Table 1) weekly at 21 locations within the River Thames catchment (Fig. 1) in England during the month of November 2009, the autumnal peak of the influenza A[H1N1]pdm09 pandemic. The aim was to quantify the pharmaceutical response to the pandemic and compare this to drug use during the late pandemic (March 2010) and the inter-pandemic periods (May 2011). One relatively large wastewater treatment plant (WWTP) in Oxford, UK employing activated sludge wastewater treatment and one relatively small WWTP in Benson, UK, employing trickle-bed wastewater treatment were sampled hourly for 24-h in November 2009 to 1) understand the differential pharmaceutical use patterns among the people within the WWTP catchments during the pandemic, 2) characterize the fate of the analytes in the two very different WWTPs prior to their entry in the receiving River Thames, and 3) examine the suitability of employing a wastewater epidemiology approach for the estimation of drug users within the Oxford and Benson WWTP populations.

7.2 EXPERIMENTAL SECTION

7.2.1 WASTEWATER TREATMENT PLANT CHARACTERISATION

The Benson WWTP serves a population of 6,230 people with a consented dry weather flow (DWF) of 2,517 m^3/d and an annual average DWF of 1,368 m^3/d (Fig. 1 and Fig. S1 in File S1). The Benson WWTP has a hydraulic retention time of 7–8 h at DWF and consists of trickling filters as the main biological treatment step. Oxford WWTP serves a population of 208,000 with a consented DWF of 50,965 m^3/d and an annual mean DWF of 38,000 m^3/d (Fig. 1 and Fig. S1 in File S1). The Oxford WWTP has a hydraulic retention time of 15–18 h, and utilizes activated sludge as the main biological treatment step. Both WWTPs have primary and secondary sedimentation steps. Both Oxford and Benson WWTPs feed into the main stem of the River Thames (Fig. 1), separated by approximately 10 miles.

7.2.2 WASTEWATER TREATMENT PLANT SAMPLING

The sampling of all analytes (Table 1) in Benson and Oxford WWTP was performed during a 24-hour period spanning 10–11 November 2009. An additional 24-h sampling was initiated on May 11, 2011 from only the Benson WWTP effluent for the primary purpose of confirming the background concentration of the antiviral, oseltamivir carboxylate (OC), during the inter-pandemic period. The pandemic officially ended on August 10, 2010, hence, the expectation was that pharmaceutical use in the study catchments in May 2011 would reflect inter-pandemic pharmaceutical usage [21]. An automated sampler was used to recover time-proportional samples (approximately 750 ml) of influent and effluent every hour for 24 hours. Samples were aliquoted into triplicate 50-ml borosilicate glass vials with PTFE-lined caps and immediately stored at −80°C until analysis.

7.2.3 RIVER SAMPLING

Grab samples were acquired within 250 ml borosilicate brown glass bottles at the end of a 1.5 m-long sampling rod at 21 river locations within the River Thames catchment (Fig. 1). Sampling was undertaken on November 3, 11, 17, and 24, 2009, as well as on March 15, 2010 (late-pandemic period) and May 11, 2011 (inter-pandemic period). These sites are part of the CEH Thames Initiative Research Platform [22]. Samples were transported from the field to the laboratory within 6 hours and transferred into 50-ml borosilicate glass vials with PTFE-lined caps, in triplicate. The samples were stored at −80°C until analysis.

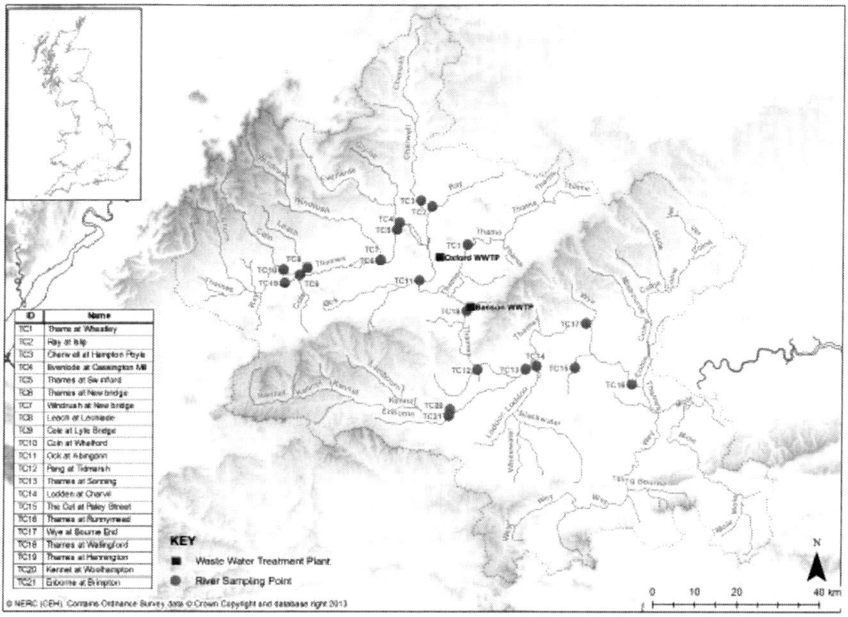

FIGURE 1: The 21 river sampling locations within the River Thames Catchment (TC) in southern England and the location of the Oxford and Benson WWTPs.

7.2.4 RIVER FLOW

River flow data was acquired from the National River Flow Archive (http://www.ceh.ac.uk/data/nrfa/) for all locations at the closest gauging station to the sampling site (Fig. S2 in File S1). In the case of Loddon at Twyford (TC14) and The Cut at Binfield (TC15), the closest active gauging station was appreciably upstream. In these two cases we employed an infilling method (equipercentile transfer) to estimate the flow at the sampling location, as previously described [23].

7.2.5 ENVIRONMENTAL CONDITIONS

Precipitation data for the 48 h before the sampling periods was used for the town of Benson, England (Fig. S3 in File S1) [24], which is geographically central to all river sampling locations and reflects the climatic conditions of all the sampling sites for the specific days of the study.

7.2.6 STUDY ANALYTES

Oseltamivir is a prodrug, which means it is metabolized in vivo to the active antiviral, oseltamivir carboxylate (OC). Eighty-percent of the parent dose is converted to OC, which is excreted and easily recorded in the environment. If only the parent compound was recorded one could not be sure that the drug wasn't flushed down the drain prior to having been consumed or the result of improper disposal from a manufacturing plant [25]. Given that oseltamivir is not manufactured in the Thames catchment, the measures reflected in this study should reflect oseltamivir consumption and not improper disposal. The antibiotics examined in this study were selected as they reflect the drugs most likely to be used during an influenza pandemic [2], [26]. beta-Lactams were not included in this study owing to their relatively high propensity to hydrolysis and biodegradation and thus low likelihood for persistence in the environment. The decongestants examined in this study represent a cross section of this class of drug, and does not reflect all or even the majority of decongestants in use in England.

7.2.7 ANALYTICAL TECHNIQUE

An on-line solid phase liquid extraction/liquid chromatography-tandem mass-spectrometry (SPE/LC-MS/MS) method was used to measure the analyte levels in pre-filtered and acidified 1 mL-samples. This on-line SPE/LC-MS/MS method used has been evaluated and described in detail previously [27]. The on-line SPE/LC system consisted of a PAL HTC auto sampler (CTC Analytics AG, Zwingen, Switzerland), a Surveyor LC-Pump (Thermo Fisher Scientific, San Jose, CA, USA), an on-line SPE Hypersil GOLD C18 column (20 mm×2.1 mm i.d.×12 μm, Thermo Fisher Scientific, Waltham, MA, USA), an Accela LC pump (Thermo Fisher Scientific, San Jose, CA, USA), and a Hypersil GOLD C18 column (50 mm×2.1 mm i.d.×3 μm particles, Thermo Fisher Scientific, San Jose, CA, USA) with a guard C18 column (2 mm×2 mm i.d.×3 μm particles, Thermo Fisher Scientific, San Jose, CA, USA). The liquid chromatography system was coupled to a heated electrospray ionization (HESI) source and a Quantum Ultra triple quadrupole mass spectrometer made by Thermo Fisher Scientific (Waltham, MA). The MS/MS parameters used are described in Table S3 in File S1. The following internal standards were obtained from Cambridge Isotope Laboratories (Andover, MA, USA): $^{13}C_2$-Trimethoprim ($^{13}C_2$-TRI) (99%), $^{13}C_3$15N–Ciprofloxacin ($^{13}C_3$-CIP) (99%), $^{13}C_2$-Erythromycin ($^{13}C_2$-ERY) and $^{13}C_6$-Sulphamethoxazole ($^{13}C_6$-SUL). Oseltamivir carboxylate labeled with deuterium (OCD3) (RO0604802-004; lot: 511-001-2197/4) was obtained from Roche (F. Hoffmann-La Roche Ltd., Basel, Switzerland).

7.2.8 FATE IN THE WWTPS AND THE THAMES RIVER CATCHMENT-CALCULATING WWTP AND RIVER LOAD AND PERCENT LOSS IN WWTP

Pharmaceutical concentrations were converted to mass loading using hourly WWTP (Fig. S1 in File S1) and river flows for the sampling period. Percent loss of analytes between WWTP influent and effluent as a result of biodegradation and sorption was calculated from the change in 24-h load of each analyte between the two sampling locations. As the influent and effluent samples were acquired simultaneously, the calculated 'percent loss' assumes negligible change in drug use between the Monday and Tuesday during which the samples were acquired and a negligible change in hydraulic retention time. Resulting from such assumptions, the interpretations of the recalcitrance of analytes were considered with caution.

7.2.9 WASTEWATER EPIDEMIOLOGY-FORWARD-CALCULATING ENVIRONMENTAL CONCENTRATIONS OF ANTIBIOTICS FROM PRESCRIPTION STATISTICS

The National Health Service Business Services Authority (NHS BSA) annual antibiotic prescriptions for England [28] was used for estimating 'background' pharmaceutical use for the population residing within the two study WWTP catchments in 2009. NHS BSA data are resolved at the national level and reflect annual prescription rates (Table S1 in File S1). More spatially resolved data was acquired from the four Primary Care Trust (PCT) [29] hospitals and clinics serving the Oxford and Benson WWTP catchments, however, this level of detail was only available from November 2011. To assess the value of this data as a proxy for 'background' antibiotic use in November, we examined total antibacterial use in general practice in England since 2007. There was a <1% change in total antibiotic prescriptions per year with the exception of penicillin where an increase of approximately 5% was seen between 2007–2011 [28]. As penicillins were not monitored in this study, we argue that the November 2011 data might serve as an adequate proxy for 'background' antibiotics prescribed during the study period. A large fluctuation from this 'background' usage might be indicative of pandemic-linked usage.

The National Pandemic Flu Service (NPFS) [30] recorded approximately 66,218 courses of Oseltamivir dispensed in Week 43 in 2009, representing 0.13% of the population of England, and 6% of all antivirals dispensed during the pandemic [30]. The national peak for the autumnal wave of the influenza pandemic was 3 weeks prior to the WWTP sampling on 10–11 November [30]. The antiviral prescription rate did not rapidly decline after the peak (see Fig. 15 in [30]), suggesting that the peak antiviral prescription rate of 0.13% might be a good proxy for antiviral use during the sampling period. The standard adult Oseltamivir dosing regime was assumed: 0.075 g per dose, consumed twice per day (0.150 g/d).

An additional dataset produced by the HPA's QSurveillance National Syndromic Surveillance System [31], was examined for estimating Oseltamivir prescription rates. The HPA dataset reports 54.2 people per 100,000 with influenza-like illness (ILI) in the Oxfordshire PCT during the week of WWTP sampling (i.e., Week 46), which was used for modelling purposes. However, the ILI reporting rate declined during November from Week 46 to Week 49, which reached as low as 33 per 100,000 [31].

We provide a general model for calculating the concentration of pharmaceuticals (ng/L) in wastewater influent (C_w) using the different data sources discussed above:

$$C_w = \left(\frac{M \cdot E \cdot F}{P \cdot L}\right) \cdot 10^9 \tag{1}$$

where the product of the population of each WWTP catchment (P) and the volume of wastewater per person (L; 230 L/capita/d [2]) was divided into the product of the mass of prescriptions (M) in grams acquired from Average Daily Quantity (ADQ) conversions (Table 1 and S1) [32], mass of parent compound excreted in its parent form (E) in grams, and correction factor (F) for adjusting for when the population served by a PCT is served by more than one WWTP.

When deriving the mass of drug prescribed using NHS BSA statistics, $M = M_a + M_s$, where M_a was the annual mass (g) of pharmaceutical prescribed (Table S1 in File S1) and M_s reflected the additional mass of drug used (g) in the winter in excess of the average monthly usage (Table S1 in File S1). Hence, $M_s = (M_a \times 0.09375)$, where 0.9375 is the additional fraction of drug used in the winter as compared to the annual mean (i.e., $M_a/12$). This adjustment was performed because it is known that the mean variation in antibiotic consumption between the summer and winter period in the UK was approximately 18.75% in 2005 [33]. Hence, the annual prescription rate provided by the NHS BSA was increased by 9.375% from the annual mean prescription rate (i.e., reflecting 50% of the total seasonal variability between summer and winter). Differences in pharmacokinetics were accounted for using factor E, which reflects the fraction of parent chemical excreted into wastewater (Table 1) [4].

The patient population of the PCT serving Benson is served by two different WWTPs at Benson and Cholsey. This difference between the population served by a PCT and the WWTP catchment were accounted for by factor F, where F = 0.389, resulting from the ratio of the population served by the Benson WWTP (6230) to the patient population of the local PCT (16,000). However, F = 1 for the Oxford WWTP as it was assumed that the populations served by the PCT in Oxford all fed into the Oxford WWTP. Lastly, F = 1 for all NHSBSA statistics, as the NHSBSA dataset used a national average prescription rate and was not stratified to the local level.

Notably, trimethoprim and sulfamethoxazole are routinely dispensed as a mixture, co-trimoxazole. Use of these drugs was calculated assuming one ADQ of co-trimoxazole contained 0.16 g of trimethoprim and 0.8 g of sulfamethoxazole.

All raw data has been made freely available at http://doi.org/10/t2x.

7.3 RESULTS

7.3.1 PHARMACEUTICALFATE IN BENSON AND OXFORD WWTPS

7.3.1.1 ANTIVIRAL IN BENSON WWTP

The concentration of OC in the influent on 10–11 November, 2009, ranged from <limit of quantification (LOQ; 1–11 ng/L, see Table 1) to 2070 ng/L (Table 2), with 18 of the 24 measures above the LOQ (Fig. S4 in File S1). The mean hourly concentration (for samples >LOQ), was 433±472 ng/L. The 24-h load was 410 mg/d, reaching a maximum load of 133 mg/h at the 18:00 sampling point, equating to 66 μg OC/capita/d.

The concentration of OC in the effluent ranged from <LOQ to 287 ng/L, with 21 of the 24 measures above the LOQ. The mean hourly concentration (for samples >LOQ), was 208±40 ng/L. The 24-h load was 206 mg/d, reaching a maximum of 16.9 mg/h at the 9:00 sampling point. The change in load from the influent to the effluent was 204 mg/d, a reduction of 50%.

TABLE 2: Mean concentration and load of pharmaceuticals measured within the Benson and Oxford WWTP on November 10–11, 2009 (n = 24).

	Oxford Load (mg/d) Mean concentration (ng/L)[a]			Benson Load (mg/d) Mean concentration (ng/L)[a]		
	Inlet	Outlet	% Loss[b]	Inlet	Outlet	% Loss[b]
Antibiotic						
	4780	492		32.3	21.8	
Azithromycin	163 ± 96	30 ± 6	90%	40 ± 20	34 ± 10	32%
		116		6.80		
Cefotaxime	<LOQ	51	<	18 ± 7	<LOQ	>99%
	57300	1840		221	1.2	
Ciprofloxacin	1090 ± 300	52 ± 50	97%	200 ± 303	14 ± 3	99%
	27800	2590		43.7	27.3	
Clarithromycin	524 ± 179	92 ± 27	91%	61 ± 62	50 ± 21	38%
	3230	6270		414	57.9	
Doxycylcine	60 ± 43	121 ± 83	−94%	345 ± 444	67 ± 150	86%
	69500	9440		847	234	
Erythromycin	1330 ± 560	236 ± 40	86%	954 ± 1610	244 ± 83	72%
	9680	332		37	13.6	
Norfloxacin	184 ± 80	21 ± 1	97%	59 ± 26	25 ± 5	63%
	4260	472		1462	9.68	
Ofloxacin	81 ± 26	23 ± 8	89%	2320 ± 3340	195	99%
	56100	697		236	51.3	
Oxytetracycline	1090 ± 229	29 ± 16	99%	171 ± 400	174 ± 83	78%
	8960	2970		75.6	32.7	
Sulfamethoxazole	169 ± 28	67 ± 18	67%	57 ± 118	34 ± 6	57%
	3550	3840		312	49.0	
Trimetoprim	70 ± 14	73 ± 7	−8%	332 ± 190	68 ± 66	84%
Decongestant						
				1460	69.2	
Naphazoline	<LOQ	<LOQ	na	1650 ± 775	696 ± 981	95%

TABLE 2: *Cont.*

	Oxford Load (mg/d) Mean concentration (ng/L)[a]			Benson Load (mg/d) Mean concentration (ng/L)[a]		
	Inlet	Outlet	% Loss[b]	Inlet	Outlet	% Loss[b]
		65.8		43	172	
Oxymetazoline	<LOQ	16 ± 1	na	67 ± 61	307 ± 110	−299%
		28.0		6.82		
Xylometazoline	<LOQ	15	na	13 ± 11	<LOQ	na
Antiviral						
Oseltamivir	18600	18900	−2%	410	206	50%
	350 ± 59	358 ± 60		433 ± 472	208 ± 40	

Percent loss calculated from change in daily pharmaceutical load between the influent and effluent. [a]Mean concentrations are followed by standard deviation for all samples > LOQ during the 24-hour sampling. [b]Negative values indicate a concentration increase from inlet to outlet. < is used to indicate concentration where the influent was below the LOQ while the effluent was > LOQ. na = not applicable.

7.3.1.2 ANTIVIRAL IN OXFORD WWTP

The concentration of OC in the influent on 10–11 November, 2009, ranged from 257 to 550 ng/L (Table 2), with all measures above the LOQ. The mean hourly concentration was 350±59 ng/L. The 24-h load was 18,600 mg/d, equating to 89 μg OC/capita/d reaching a maximum load of 1,330 mg/h at the 10:00 sampling point.

The concentration of OC in the effluent ranged from 474 to 1,130 ng/L, with all measures above the LOQ. The mean hourly concentration was 358±60 ng/L. The 24-h load in the effluent was 18,900 mg/d, reaching a maximum of 1130 mg/h at the 8:00 sampling point. The change in 24-h load from the influent to the effluent was −300 mg/d, a trivial increase of 2% in the effluent, suggestive of a fully conservative chemical.

7.3.1.3 ANTIBIOTICS IN BENSON WWTP

Erythromycin showed the highest average antibiotic concentration in the influent, 954 ng/L (n = 17, 134 μg/capita/d) and effluent 244 ng/L (n = 20, 37 μg/capita/d; Table 2) and reached as high in concentration as 6,870 ng/L in the influent. Three antibiotics, at least once in the 24-h sampling, exceeded the average inlet concentration of erythromycin: ofloxacin (max = 11,000 ng/L, 235 μg/cap/d, n = 12), doxycycline (max = 1,550 ng/L, 66 μg/cap/d, n = 23) and oxytetracycline (max = 1,700 ng/L, 38 μg/cap/d, n = 20). Notably, ciprofloxacin was recorded as high as 917 ng/L (35 μg/cap/d, n = 16) and trimethoprim as high as 780 ng/L (50 μg/cap/d, n = 21).

With the exception of cefotaxime, which was found in only 7 of 24 samples from Benson influent, all antibiotics were found in at least half of the 24 samples, with 6 antibiotics found at each of the 24 hourly measurements (Fig. S4 in File S1).

The load of each of the 11 antibiotics in the Benson WWTP effluent was reduced by 32% to a maximum of 99%+ after treatment (Table 2), the most persistent being azithromycin (32% loss, n = 18 and n = 15 for inlet and outlet, respectively) and clarithromycin (38% loss, n = 13 and n = 11, respectively).

7.3.1.4 ANTIBIOTICS IN OXFORD WWTP

Erythromycin showed the highest average antibiotic concentration in the influent (1,330 ng/L, 341 μg/cap/d, n = 24) and reached a maximum concentration of 2,930 ng/L (Table 2). Two other antibiotics recorded mean concentrations in the influent above 1,000 ng/L, oxytetracycline (1,090 ng/L, 275 μg/cap/d, n = 24) and ciprofloxacin (1,090 ng/L, 281 μg/cap/d, n = 24). Two antibiotics achieved maximum concentration that exceeded the average inlet concentration set by erythromycin: oxytetracycline (1,430 ng/L) and ciprofloxacin (1,530 ng/L; Table 2). Notably, clarithromycin achieved a maximum of 980 ng/L (136 μg/cap/d), much higher than the maximum in Benson (243 ng/L, 7 μg/cap/d).

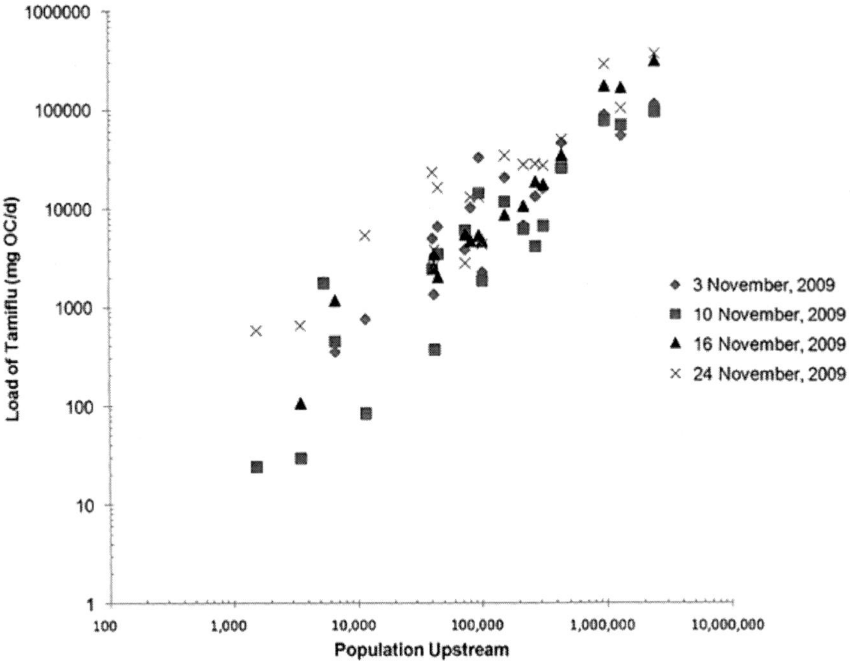

FIGURE 2: Correlation between population upstream and daily load of OC in river (mg OC/d) for each of the four sampling points: November 3 (diamond), November 10 (square), November 16 (triangle) and November 24 ('x'), 2009.

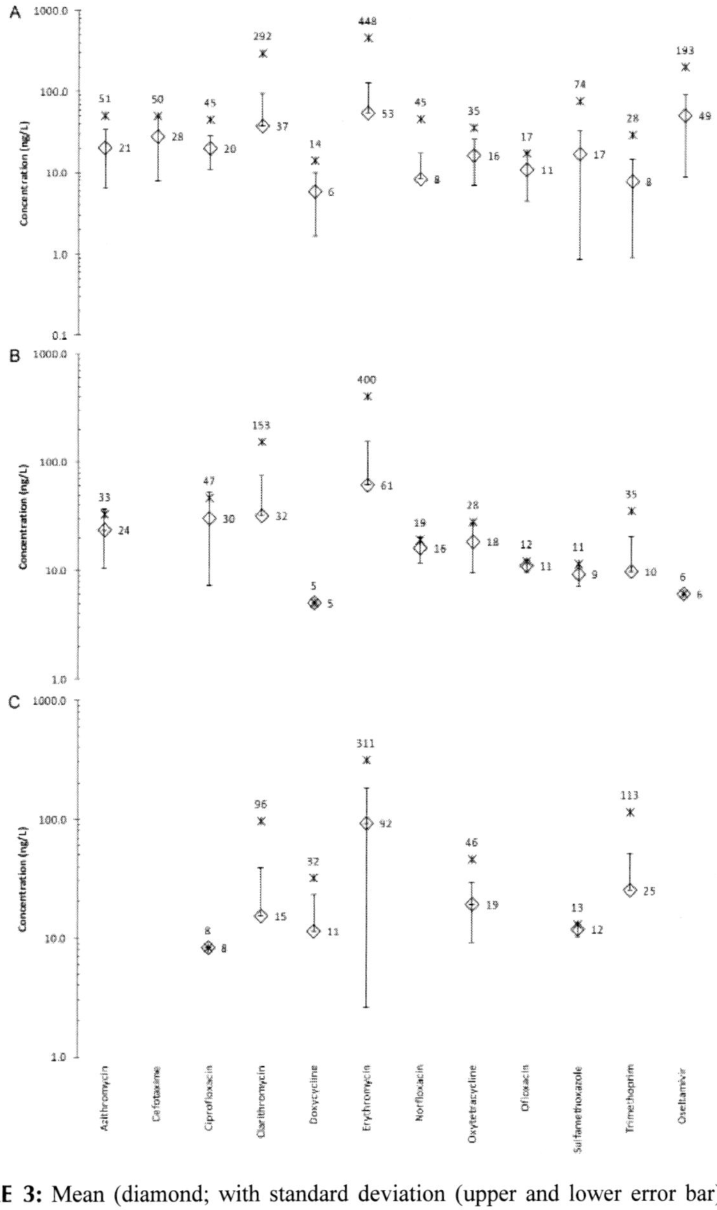

FIGURE 3: Mean (diamond; with standard deviation (upper and lower error bar)) and maximum (asterisk) concentration (ng/L) of pharmaceuticals across the 21 River Thames locations during (A) the four sampling occasions in November 2009 (intra-pandemic, n = 84); (B) 15 March, 2010 (late-pandemic. n = 21); and (C) 11 May, 2011 (inter-pandemic, n = 21). 'Maximum' data label is above asterisk, while 'mean' data label is to the right of the diamond. Decongestants were omitted as none were found >LOQ.

With the exception of cefotaxime (0/24), azithromycin (13/24) and trimethoprim (23/24), all antibiotics were found at each of the 24 hourly measurements from the Oxford influent.

The antibiotic load in the Oxford WWTP effluent was reduced by 67% to <LOD after treatment for many antibiotics (Table 2), however, unlike Benson, several demonstrated persistence and concentration, including: cefotaxime (increase from <LOQ to 51 ng/L (n = 1) for inlet and outlet, respectively), doxycycline (increase by 94% from 3230 ng/L (n = 24) to 6270 ng/L (n = 24), respectively) and trimethoprim (increase by 8% from 70 ng/L (n = 24) to 73 ng/L (n = 24), respectively).

7.3.1.5 DECONGESTANTS IN BENSON WWTP

Approximately 1.56 g/d of decongestant was quantified in the Benson WWTP influent (0.25 mg/capita/d), 97% of which was a single decongestant, naphazoline. The most frequently quantified decongestant in the influent was naphazoline (n = 19), however, oxymetazoline was the most frequently found decongestant in the effluent (n = 18; Table 2). Oxymetazoline achieved a mean influent and effluent concentration of 67 ng/L and 307 ng/L, respectively. Unlike oxymetazoline, naphazoline averaged higher concentrations in the influent (1,650 ng/L) than the effluent (696 ng/L). Naphazoline reached a higher maximum concentration in the influent and effluent (3,070 ng/L, 1,390 ng/L, respectively) than oxymetazoline (177 ng/L, 440 ng/L, respectively). The influent load (1,460 mg/d) for naphazoline, was reduced by 95% as compared to the effluent (69 mg/d, n = 2), indicating it is unlikely to be a persistent environmental pollutant. Xylometazoline averaged concentrations only marginally above the LOQ (11 ng/L) in the influent (13 ng/L; 7 mg/d, n = 10), and was <LOQ in all of the effluent samples.

7.3.1.6 DECONGESTANTS IN OXFORD WWTP

Unlike at Benson WWTP, no decongestant was found from the Oxford WWTP influent, within the limits of quantification. However, decongestants were quantified from the Oxford WWTP effluent during the hours 2:00 and 3:00–notably the time when the Oxford WWTP influent flow was at its minimum (Fig. S1 in File S1); these were xylometazoline (15 ng/L, 28 mg/d, n = 1) and oxymetazoline (16 ng/L, 66 mg/d, n = 2), the latter of which indicated a propensity for concentration within the Benson WWTP, while the former did not.

7.3.2 PHARMACEUTICAL OCCURRENCE IN THE THAMES RIVER CATCHMENT

7.3.2.1 ANTIVIRAL

OC was the most frequently measured analyte with concentrations >LOQ at 73% of the river sampling locations during the month of November, 2009. The mean concentration of OC across the Thames catchment was 65, 61, 33 and 33 ng/L for November 3, 10, 16, and 24, respectively (Fig. 2 & 3). A maximum OC concentration of 193 ng/L was recorded at The Cut at Paley Street (T15) on Nov 10 (Figs. 3, S5b in File S1), a site known to be among the more severely impacted by sewage [22] with relatively low dilution per capita (Fig. S2 in File S1). The mean load of OC across all 21 sites was 24, 18, 46 and 55 g OC/d for November 3, 10, 16 and 24, respectively, with a maximum load on November 24 at the Thames at Runnymead (TC16) site, the location with the highest upstream population (2.3 million; Fig. S2 in File S1). A strong positive correlation ($R^2 = 0.82$ to 0.96) between the population upstream of a sampling location and the load of OC further confirms this expected relationship (Fig. 2). The Thames at Runnymead (TC16; Fig. 1), recorded 117, 98, 319, and 377 g OC/d (49, 41, 134, 158 μg/capita/d, respectively) for the same time points in November (Fig. 4). With only a few exceptions, the per capita usage at TC16 (the most downstream sampling point on the River Thames) was consistent with estimates generated from many of the upstream sampling sites (Fig. 4), indicating that OC was relatively conserved within the river environment.

OC was not found in any river sampling location in March 2010 or May 2011 except for one location in March, TC17 (River Wye at Bourne End; Fig. S5e, f in File S1). In this location a concentration of 6 ng/L was recorded with an estimated load/d of 622 mg/d or 8.48 μg/capita/d, approximately 13% that achieved during the month of November (63 μg/capita/d) at this site.

7.3.2.2 ANTIBIOTICS

The concentration range for antibiotics largely fell within the low-ng/L range (17–74 ng/L; Figs. 3, S5a–e in File S1), with the exception of clarithromycin (max = 292 ng/L) and erythromycin (max = 448 ng/L), at TC15 and TC14, respectively. Much like OC, the load of antibiotics across all time points was positively correlated with the number of people upstream of the sampling location. Erythromycin was by far the most frequently recorded antibiotic, >LOQ in 87% of samples, equally as abundant as OC (Fig. 3 & 5). Clarithromycin and trimethoprim were recorded in approximately 50% of samples. Ciprofloxacin and norfloxacin were measured in 39 and 33% of samples. Sulfamethoxazole was found in 30% of the river samples, while they were recorded in 79–100% of WWTP inlet and 68–83% of outlet samples. Similarly, doxycycline was frequently found in WWTP inlet and outlet samples (96–100% and 71–100%, respectively), but only measured >LOQ in 18% of river samples.

The mean load of antibiotics (μg) per capita (upstream population) per day was calculated for each of the analytes across the four November sampling times. Estimates range between 84 μg/cap/d for TC6, where only 4 antibiotics were recorded (ciprofloxacin, clarithromycin, erythromycin and norfloxacin), to 893 μg/cap/d for TC4 where six antibiotics were recorded. Despite all 11 antibiotics being recorded at TC15, it was among the lower yielding sites at 119 μg/cap/d. The low levels of antibiotics per person in The Cut at Paley Street could be down to the fact that The Cut is impacted by a very small number (3) of relatively large WWTPs (Bracknell = 77600 person equivalents (PE); Ascot = 26000 PE; White Waltham = 5150 PE), and therefore all the effluent has a relatively good level of treatment. There are no small WWTPs in this catchment, unlike all the other sites. The Cut was also one of only three sites to record cefotaxime. The sites that reflect the highest population upstream, TC18, TC13 and TC16, yielded estimates of 633, 199 and 263 μg/cap/d and recorded 8 or 9 of the 11 antibiotics.

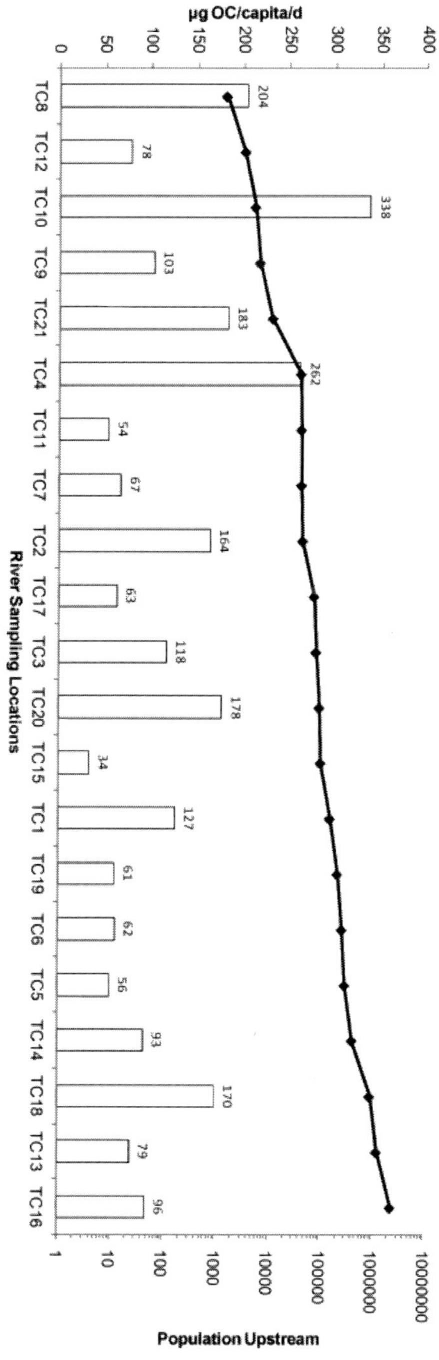

FIGURE 4: Mean load of oseltamivir (OC) per capita per day (µg/cap/d) across all river sampling locations throughout November 2009 (intra-pandemic period). Triangles indicate the population upstream at that location.

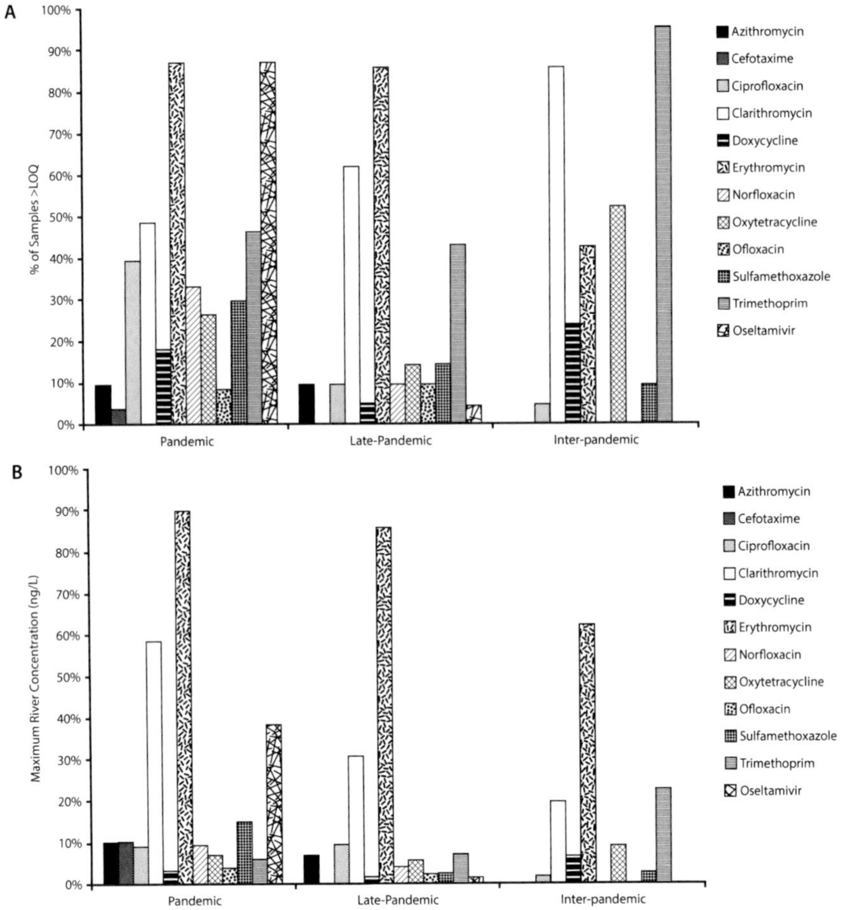

FIGURE 5: Comparison of pharmaceutical (A) abundance (% of samples above the limit of quantification (LOQ)) and (B) the maximum river concentration (ng/L) during the pandemic period (November, 2009), late pandemic period (March 2010) and the inter-pandemic period (May 2011).

7.3.2.3 LATE-PANDEMIC/INTER-PANDEMIC PERIOD

Fewer antibiotics were found at the late-pandemic period (mean 2.6±2.6; March 15, 2010) and inter-pandemic period (mean 3.1±1.7; May 11, 2011) than in the pandemic period (mean 6.9±2.3; Figs. 5 and S5a–e in File S1). The load of antibiotic found at each river location during the inter-pandemic period (May 2011) was less than that found during the pandemic (Figs. S5a–e in File S1). In all but three cases (TC11, TC15 and TC17), the same was true for the late-pandemic sampling points. Notably during the late-pandemic period, erythromycin was found at a concentration of 400 ng/L in TC15, a highly sewage impacted site; a site that achieved nearly 750 ng/L in total antibiotics. The high erythromycin level was again found in the inter-pandemic sampling (May, 2011) at TC17, achieving 311 ng/L (Fig. S5e in File S1). At site TC15, the level of trimethoprim was also exceptionally high (113 ng/L), far exceeding the maximum concentration found during the pandemic (November, 2009; 28 ng/L). It is unclear whether these levels are the result of a local outbreak, normal fluctuation, or evidence of improper disposal.

7.3.2.4 DECONGESTANTS

Naphazoline, the most frequently found decongestant in the Benson WWTP was not found in the river samples. Oxymetazoline in the Benson WWTP effluent, along with xylometazoline, was also not found in the river samples. Decongestants were also not found in the river samples collected during the late-pandemic or the inter-pandemic period.

7.3.3 DRUG USE BY WASTEWATER EPIDEMIOLOGY

Estimation of oseltamivir compliance, i.e., use of the drug as prescribed, was approximately 45–60% as previously reported [34]. Estimation of antibiotic use were derived from government statistics (mg/excreted/d; Table 3). Overall, neither of the government statistics (PCT or NHS BSA) were accurate predictors of MELs.

TABLE 3: Comparison of measured environmental load (MEL) of antibiotics in WWTP influent to MEL estimates generated from PCT- and NHS BSA-derived prescriptions.

	Benson			Oxford		
	mg/excreted/d[a]		mg/d	mg/excreted/d[a]		mg/d
	PCT[b]	NHS BSA[c]	MEL[d]	PCT[b]	NHS BSA[c]	MEL[d]
Azithromycin	55.2	75.7	32.3	652	6493	4782
Cefotaxime	na	na	7.00	na	na	<LOQ
Ciprofloxacin	389	1229	221	7150	105409	57258
Clarithromycin	89.2	441	43.7	2035	37810	27846
Doxycycline	54.0	34.2	414	757	2933	3234
Erythromycin	1259	3920	847	18833	336086	69475
Norfloxacin	0	0	37	0	0	9677
Ofloxacin	0	6	1462	26.1	479	4257
Oxytetracycline	195	187	236	3080	16040	56136
Sulfamethoxazole	51.9	103	75.6	693	8799	8956
Trimethoprim	800	1120	312	6693	96060	3548

[a]Predicted amount of drug excreted from the daily doses (ADQ) of antibiotic consumed during 24-h sampling from Benson and Oxford WWTP influent as per: [b]Primary Care Trust (PCT) and [c]National (NHS BSA) statistics for antibiotic us. [d]Measured environmental load (MEL) from 24-hourly wastewater inlet samples (sum of each antibiotic/24 h). No biodegradation other than pharmacokinetics (Table 1) was assumed. na = prescription statistics were unavailable.

7.4 DISCUSSION

7.4.1 PHARMACEUTICAL FATE AND OCCURRENCE IN OXFORD AND BENSON WWTPS, AND THE THAMES RIVER CATCHMENT

7.4.1.1 ANTIVIRAL

Mean concentrations of OC in the WWTP influent (350, 443 ng/L) and effluent (208, 358 ng/L) were consistent with the range previously reported in the literature 11–1450 ng/L (Table S2 in File S1) [1], [25], [35]–[39]. The maximum concentration found in this study (2,070 ng/L, Benson influent) was the highest reported in the literature during the pandemic (827 ng/L [35]). The persistence of OC in the Oxford WWTP is consistent with many reports in the literature that document minimal loss of OC in laboratory and field studies [13], [35], [37], [38], [40]. However, the extent of OC loss seen in the Benson WWTP (approximately 50%), was well above the range reported for WWTP not using ozonation, but it is consistent with the removal efficiency reported in WWTPs with ozonation in Germany, where elimination of OC was reported to be 59% [25], and Japan, where loss was between 30–40% [38].

Given the mild nature of the pandemic, measured environmental concentrations (MECs) of OC (33–62 ng/L) were 2–3 orders of magnitude lower than predicted environmental concentrations (PECs) in this catchment during a severe pandemic, i.e., R_0>2.0 [2], [7], but is consistent with the lower end of PECs for a mild pandemic within the Thames catchment (27–11,000 ng/L; [2]). As expected, no OC was found during the inter-pandemic period (May 2011) from either the WWTP effluent or river sampling locations, reinforcing the assumption that Oseltamivir was used in negligible amounts during the inter-pandemic period in the UK (Fig. 5).

Previous studies have demonstrated OC resistance development where influenza-infected mallard ducks had been exposed to water containing OC at concentrations ≥1000 ng/L [15], [18], which closely approximates concentrations recorded in Japan [36], [41] and within an order of magnitude of many measured concentrations in the literature and this study (Table S2 in File S1). The evidence in this study and the literature further lend support to the likelihood of OC-resistance generation in wildfowl influenza viruses; owing to its significance to human health, this is worthy of additional study.

7.4.1.2 ANTIBIOTICS

In general, the antibiotics were more labile in the Oxford WWTP as compared with Benson, mostly likely due to the fact that Oxford uses activated sludge treatment while Benson relies on a trickling filter to treat wastewater. Fewer different antibiotics were recovered in the river during the late-pandemic (March 15, 2010) and the inter-pandemic (May 11, 2011) period than in the peak pandemic period (November 2009; Figure 5). However, some antibiotics did not show any appreciable decline in concentration (erythromycin, oxytetracycline and trimethoprim). It is unclear to what extent the persistence of these antibiotics in the river is related to the average river temperatures; it could be expected that higher river temperatures would facilitate the biotransformation of the antibiotics. The water temperature during the March 15, 2010 sampling was the lowest (8.2°C), followed by the four November sampling points (12.7, 9.7, 9.9 and 10.3°C respectively), while the water temperature in May 11, 2011 was the highest, 15.6°C. Evidence for an exponential relationship between the biodegradation rate and temperature suggest the temperature range in this study is likely to be an important variable in explaining the frequency and concentration of analytes recovered in different seasons [42], [43].

7.4.1.3 DECONGESTANTS

The detection of decongestants in Oxford WWTP effluent (notably in the lowest flowing period of the day), and not in the influent, suggests a degree of concentration within the WWTP. This is likely due to the relatively high lipophilicity of the decongestants (predicted logP: naphazoline 3.8, xylometazoline 5.2, and oxymetazoline 4.5; www.chemspider.com). High logP would facilitate partitioning of the decongestant onto suspended organic matter which is recycled in the Oxford WWTP. Notably, the trickling filter treatment of Benson WWTP also appeared to concentrate oxymetazoline, with negligible change in naphazoline and a significant decline in xylometazoline. Although all the decongestants are found in the WWTPs, they do not appear to persist in rivers, or if they do, they persist in an adsorbed state (i.e., to sediment), which was not measured as part of this study. Future studies will need to examine how representative this study is across a wider range of WWTPs and over a longer period of monitoring. Furthermore, future research should target different environmental matrices to ensure these pharmaceuticals are not accumulating in river sediment.

7.4.2 DRUG USE BY WASTEWATER EPIDEMIOLOGY

7.4.2.1 ANTIVIRAL

The potential for wastewater epidemiology is highest when studying a recalcitrant, water soluble pollutant. This is one of the major reasons why the wastewater epidemiology approach has so much potential when applied to OC [34]. The physico-chemical benefits of OC are further enhanced because it is consumed as a prodrug. As such, improperly disposed Oseltamivir would be found in wastewater as the parent compound oseltamivir not OC. Hence, the difference between consumption and improper disposal can be relatively easily illuminated.

7.4.2.2 ANTIBIOTICS

To our knowledge the antibiotics within this study are not provided in a prodrug form, making estimates of their usage susceptible to misinterpretation owing to potential improper disposal. The recalcitrance of antibiotics is known to vary greatly in wastewater [44], [45] making it considerably more difficult to accurately predict environmental concentrations, as can been seen in Table 3. The variability in the recalcitrance of antibiotics in WWTPs and within the same WWTP over time and between differing WWTPs of different size and treatment technologies are likely among the main sources of error in the wastewater epidemiology approach when applied to antibiotics. Typically, the measured environmental load for each antibiotic (Table 3), was lower than the forward-calculated values from PCT or NHS BSA statistics, hence, the projected antibiotic users were frequently over estimated. Efforts to improve the wastewater epidemiology approach for antibiotics will need to address: (a) heterogeneity in the temporal distribution of prescriptions over time; (b) heterogeneity in the spatial distribution of prescriptions (across the UK) over time; (c) heterogeneity in in vivo and environmental stability of the antibiotic, including sewage pipes prior to reaching the WWTP inlet [46]; and (d) variability in compliance rate. It has been shown that the compliance rate for antibiotics can depend on the number of doses per day and age [47]–[49]. Further consideration should be given to (e) the sample size and sampling method. The relatively low number of pharmaceutical users in the two WWTP catchments, and the Benson WWTP catchment in particular, leaves model estimates of antibiotic users highly susceptible to systematic errors, as previously described [50]. The heterogeneity in the content of wastewater associated with low flush events, typical of low flow periods in the middle of the night, are a major factor influencing variations in analyte recovery over much of the sampling period [51]–[53]. This higher variability can be witnessed by the higher standard deviation in hourly measures of OC in Benson (433±472 ng/L) as compared to Oxford (358±60 ng/L). The Oxford and Benson sewer systems receive flow from a number of pumping stations that contribute to the mixing of discrete flushing events, however, the problems associated with sampling small populations would be more effectively alleviated with more intensive sampling (every 5–15 minutes) [50]. And finally, the wastewater epidemiology approach for antibiotics will likely be highly sensitive to (f) variability in environmental temperatures and precipitation, where low temperatures will likely retard biodegradation and high precipitation will dilute potentially inhibitory levels of drug while also resuspending sediment that can subsequently influence the drug's fate.

7.4.2.3 DECONGESTANTS

The ability to predict decongestant users from measured concentrations in WWTP influent was constrained by the same systematic problems discussed earlier for the antibiotics, but might be further constrained by: 1) their apparent susceptibility to biodegradation; 2) their high rate of non-prescription use (i.e., over-the-counter), thereby hindering the acquisition of spatially and temporally resolved use data to confirm model projections; and 3) their more sporadic use pattern than antibiotics, the latter of which has a typical course of two to four tables per day for 7 to 10 days, whereas decongestants are only used as and when required. Given these many limitations, there was no ability to predict decongestant user numbers from measured environmental concentrations.

7.5 CONCLUSIONS

In hindsight, the 2009 influenza A(H1N1) pdm09a virus generated a relatively small number of fatalities as compared to severe pandemics like the 1918 'Spanish flu', which meant that the medical response was proportionately lower than would have been expected in a moderate or severe influenza pandemic. Hence, the potential negative effects to WWTP operation [13] and the environment proposed to occur in a moderate and severe pandemic [2], [14] were not reported. This study provides the first evidence that antibiotic and antiviral use was elevated during the pandemic. Theoretically, the antiviral recorded in the River Thames was of sufficient concentration to select for antiviral resistance in wildfowl [18], [54]. However, it remains to be demonstrated whether this had occurred.

There remains a great deal of uncertainty with regard to pharmaceutical use patterns during a pandemic, as a result of poor adherence to prescribed drugs [34] and the widespread use of over-the-counter medications. The focus on Oseltamivir here and in the literature is unlikely to reflect antiviral practices beyond 2020 owing to an increasing number of influenza antivirals in the pipeline [55]–[57]. However, for the time being, Oseltamivir remains one of the few antivirals within national stockpiles and as such, remains an important medical tool and potentially significant environmental pollutant [58]. Future influenza pandemics might, in fact, employ a combination therapy of two or more antivirals in an effort to combat resistance [59], [60].

Opportunities to ground truth model predictions for 'black swan events' such as influenza pandemics are, by definition, very rare (every 30 years), making this study conducted during the last influenza pandemic a unique window onto public health practice, human behavior, and drug adherence in the UK. It represents the first study to measure antibiotics and decongestants in influent and effluent and receiving rivers during a public health emergency, thereby establishing a baseline from which future modeling and risk assessments can be built in preparation for more severe public health emergencies.

REFERENCES

1. Azuma T, Nakada N, Yamashita N, Tanaka H (2013) Mass balance of anti-influenza drugs discharged into the Yodo River system, Japan, under an influenza outbreak. Chemosphere 93: 1672–1677. doi: 10.1016/j.chemosphere.2013.05.025
2. Singer AC, Colizza V, Schmitt H, Andrews J, Balcan D, et al. (2011) Assessing the ecotoxicologic hazards of a pandemic influenza medical response. Environ Health Perspect 119: 1084–1090. doi: 10.1289/ehp.1002757
3. Chang CC, Cheng AC, Chang AB (2014) Over-the-counter (OTC) medications to reduce cough as an adjunct to antibiotics for acute pneumonia in children and adults. Cochrane Database Syst Rev 3: CD006088. doi: 10.1002/14651858.cd006088.pub4
4. Dollery C (1999) Therapeutic Drugs. London: Harcourt Brace and Company Limited.
5. Knox C, Law V, Jewison T, Liu P, Ly S, et al. (2011) DrugBank 3.0: a comprehensive resource for 'omics' research on drugs. Nucleic Acids Res 39: D1035–1041. doi: 10.1093/nar/gkq1126
6. Edlund A, Ek K, Breitholtz M, Gorokhova E (2012) Antibiotic-induced change of bacterial communities associated with the copepod Nitocra spinipes. PLoS One 7: e33107. doi: 10.1371/journal.pone.0033107
7. Singer AC, Nunn MA, Gould EA, Johnson AC (2007) Potential risks associated with the proposed widespread use of Tamiflu. Environmental Health Perspectives 115: 102–106. doi: 10.1289/ehp.9574
8. Escher BI, Bramaz N, Lienert J, Neuwoehner J, Straub JO (2010) Mixture toxicity of the antiviral drug Tamiflu) (oseltamivir ethylester) and its active metabolite oseltamivir acid. Aquat Toxicol 96: 194–202. doi: 10.1016/j.aquatox.2009.10.020
9. Straub JO (2009) An environmental risk assessment for oseltamivir (Tamiflu) for sewage works and surface waters under seasonal-influenza- and pandemic-use conditions. Ecotoxicol Environ Saf 72: 1625–1634. doi: 10.1016/j.ecoenv.2008.09.011
10. Hutchinson TH, Beesley A, Frickers PE, Readman JW, Shaw JP, et al. (2009) Extending the environmental risk assessment for oseltamivir (Tamiflu) under pandemic use conditions to the coastal marine compartment. Environ Int 35: 931–936. doi: 10.1016/j.envint.2009.04.001
11. Chen WY, Lin CJ, Liao CM (2014) Assessing exposure risks for aquatic organisms posed by Tamiflu use under seasonal influenza and pandemic conditions. Environ Pollut 184: 377–384. doi: 10.1016/j.envpol.2013.09.019
12. Sacca ML, Accinelli C, Fick J, Lindberg R, Olsen B (2009) Environmental fate of the antiviral drug Tamiflu in two aquatic ecosystems. Chemosphere 75: 28–33. doi: 10.1016/j.chemosphere.2008.11.060
13. Slater FR, Singer AC, Turner S, Barr JJ, Bond PL (2011) Pandemic pharmaceutical dosing effects on wastewater treatment: no adaptation of activated sludge bacteria to degrade the antiviral drug oseltamivir (Tamiflu) and loss of nutrient removal performance. FEMS Microbiol Lett 315: 17–22. doi: 10.1111/j.1574-6968.2010.02163.x
14. Singer AC, Howard BM, Johnson AC, Knowles CJ, Jackman S, et al. (2008) Meeting report: risk assessment of tamiflu use under pandemic conditions. Environ Health Perspect 116: 1563–1567. doi: 10.1289/ehp.11310

15. Achenbach JE, Bowen RA (2013) Effect of oseltamivir carboxylate consumption on emergence of drug-resistant H5N2 avian influenza virus in Mallard ducks. Antimicrob Agents Chemother 57: 2171–2181. doi: 10.1128/aac.02126-12
16. Jarhult JD (2012) Oseltamivir (Tamiflu) in the environment, resistance development in influenza A viruses of dabbling ducks and the risk of transmission of an oseltamivir-resistant virus to humans - a review. Infect Ecol Epidemiol 2.
17. Gillman A, Muradrasoli S, Söderström H, Nordh J, Bröjer C, et al. (2013) Resistance Mutation R292K Is Induced in Influenza A(H6N2) Virus by Exposure of Infected Mallards to Low Levels of Oseltamivir. PLoS ONE 8: e71230. doi: 10.1371/journal.pone.0071230
18. Jarhult JD, Muradrasoli S, Wahlgren J, Soderstrom H, Orozovic G, et al. (2011) Environmental levels of the antiviral oseltamivir induce development of resistance mutation H274Y in influenza A/H1N1 virus in mallards. PLoS One 6: e24742. doi: 10.1371/journal.pone.0024742
19. Wellington EM, Boxall AB, Cross P, Feil EJ, Gaze WH, et al. (2013) The role of the natural environment in the emergence of antibiotic resistance in gram-negative bacteria. Lancet Infect Dis 13: 155–165. doi: 10.1016/s1473-3099(12)70317-1
20. Gaze WH, Krone SM, Larsson DG, Li XZ, Robinson JA, et al. (2013) Influence of humans on evolution and mobilization of environmental antibiotic resistome. Emerg Infect Dis 19.
21. World Health Organization (2010) WHO recommendations for the post-pandemic period: Pandemic (H1N1) 2009 briefing note 23. Available: http://www.who.int/csr/disease/swineflu/notes/briefing_20100810/en/. Accessed 2012 Jun 1.
22. Bowes MJ, Jarvie HP, Naden PS, Old GH, Scarlett PM, et al. (2014) Identifying priorities for nutrient mitigation using river concentration–flow relationships: The Thames basin, UK. Journal of Hydrology 517: 1–12. doi: 10.1016/j.jhydrol.2014.03.063
23. Harvey CL, Dixon H, Hannaford J (2012) An appraisal of the performance of data-infilling methods for application to daily mean river flow records in the UK. Hydrology Research 43: 618–636. doi: 10.2166/nh.2012.110
24. TuTiempo.net (2009) Climate: Benson (November, 2009). Available: http://www.tutiempo.net/en/Climate/Benson/11-2009/36580.htm. Accessed 2012 Oct 26.
25. Prasse C, Schlusener MP, Schulz R, Ternes TA (2010) Antiviral drugs in wastewater and surface waters: a new pharmaceutical class of environmental relevance? Environ Sci Technol 44: 1728–1735. doi: 10.1021/es903216p
26. Balcan D, Colizza V, Singer AC, Chouaid C, Hu H, et al. (2009) Modeling the critical care demand and antibiotics resources needed during the Fall 2009 wave of influenza A(H1N1) pandemic. PLoS Curr 1: RRN1133. doi: 10.1371/currents.rrn1133
27. Khan GA, Lindberg R, Grabic R, Fick J (2012) The development and application of a system for simultaneously determining anti-infectives and nasal decongestants using on-line solid-phase extraction and liquid chromatography-tandem mass spectrometry. J Pharm Biomed Anal 66: 24–32. doi: 10.1016/j.jpba.2012.02.011
28. NHS BSA (2011) Prescribing Analysis Charts: National Antibiotics Charts. Available: http://www.nhsbsa.nhs.uk/PrescriptionServices/2587.aspx. Accessed 2012 Oct.

29. NHS Information Centre for Health and Social Care (2011) Number of GP Prescriptions by drug by postcode. Available: http://data.gov.uk/apps/number-of-gp-pre scriptions-by-drug-by-postcode.
30. Health Protection Agency (2010) Epidemiological report of pandemic (H1N1) 2009 in the UK (April 2009 - May 2010). Available: http://www.hpa.org.uk/webc/HPAwebFile/HPAweb_C/1284475321350. Accessed 2012 Jun 1.
31. Rogers S, Sedghi A (2011) Swine flu data in the UK: see how bad cases are where you live. Guardian Available: http://www.guardian.co.uk/news/datablog/2011/jan/06/swine-flu-data-uk#data. Accessed 2012 Jun 1.
32. The Health and Social Care Information Centre PSU (2012) Average Daily Quantities. Available: http://www.ic.nhs.uk/services/prescribing-support-unit-psu/using-the-service/reference/measures/volume-measures/average-daily-quantities-adq. Accessed 2012 Oct 19.
33. Davey P, Ferech M, Ansari F, Muller A, Goossens H (2008) Outpatient antibiotic use in the four administrations of the UK: cross-sectional and longitudinal analysis. Journal of Antimicrobial Chemotherapy 62: 1441–1447. doi: 10.1093/jac/dkn386
34. Singer AC, Jarhult JD, Grabic R, Khan GA, Fedorova G, et al. (2013) Compliance to oseltamivir among two populations in Oxfordshire, United Kingdom affected by influenza A(H1N1) pdm09, November 2009–a waste water epidemiology study. PLoS One 8: e60221. doi: 10.1371/journal.pone.0060221
35. Azuma T, Nakada N, Yamashita N, Tanaka H (2012) Synchronous dynamics of observed and predicted values of anti-influenza drugs in environmental waters during a seasonal influenza outbreak. Environ Sci Technol 46: 12873–12881. doi: 10.1021/es303203c
36. Takanami R, Ozaki H, Giri RR, Taniguchi S, Hayashi S (2012) Antiviral Drugs Zanamivir and Oseltamivir Found in Wastewater and Surface Water in Osaka, Japan. Journal of Water and Environment Technology 10: 57–68. doi: 10.2965/jwet.2012.57
37. Leknes H, Sturtzel IE, Dye C (2012) Environmental release of oseltamivir from a Norwegian sewage treatment plant during the 2009 influenza A (H1N1) pandemic. Sci Total Environ 414: 632–638. doi: 10.1016/j.scitotenv.2011.11.004
38. Ghosh GC, Nakada N, Yamashita N, Tanaka H (2010) Occurrence and fate of oseltamivir carboxylate (Tamiflu) and amantadine in sewage treatment plants. Chemosphere 81: 13–17. doi: 10.1016/j.chemosphere.2010.07.023
39. Ghosh GC, Nakada N, Yamashita N, Tanaka H (2010) Oseltamivir carboxylate, the active metabolite of oseltamivir phosphate (Tamiflu), detected in sewage discharge and river water in Japan. Environ Health Perspect 118: 103–107. doi: 10.1289/ehp.0900930
40. Fick J, Lindberg RH, Tysklind M, Haemig PD, Waldenstrom J, et al. (2007) Antiviral oseltamivir is not removed or degraded in normal sewage water treatment: implications for development of resistance by influenza A virus. PLoS One 2: e986. doi: 10.1371/journal.pone.0000986
41. Takanami R, Ozaki H, Giri RR, Taniguchi S, Hayashi S (2010) Detection of antiviral drugs oseltamivir phosphate and oseltamivir carboxylate in Neya River, Osaka Japan. Journal of Water and Environment Technology 8: 363–372. doi: 10.2965/jwet.2010.363

42. Kohring GW, Rogers JE, Wiegel J (1989) Anaerobic biodegradation of 2,4-dichlorophenol in freshwater lake sediments at different temperatures. Applied and Environmental Microbiology 55: 348–353. doi: 10.1007/978-1-4899-0824-7_45
43. Atlas RM (1975) Effects of Temperature and Crude Oil Composition on Petroleum Biodegradation. Applied Microbiology 30: 396–403.
44. Michael I, Rizzo L, McArdell CS, Manaia CM, Merlin C, et al. (2013) Urban wastewater treatment plants as hotspots for the release of antibiotics in the environment: a review. Water Res 47: 957–995. doi: 10.1016/j.watres.2012.11.027
45. Luo Y, Guo W, Ngo HH, Nghiem LD, Hai FI, et al. (2014) A review on the occurrence of micropollutants in the aquatic environment and their fate and removal during wastewater treatment. Sci Total Environ 473–474: 619–641. doi: 10.1016/j.scitotenv.2013.12.065
46. Lindberg RH, Ostman M, Olofsson U, Grabic R, Fick J (2014) Occurrence and behaviour of 105 active pharmaceutical ingredients in sewage waters of a municipal sewer collection system. Water Res 58C: 221–229. doi: 10.1016/j.watres.2014.03.076
47. Kardas P (2002) Patient compliance with antibiotic treatment for respiratory tract infections. J Antimicrob Chemother 49: 897–903. doi: 10.1093/jac/dkf046
48. Llor C, Hernandez S, Bayona C, Moragas A, Sierra N, et al. (2013) A study of adherence to antibiotic treatment in ambulatory respiratory infections. Int J Infect Dis 17: e168–172. doi: 10.1016/j.ijid.2012.09.012
49. McNulty CA, Boyle P, Nichols T, Clappison P, Davey P (2007) The public's attitudes to and compliance with antibiotics. J Antimicrob Chemother 60 Suppl 1: i63–68. doi: 10.1093/jac/dkm161
50. Ort C, Lawrence MG, Reungoat J, Mueller JF (2010) Sampling for PPCPs in wastewater systems: comparison of different sampling modes and optimization strategies. Environ Sci Technol 44: 6289–6296. doi: 10.1021/es100778d
51. Ort C, Gujer W (2006) Sampling for representative micropollutant loads in sewer systems. Water Sci Technol 54: 169–176. doi: 10.2166/wst.2006.591
52. Ort C, Schaffner C, Giger W, Gujer W (2005) Modeling stochastic load variations in sewer systems. Water Sci Technol 52: 113–122.
53. Ort C, Lawrence MG, Rieckermann J, Joss A (2010) Sampling for pharmaceuticals and personal care products (PPCPs) and illicit drugs in wastewater systems: are your conclusions valid? A critical review. Environ Sci Technol 44: 6024–6035. doi: 10.1021/es100779n
54. Gillman A, Muradrasoli S, Soderstrom H, Nordh J, Brojer C, et al. (2013) Resistance mutation R292K is induced in influenza A(H6N2) virus by exposure of infected mallards to low levels of oseltamivir. PLoS One 8: e71230. doi: 10.1371/journal.pone.0071230
55. Ivachtchenko AV, Ivanenkov YA, Mitkin OD, Yamanushkin PM, Bichko VV, et al. (2013) A novel influenza virus neuraminidase inhibitor AV5027. Antiviral Res 100: 698–708. doi: 10.1016/j.antiviral.2013.10.008
56. Wathen MW, Barro M, Bright RA (2013) Antivirals in seasonal and pandemic influenza–future perspectives. Influenza Other Respir Viruses 7 Suppl 1: 76–80. doi: 10.1111/irv.12049

57. Jefferson T, Jones MA, Doshi P, Del Mar CB, Hama R, et al. (2014) Neuraminidase inhibitors for preventing and treating influenza in healthy adults and children. Cochrane Database Syst Rev 4: CD008965. doi: 10.1002/14651858.cd008965.pub3
58. Drinking Water Inspectorate (2007) Desk based review of current knowledge on pharmaceuticals in drinking water and estimation of potential levels (Defra Project Code: CSA 7184/WT02046/DWI70/2/213). Available: http://googl/nkBVlP.
59. Perelson AS, Rong L, Hayden FG (2012) Combination Antiviral Therapy for Influenza: Predictions From Modeling of Human Infections. Journal of Infectious Diseases 205: 1642–1645. doi: 10.1093/infdis/jis265
60. Kim W-Y, Young Suh G, Huh JW, Kim S-H, Kim M-j, et al. (2011) Triple-Combination Antiviral Drug for Pandemic H1N1 Influenza Virus Infection in Critically Ill Patients on Mechanical Ventilation. Antimicrobial Agents and Chemotherapy 55: 5703–5709. doi: 10.1128/aac.05529-11

There are several supplemental files that are not available in this version of the article. To view this additional information, please use the citation on the first page of this chapter.

CHAPTER 8

Wastewater Recycling in Greece: The Case of Thessaloniki

ANDREAS ILIAS, ATHANASIOS PANORAS, AND ANDREAS ANGELAKIS

8.1 INTRODUCTION

Treated wastewater has been increasingly used around the world for irrigation, environmental applications, industrial use, groundwater recharge, urban use, indirect potable use and in some rare cases, direct potable use [1,2]. The principal causes preventing the expansion of effluent reuse worldwide are public health and environmental concerns [2]. To reduce the potential risks to acceptable levels, many countries have set regulations or guidelines governing effluent reuse [3,4].

The climate of Greece is sub-humid Mediterranean with humid and relatively cold winters and dry and warm summers, with an average rainfall of 874 mm/year (115,937 Mm^3/year). Greece is characterized by high temporal and spatial precipitation imbalance, with low precipitation and increased water demands for irrigation and tourism during the summer. Annual rainfall ranges from 300 to 500 mm in southeastern Greece and from 800 to 1200 mm in the northwestern plains of the mainland, while in some mountainous areas it may be above 2000 mm. It is worth mentioning that during dry years, and at the most hot and arid areas of the country, the irrigation period may be as long as 6 to 7 months. Effluent from WWTPs is an alternative water resource that should be considered in order to catch up with the demand [5,6].

A high potential of reusing treated wastewater for crop and landscape irrigation has been reported in Greece by several researchers ([7,8,9,10] and others). In 2009, it was estimated that more than 75% of the Greek population was connected to WWTPs with a total capacity of over 1.50 mm^3/day [10]. An analysis of data concerning the water balance of the areas served by WWTPs has demonstrated that more than 83% of the treated effluents are produced in regions with a negative water balance [9].

In this paper, the wastewater treatment and reuse status in Greece, in terms of treatment technology, effluent quality, legislation, and future perspectives is presented. Emphasis is given to the major water recycling project in Greece, which is the reuse of the Thessaloniki WWTP effluent for crop irrigation.

8.2 WASTEWATER TREATMENT TECHNOLOGIES AND STATUS

Greece, with a population of approximately 11 million inhabitants, has to comply with the EU Urban Wastewater Treatment Directive (271/91) treatment [11]. According to this Directive, Greece as a member of EU, was bound to connect all urban agglomerations with more than 2000 PE (population equivalents) to WWTPs by the end of 2005. In 2005, it was reported by Tsagarakis that about 350 WWTPs served over 75% of the country's permanent population [9]. In 2010, the Greek Ministry of Environment, Energy and Climate Change reported that 100%, 93%, 74% and 32% of the population living in agglomerations of over 150,000 EP, 15,000–150,000 EP, 10,000–15,000 EP and 2000–10,000 EP, respectively, were covered by complying to the EU regulations for WWTPs ([12]. Furthermore, it is noted that 10% of the population live in villages of less than 500 PE, for which on-site sanitation technologies should be applied.

Thus far, a number of different technologies for municipal wastewater treatment have been adopted in Greece. Out of them, 88% are activated sludge systems, 10% are natural systems and 2% are attached growth systems. Activated sludge systems are comprised of 85% extended aeration systems, 10% conventional systems and 5% sequencing batch reactors. It is obvious that extended aeration is the dominant system, as it provides significant advantages for the Mediterranean climatic conditions. Out of the activated sludge systems, 44% remove nitrogen and 15% remove phosphorous [13].

In the early years of wastewater treatment implementation, Greece was lacking the necessary knowledge and its own regulations, and therefore the effluent was either tossed away or reused following other countries' regulations. Until recently, most of the country's WWTPs outflows were driven to sea, or to permanent or ephemeral water bodies, such as lakes, rivers or torrents. Occasionally treated wastewater was applied at agricultural or forest land. However, during the last years, water recycling has increasingly been adopted instead of disposal, mainly for crop, forestry or landscape irrigation [9].

8.3 WATER EFFLUENT QUALITY

An early evaluation of the performance of WWTPs, taking into account physicochemical and biological parameters, was performed by Tsagarakis et al. [13]. According to that study, 42% of the operating WWTPs performed well, 41% moderately and 17% poorly. These findings also revealed a positive correlation between the size of WWTP and effluent quality. Furthermore, Mamais et al. [14] investigating the quality of effluent from fifteen large WWTPs, representing the 60% of effluent volume in Greece, found that most of these WWTPs appeared to comply with EC Directive 91/271 for the produced effluent. Most of the reported WWTPs performed well in terms of BOD_5 and TSS removal. However, a few higher BOD_5 values have been reported for some WWTPs due to insufficient aeration, or foaming incidents, or to overloading, especially during the summer [10,15]. Increased levels of salts in the effluent of a number of

WWTPs suggest that appropriate management practices should be implemented when it is to be used for agricultural reuse, to ensure land sustainability and productivity. Salinity problems are more frequent in WWTPs located at coastal areas, likely due to seawater intrusion [8], or intrusion from shallow saline aquifers into the municipal sewerage systems. In terms of trace elements, limited data are available, but it appears that they do not pose a risk for the environment or public health [2,16]. Finally, the microbiological parameters fecal coliforms (FC) and total coliforms (TC) range from 0 to 500 and 3,0 to 1500 MPN/100 mL respectively, in the major WWTPs of Greece [10].

8.4 HISTORY OF WATER RECYCLING AND REUSE IN GREECE

Land application of wastewater is an old and common practice, which has gone through different development stages with time, knowledge, treatment technology, and regulation evolution [17]. Greece, as is the case with other European regions, has a long history on wastewater reuse, as it has been practiced since the Ancient Greek and Roman civilizations [18]. Wastewater was used for irrigation in Mediterranean cities in the 14th and 15th centuries in, for example, the Milanese Marcites and the Valencia huerta, in addition to central and northern European countries like Great Britain, Germany, France, and Poland [19]. Land application of wastewater was practiced in modern Greece in the late 1970s, just before the establishment of the first WWTP.

8.5 EFFLUENT RECYCLING AND REUSE IN GREECE: MAJOR PROJECTS

Although many regions of Greece, especially its southeastern regions, are under water stress, effluent reuse is not practiced widely. Tsagarakis et al. [20] estimated that by reusing effluent of existing WWTPs, in order to increase water availability for crop irrigation and ensure environmental protection, 3.2% of the total water currently used for irrigation would be saved. This percentage has actually been exceeded to 4.5% [10]. However, the lack of a recycling context and the complex approval processes have inhibited the development of well-organized water recycling projects. Over the last few years, however, the concern for effluent reuse has arisen and many recycling projects are being implemented or planned in Greece, mainly for crop or landscape irrigation.

In Thessaloniki, the secondary effluent of the WWTP (165,000 m3/day), is used for agricultural irrigation of the Halastra-Kalohori area, after mixing with freshwater from the Axios River at a 1:5 ratio. Approximately 2500 ha of spring crops are irrigated in the Halastra-Kalohori area, with the freshwater—effluent mixture. A recycling scheme has been in operation at the surrounding area of Chalkida since 1998 for landscape irrigation [21]. In the WWTP that is located on the island of Ioannina, the effluent receives tertiary treatment which includes coagulation, filtration, UV disinfection and chlorination. The infrastructure includes four underwater pipes of a total length of 2500 m for effluent transportation, an effluent storage tank of 200 m^3, a distribution network, and a supervision control and data acquisition system for the treatment and irrigation procedures. In this project, approximately 4000 m^3/day are recycled for tree and shrub irrigation. A number of small wastewater reuse projects are in operation around the country, such as those at Chersonisos, Crete (4500 m^3/day) mainly for agricultural irrigation, and secondarily for fire protection and landscape irrigation, at Malia (2500 m^3/day), at Levadia (3500 m^3/day), at the island of Kos (3500 m^3/day), at Amfissa (400 m^3/day), and at Nea Kalikratia (800 m^3/day). A more recent water reuse project has been in operation in the city of Iraklion (Crete) since 2012. The effluent of a tertiary treatment plant (9500 m^3/day), which includes coagulation, filtration, and UV disinfection, is used for grape and olive tree irrigation, in the southwestern area of the city. Furthermore a membrane bioreactor (MBR) treatment and reuse plant is under construction at the same area.

A few other projects are under planning, such as the one at Agios Nikolaos in eastern Crete, at Lamia in central Greece, and on several Aegean islands. Finally, it should be mentioned that there are several cases of indirect reuse, especially in central Greece, which occur after the disposal of the treated effluents in rivers and the downstream uptake of the mixed water for irrigation (Larissa, Trikala, Karditsa, Lamia, and Tripolis). The reuse potential in Greece, however, is restricted since effluent recycling from Athens's WWTP, which serves approximately 35% of the country's population, is most likely not economically feasible, due to the location of the treatment plant (the small island of Psitalia). Assessment of the possibility to transport 20,000 m^3/day of treated effluent from the island back to the city areas for landscape irrigation and industrial use at an estimated cost of 0.40€/m^3, was shown to be not cost effective, at least for the time being [22].

In Table 1, the major projects of treated wastewater reuse in Greece are presented, along with the capacity of the WWTPs, the irrigated areas, and the main crops irrigated with recycled water.

TABLE 1: Major wastewater reuse sites (adapted from [10,21,23]).

Project	Region	Capacity (m^3/day)[a]	Irrigated area (ha)	Crops
Irrigation of agricultural land				
Thessaloniki (Sindos)	Central Macedonia	165,000	2500	Corn, sugar beet, rice, etc.
Iraklion	Crete	9500	570	Grapes and Olive trees
Levadia	Central Greece	3500		Cotton, corn
Amfissa	Central Greece	400		Olive trees
Nea Kalikratia	Central Macedonia	800	150	Olive trees
Chersonissos	Crete	4500	270	Olive trees
Malia	Crete	2500	150	
Archanes	Crete	550	33	Grapes and olive trees
Kos	North Aegean	3500	210	Olive trees, citrus, etc.
Others		10,000		Various
Irrigation of other land (parks, forest, etc.)				
Chalkida	Central Greece	4000	50	
Chersonisssos	Crete	500	8	
Agios Costantinos	North Aegean	200	10	
Kentarchos	North Aegean	100	5	
Kos	North Aegean	500	10	
Karistos	North Aegean	1450	30	
Ierissos	South Aegean	1500	25	
Others		2000		

TABLE 1: *Cont.*

Project	Region	Capacity (m³/day)[a]	Irrigated area (ha)	Crops
Indirect reuse				
Larissa	Thessaly	25,000		Cotton, corn, etc
Karditsa	Thessaly	15,000		Cotton, corn, etc.
Lamia	Central Greece	15,000		Cotton, olive trees, corn, etc.
Tripoli	Peloponissos	18,000		
Others		35,000		
Total		318,500		

[a]*The effluent is used only during the dry period of the year, ranging from 3 to 6 months/year depending on climate, agronomical and other local conditions.*

8.6 THESSALONIKI WATER REUSE PROJECT

The Thessaloniki plain is one of the biggest agricultural areas of Greece (100,000 ha), located in the country's northern part. Extended open canal irrigation networks have been constructed to transport water from the Axios and Aliakmon Rivers to the fields. The main irrigated crops are rice, corn, cotton, sugar beet, alfalfa and orchards. During dry years, at the peak of the irrigation period (July–August) the flow of the two rivers that supply the network, especially the one of the Axios, is getting too low and additional water resources are needed in order to meet the demand.

A WWTP effluent reuse project has been built upon the findings of long (more than 10 years) experimental work carried out by the Land Reclamation Institute of the National Agricultural Research Foundation (LRI-NAGREF) in the area of the Thessaloniki WWTP [24]. The location of the WWTP and the recycling scheme of Thessaloniki area are shown in Figure 1.

8.7 EXPERIMENTAL WORK

In 1995, the Land Reclamation Institute (LRI-NAGREF) began to study and experiment on the utilization of Thessaloniki's WWTP effluent for crop irrigation. The appropriate practices to irrigate sugar beet and cotton (Figure 2), corn and rice (Figure 3), which are the main crops of the neighboring Halastra–Kalohori area, were investigated. During the experimentation, two water qualities, treated wastewater and freshwater (control), were tested in relation to the effects on crop production, soil properties, irrigation equipment and health risk. This important agricultural area with an open-canal collective irrigation network is located nearby Thessaloniki's WWTP (Figure 1), and the farmers often expressed their willingness to use the treated effluent on their fields.

Extended field experiments and research concerning the reuse of the treated effluent for crop irrigation (conducted by LRI-NAGREF) have resulted in the following conclusions.

Irrigation water evaluation. The evaluation of the suitability of the WWTP effluent for irrigation, taking into account its composition and the established agronomical standards, showed that if the recommended practices are followed, it could be safely used without posing high risk to the soil, crop, irrigation systems and human health. However, only rational use of the effluent and systematic surveillance of the system could ensure sustainability and long-term safety.

Soil. As mentioned previously, effluents from near-coast cities and WWTPs often have a high salt content. The use of such effluent increases the soil EC regardless of the irrigation method and, to be on the safe side, the level of the soil salinity and ESP values related to permeability issues have to be regularly monitored. Furthermore, taking into account the property of the soil to accumulate substances, long-term soil monitoring should be undertaken.

Trace elements. The trace elements concentration in the soil and plant tissues was quite low compared to the international criteria. This was due to the low concentration of trace elements in the effluent, ensuring a safe long-term use for crop irrigation [16].

Microorganisms. There is no significant health risk from the use of this effluent, because pathogenic microorganisms in the chlorinated effluent were well within the limits, according to various health criteria. Both the level furrows with blocked ends and the drip irrigation system satisfactorily protect the farmers from having contact with the water. Nevertheless, every regulation related to health protection should be strictly followed, as using treated wastewater always involves a potential health risk.

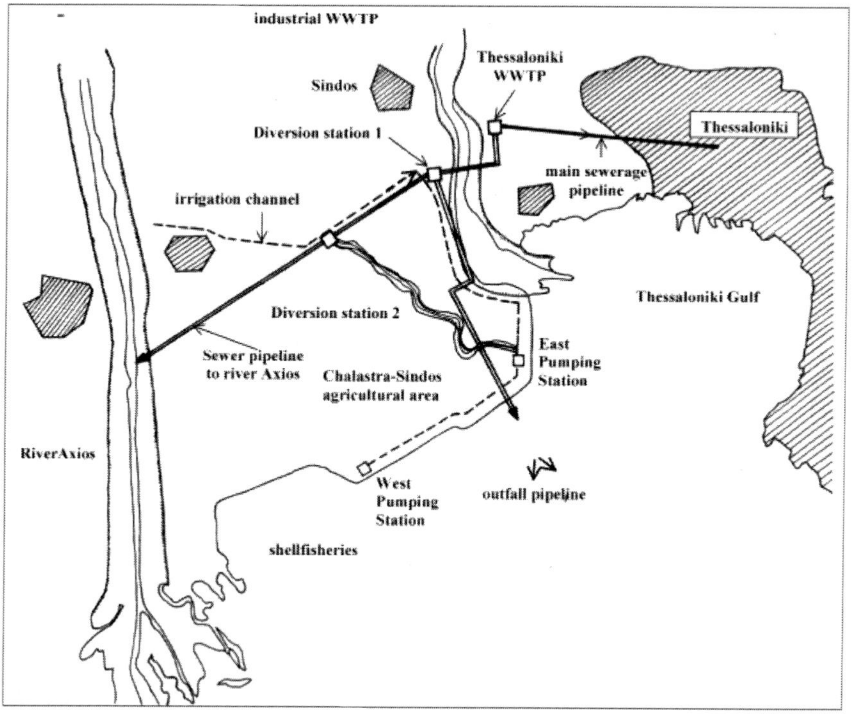

FIGURE 1: The WWTP and the recycling scheme of Thessaloniki [23].

FIGURE 2: Experimental fields irrigated with treated wastewater at Sindos, Thessaloniki: (a) sugar beet and (b) cotton.

FIGURE 3: Experimental fields irrigated with treated wastewater at Sindos, Thessaloniki: (a) corn and (b) rice.

FIGURE 4: Irrigation canal and mixing point of Thessaloniki's WWTP effluent with the river water from the Axios.

Drip irrigation/Clogging problems. All the emitter tests for the drip irrigation system have shown no emitter clogging even at the far-end of the laterals, after three years of continuous use of the effluent [25]. Furthermore, the test for deposits within the emitters have proved that the cylinder of the labyrinth and the inlet were clean. Only negligible amounts of deposits have been detected on the teeth of the labyrinth. However, there is still need for longer-term experiments.

Crop yields. Effects of the use of treated wastewater on crop yield were as follows:

- Sugar beet. No statistically significant differences either between the two water qualities (effluent and control water) or between the two irrigation methods (furrows and drip irrigation) have been detected, with relation to sugar content and white sugar yield [26].

- Cotton. Statistically significant differences in seed cotton yield and lint have been observed between the treated effluent and the control water. The wastewater has resulted in higher yield than the freshwater. There have been no significant differences in cotton quality characteristics between the two water quality treatments. Furthermore, no significant differences in seed cotton yield have been observed between furrow and drip irrigation, unlike the significant differences in lint, fiber length, micronaire and mean maturation date [27].
- Corn. The statistical analysis has shown that no significant differences in relation to corn yield have been observed between the two water qualities. On the other hand, there have been significant differences in terms of crop height between the two water qualities, in favor of the corn irrigated with the WWTP effluent. Significant differences in corn yield and crop height have been observed between furrow and drip irrigation. Furrow irrigation has produced 10% more corn yield than drip irrigation. The statistical analysis of the corn yield and crop height in relation to water quality or irrigation method has shown that the higher the plants, the less seed that is produced. Although the corn breeders have reported a positive correlation between plant height and yield, there are experimental results where no or negative correlation has been observed [28].
- Rice. The reuse of the wastewater for rice irrigation has not affected the yield and the quality characteristics in relation to irrigation with freshwater [29].

Conclusions. The evaluation of the results of crop yield and soil data in relation to the water quality and the irrigation method shows that the municipal wastewater of Thessaloniki city treated by activated sludge and chlorination can reinforce the water resources of the Thessaloniki plain. The treated effluent is discharged to the neighboring Halastra-Kalohori irrigation network, and reused after mixing with river water from the Axios. The mixing of the two waters is essential because of the effluent's increased salinity due to seawater intrusion into the sewerage system. Reuse of the treated effluent of the WWTP of Thessaloniki for crop irrigation will save freshwater for other uses and, furthermore, will protect the Gulf of Thessaloniki from pollution [24].

8.8 PROJECT IMPLEMENTATION

The aforementioned findings helped to implement a wastewater reuse scheme to the Halastra-Kalohori irrigation network since 2007 (Figure 4). Part of a wastewater effluent of 165,000 m³/day from the Thessaloniki

WWTP mixed with river water from the Axios is reused to irrigate approximately 2500 ha of rice, corn, alfalfa, sugar beet and cotton. The distribution of the major crops in the irrigated area is shown in Table 2. Water requirements varied from 60 to 111 m^3/ha/day, for cotton and rice, respectively. The minimum and the maximum water requirements were determined upon the irrigation method and the water application efficiency. In the traditional surface water irrigation methods, the efficiency is less than 70% [24].

TABLE 2: Distribution of major crops in the irrigated area of the water reuse project [30].

Crops	Area (ha)	Percentage (%)	Water requirements (max and min in m^3/ha/day)
Cotton	104	4.26	54–66
Corn	559	22.8	62–76
Alfalfa	111	4.5	70–86
Rice	1677	68.5	100–222

In this implementation, the effluent is mixed with freshwater from the Axios River at a maximum ratio of 1:5 (Figure 4). The exact percentage of wastewater in the mixture depends on the quality of the effluent and especially its salinity, and is decreased if higher than usual salinity values are measured [24]. The Land Reclamation Institute is in charge of the surveillance of the system, checking the quality characteristics of the effluent delivered to the irrigation network and the possible effects of its use. During the early years of the project, the mixed effluent (WWTP effluent mixed with the Axios's river water) should follow the quality criteria: pH 6.5-8.5, EC < 3.0 dS/m, $BOD_5 < 20$ mg/L, COD < 80 mg/L, SS < 30 mg/L, Residual chlorine < 0.5 mg/L, FC $< 1000/100$ mL, Helminths eggs < 1 egg/L, B < 2 mg/L, TKN < 30 mg/L according to the 141937 Common Ministerial Decision [31]. The actual concentration of pathogens in WWTP effluent, the Axios's freshwater and the mixed water are shown in Table 3. As indicated from those measurements, pathogen concentrations in the mixed water used for irrigation, were lower than those indicated by the 141937 Common Ministerial Decision (2005) as well as those by WHO [32]. In addition, concentrations of TC, *E. coli,* and *Enterococcus* spp in the wastewater effluent were significantly lower than those of the Axios' river water (Table 3).

TABLE 3: Concentration of pathogens in wastewater effluent, the Axios's water and mixed water.

Type of water	TC (MPN/100 mL)	*E. coli* (MPN/100 mL)	*Enterococcus* spp. (MPN/100 mL)	*Salmonella* spp. (MPN/100 mL)	Intestinal Nematodes (No of eggs/L)
WWTP effluent	<3	<3	<3	none	none
Axios	23	23	<3	none	none
Mixed	21	15	<3	none	none

TABLE 4: Chemical and physical parameters of the WWTP effluent, the Axios's water and mixed water (adapted from [33]).

Parameter	Units	Effluent	Axios	Mixed
SS	mg/L	11	16	12
COD	mg/L	60	16	18
BOD_5	mg/L	3	2	2
pH		7.27	7.71	7.6
EC	dS/m	5.13	0.56	0.7
Ca	meq/L	5.33	3.07	3.09
Mg	meq/L	7.84	0.73	0.75
Na	meq/L	49.74	1.35	1.61
SAR		19.25	0.98	1.16
Cl^-	mg/L	1250	50	75
$N\text{-}NH_4$	mg/L	0.33	0.18	0.26
$N\text{-}NO_3$	mg/L	9.66	1.1	1.2
TKN	mg/L	10.6	1.4	1.4
$P\text{-}PO_4$	mg/L	3.6	<0.5	0.81
K	meq/L	0.35	0.06	0.08
HCO_3	meq/L	5.46	3.32	3.82
SO_4	meq/L	2.5	0.3	1.0

Concentrations of several chemical compounds and heavy metals in the WWTP effluent, the Axios's fresh water and the mixed water are given in Table 4 and Table 5. The mixed water values of SS, COD and BOD_5 were less than 12, 18, and 2 mg/L, respectively, thus much less than those deemed satisfactory by several regulations and guidelines around the world [2,32] as well as of those considered safe by the specific regulations for this project.

TABLE 5: Concentrations of heavy metals in the WWTP effluent, the Axios's river water and mixed water in mg/L (adapted from [33]).

Parameter	Effluent	Axios water	Mixed water
Fe	0.27	0.56	0.36
Cu	0.024	0.03	0.04
Zn	0.215	0.12	0.15
Ni	0.006	0.62	0.48
Cd	<0.0001	<0.0001	<0.0001
Pb	0.002	0.069	0.061
Hg	<0.001	0.0027	0.003
As	0.006	0.004	0.003
Mn	0.06	0.04	0.03
Cr (total)	0.001	0.05	0.02
B	0.52	0.11	0.1

The EC of the treated wastewater effluent was relatively high, with a value of 5.13 dS/m (Table 4). However, the EC of the irrigation water (mixed water) was within acceptable agronomical limits. In the past, several guidelines were proposed in order to protect soil and plants from irrigation water salinity (EC). Such guidelines were organized, modified and presented (among others) by Ayers and Westcot [34]. In order to eliminate the impacts of salinity on the productivity of agricultural land, appropriate management practices should be adopted, including the selection of suitable irrigation methods, soil cultural practices, soil drainage, and selection of salt-tolerant plant species. In the Thessaloniki case study, no detrimental impact on crops growth, production and soil was observed [24].

Two of the substances of concern in recycled water are nitrogen (N) and phosphorus (P). Quite often they occur in recycled water at higher concentrations than those required for optimum crop growth. Excessive application of nutrients to the land may result in the leaching of nitrates to underground aquifer and runoff of both phosphorous and nitrogen to surface waters. Nutrient transport to water bodies may degrade quality, reduce biodiversity, cause eutrophication and result in health risks if these supplies are used as potable water. Furthermore, increased concentrations of nutrients, and particularly of nitrogen, may significantly affect the yield of the irrigated crops through excessive vegetative growth at the expense of yield and lower accumulation of carbohydrates in sugar beets and fruit [2,26]. On the other hand, if the nutrient content of WWTP effluents is utilized with proper practices, significant quantities of fertilizers could be saved. In the Thessaloniki WWTP effluent reuse scheme, amounts of 70–125 kg N/ha and 140–320 kg P/ha (Table 4) were applied with the irrigation water in the rice fields and 17–30 kg N/ha and 35–80 kg P/ha to other crops. Further work is required to investigate the potential effects of nutrient content of the treated effluent to crops, groundwater and the environment.

Finally, other nutrients and heavy metals were within the limits reported by Ayers and Westcot [34], Panoras and Ilias [35], Paranychianakis et al. [2], and others. Additional measures in the Thessaloniki reuse project include the restriction of effluent application to crops that are consumed raw by humans while, in the case of horticultural crops, irrigation must cease at least two weeks before harvesting. Farmers are encouraged to use trickle irrigation to minimize the contact of crop and humans with recycled water, while the use of sprinklers is prohibited. Moreover, notices should be placed in the effluent-irrigated fields and effluent (mixed water) conveying canals. Educational programs are to be implemented to inform farmers of the safe use of the recycled water, to promote public acceptance and familiarize farmers and public with the idea of treated wastewater reuse in crop irrigation.

8.9 CURRENT QUALITY CRITERIA

A preliminary study about WWTP effluent quality criteria was implemented in 2000 by the Hellenic Union of Water Supply and Sewerage

Services Association and was mainly influenced by the WHO guidelines [36]. This study was updated in 2009 taking into account the scientific standards of that time and suggested criteria based on the revised WHO and Australian guidelines [10], but these criteria were not adopted by Greek authorities. In March 2011, the 145116 Common Ministerial Decision [37] was published and set effluent reuse criteria that were applicable to any project using recycled water in Greece. These criteria define treatment processes, water quality limits and, in some applications, additional measures to protect consumers, workers and, in general, public health and the environment (Table 6 and Table 7).

Current Greek regulations [37] distinguish the following applications: (a) urban uses including landscape irrigation, recreational uses, car washing, and fire protection, (b) irrigation of agricultural crops and commercial nurseries with or without restrictions, (c) industrial uses such as cooling, boiler feeding, processes that require water use, etc., and (d) recharge of aquifers not used for potable water uptake. The regulations distinguish three basic effluent qualities with respect to microbiological criteria [2]:

- The highest quality which refers to urban uses and direct injection for groundwater recharge, with a threshold of 2 cfu/100 mL for TC (Table 7).
- The medium quality refers to unrestricted irrigation and to industrial uses except that of disposable cooling water, with a threshold of 5 cfu/100 mL for *E. coli* (Table 6).
- The lowest effluent quality refers to restricted irrigation, aquifer recharge by percolation and to disposable industrial use (cooling), with a threshold for *E. coli* of 200 cfu/100 mL (Table 6).

Additionally to the effluents' quality limits and treatment processes described in Table 6 and Table 7, the latest Greek regulations include maximum concentrations for 19 heavy metals and metalloids, as well as numerous barriers for each application, which are expected to discourage the development of water recycling projects. For example, monitoring of heavy metals and metalloids is required and varies with the capacity of the WWTP from 2 (<10,000 EP) to 12 (200,000 EP and bigger) times per year. In addition, a set of 40 organic substances should be monitored two times per year in WWTPs serving more than 100,000 EP [2].

Furthermore, in order to use the effluent for irrigation, the regulations define agronomical thresholds for various parameters based on the Ayers and Westcot [34] guidelines and maximum allowable concentrations of priority substances and toxicity in the reused water.

8.10 CONCLUSIONS

In Greece, because of the restricted water resources and of the high demand for freshwater with a peak during the hot period, it is essential that every possible water resource is exploited. Crop irrigation is one of the most demanding water consumers, especially during summer. At the eastern and southern regions of the country, the integration of treated wastewater into water resources management is of paramount importance to meet current and future demands. Despite this need, only a few projects of effluent reuse have been implemented, and most of them are pilot projects dealing with crop or landscape irrigation. The most important projects which are currently in practice include those of Thessaloniki, Chalkida, Malia, Livadia, Amfisa, Kalikratia, and Chersonissos. One of the most significant wastewater reuse projects is that of Thessaloniki, in northern Greece. In this project, the treated wastewater is mixed with freshwater at a ratio of 1:5, mostly because of the high salinity of the effluent. A fraction of the total WWTP discharge of 165,000 m^3/day is mixed with the Axios's river water to irrigate approximately 2500 ha of rice, corn, alfalfa, sugar beet, and cotton.

A few other projects are under planning, such as that in Iraklion, in Agios Nikolaos, and on several Aegean islands. It should be mentioned that there are several cases of indirect reuse of WWTPs effluent in central Greece. However the reuse potential in Greece is limited, since effluent recycling from Athens's WWTP, which serves approximately half of the Greek population, is not economically feasible due to the location of the treatment plant.

The increasing pressure on water resources in Greece—especially in southern and eastern Greece—can only be met by the adoption of integrated water management schemes. Emphasis should be given to the efficient use of existing water supplies, to the protection of their quality, as well as to the use of marginal waters, such as treated effluents. A saving of 4.5% to 5% of the total water use could be achieved by proper management and reuse of the effluents from the existing WWTPs. This percentage may increase in the near future as the number and the capacity of WWTPs increases and as new plants will be incorporated in effluent reuse schemes. In order to achieve this, national water policy should be improved and extended to encourage the safe use of recycled water for various uses.

Recently, quality criteria for treated wastewater reuse have been established in Greece, for various reuse options. Projects that have already been implemented are required to conform to the new regulations [37].

REFERENCES

1. Asano, T.; Burton, F.L.; Leverenz, H.L.; Tsuchinhashi, R.; Tchobanoglous, G. Water Reuse: Issues, Technologies, and Applications; McGraw-Hill: New York, NY, USA, 2007.
2. Paranychianakis, N.V.; Salgot, M.; Angelakis, A.N. Irrigation with recycled water: Guidelines and regulations. In Treated Wastewater in Agriculture: Use and Impacts on the Soil Environments and Crops; Levy, G., Fine, A., Bar-Tal, A., Eds.; Wiley Knowledge for Generations: Hoboken, NJ, USA; Oxford, UK, 2011; Chapter 3. pp. 77–111.
3. Marecos do Monte, M.H.F.; Angelakis, A.N.; Asano, T. Necessity and basis for the establishment of European guidelines on wastewater reclamation and reuse in the Mediterranean region. Water Sci. Technol. 1996, 33, 303–316.
4. Angelakis, A.N.; Marecos do Monte, M.H.; Bontoux, L.; Asano, T. The status of wastewater reuse practice in the Mediterranean Basin. Water Res. 1999, 33, 2201–2217.
5. Angelakis, A.N.; Diamadopoulos, E. Water resources management in Greece: Current status and prospective outlook. Water Sci. Technol. 1995, 32, 267–272.
6. Kalavrouziotis, I.K.; Kokkinos, P.; Oron, G.; Fatone, F.; Bolzonella, D.; Vatyliotou, M.; Fatta-Kassinos, D.; Koukoulakis, P.H.; Varnavas, S.P. Current Status in Wastewater Treatment, Reuse and Research in Some Mediterranean Countries. Desalin. Water Treat. 2013.
7. Chartzoulakis, K.S.; Paranychianakis, N.V.; Angelakis, A.N. Water resources management in the island of Crete, Greece: With emphasis the agricultural use. Water Policy 2001, 3, 193–205.
8. Panoras, A.; Ilias, A. Reclaimed municipal wastewater reuse in Thessaloniki. In Proceedings of the International Symposium on Wastewater Reclamation and Reuse, Thessalonki, Greece, 13–14 February 2003; pp. 137–145.
9. Tsagarakis, K.P.; Dialynas, G.E.; Angelakis, A.N. Water resources management in Crete (Greece) including water recycling and reuse and proposed quality criteria. Agric. Water Manag. 2004, 66, 35–47.
10. Paranychianakis, N.V.; Kotselidou, O.; Vardakou, E.; Angelakis, A.N. Greek Regulations on Wastewater Reclamation and Reuse; Hellenic Union of Municipal Enterprises for Water Supply and Sewage: Larissa, Greece, 2009; p. xxiv, 160.(In Greek).

11. EU. Council Directive 91/271/EEC concerning urban waste water treatment. Offic. J. EU: Legislation 1991, 56, 40–52.
12. MEECC (Ministry of Environment, Energy and Climate Change). Greek State Conformation to the EU 271/91 Directive; MEECC: Athens, Greece, 2010.
13. Tsagarakis, K.P.; Mara, D.D.; Angelakis, A.N. Evaluation of municipal wastewater treatment plants in Greece. Tech. Chron. 1998, 18, 97–109.(In Greek).
14. Mamais, D.; Andreadakis, A.D.; Gavalaki, E. Assessment of secondary and tertiary treatment processes for wastewater reclamation in Greece. In Proceedings of the IWA Regional Symposium on Water Recycling and Reuse, Iraklion, Greece, 26–29 September 2002.
15. Andreadakis, A.; Mamais, D.; Gavalakis, E.; Panagiotopoulou, V. Evaluation of treatment schemes appropriate for wastewater reuse in Greece. Glob. Nest: Int. J. 2004, 5, 1–8.
16. Panoras, G. Heavy Metal Accumulation in Soil Irrigated for Six Years by Treated Municipal Wastewater. Master's Thesis, Aristotle University of Thessaloniki, Thessaloniki, Greece, 2011 (In Greek).
17. Angelakis, A.N.; Koutsoyiannis, D.; Tchobanoglous, G. Water resources technologies in the ancient Greece. Water Res. 2005, 39, 210–216.
18. Koutsoyiannis, D.; Zarkadoulas, N.; Angelakis, A.N.; Tchobanoglous, G. Urban water management in ancient Greece: Legacies and lessons. ASCE J. Water Resour. Plan. Manag. 2008, 134, 45–54.
19. Tzanakakis, V.E.; Koo-Oshima, S.; Haddad, M.; Apostolidis, N.; Angelakis, A.N. The history of land application and hydroponic systems for wastewater treatment and reuse. In Evolution of Sanitation and Wastewater Management through the Centuries; Angelakis, A.N., Wilderer, P.A., Rose, J.B., Eds.; IWA Publishing: London, UK, 2014. accepted.
20. Tsagarakis, K.P.; Tsoumanis, P.; Chartzoulakis, K.; Angelakis, A.N. Water resources status including wastewater treatment and reuse in Greece: Related problems and prospectives. Water Int. 2001, 26, 252–258.
21. Sbirilis, N.; Kanaris, S. Wastewater reclamation and reuse in city of Chalkis, Greece. In Proceedings of the IWA Regional Symposium on Water Recycling in Mediterranean Region, Iraklion, Greece, 26–29 September 2002; pp. 63–68.
22. EYDAP (Athens Water Supply and Sewage Company). Personal communication, 2013.
23. Soupilas, A.; Papastergiou, F. Preparations for large scale reuse of treated domestic wastewater for irrigation purposes in Thesaloniki. In Proceedings of the IWA Regional Symposium on Water Recycling in Mediterranean Region, Iraklion, Greece, 26–29 September 2002; pp. 465–471.
24. Ilias, A.; Panoras, G.; Karamoutzis, D.; Angelakis, A.N. Reuse of treated wastewater in Greece: The project of Thessaloniki. In Proceedings of the IWA Regional Conference on Wastewater Purification & Reuse, Iraklion, Greece, 28–30 March 2012; p. 183.
25. Papayiannopoulou, A.; Parissopoulos, G.; Panoras, A.; Kampeli, S.; Papadopoulos, F.; Papadopoulos, A.; Ilias, A. Emitter performance in conditions of treated municipal wastewater. In Proceedings of the 2nd International Conference Entitled "Advanced Wastewater Treatment, Recycling and Reuse. AWT98", Fiera, Milano, Italy, 14–16 September 1998; pp. 1011–1014.

26. Panoras, A.; Ilias, A.; Skarakis, G.; Zdragas, A. Suitability of reclaimed municipal wastewater for sugar beet irrigation. In Proceedings of the 5th International Congress on Environmental Pollution, Thessaloniki, Greece, 28 August–1 September 2000; pp. 221–232.
27. Panoras, A.; Kehagia, U.; Xanthopoulos, F.; Doitsinis, A.; Samaras, I. The use of municipal wastewater in cotton irrigation. In Proceedings of the Inter-Regional Research Network Meeting on Cotton, Chania, Greece, 27–30 September 2001.
28. Panoras, A.; Evgenidis, G.; Bladenopoulou, S.; Melidis, B.; Doitsinis, A.; Samaras, I.; Sdragkas, A.; Matsi, T. Corn irrigation with reclaimed municipal wastewater. Glob. Nest: Int. J. 2005, 5, 47–54.
29. Papadopoulos, F.; Parissopoulos, G.; Papadopoulos, A.; Zdragas, A.; Ntanos, D.; Prochaska, C.; Metaxa, I. Assessment of reclaimed municipal wastewater application on rice cultivation. Environ. Manag. 2009, 43, 135–143.
30. Panoras, A.; Ilias, A.; Hatzigiannakis, E.; Arampatzis, G.; Panagopoulos, A.; Stathaki, S.; Diamantidis, I.; Tsekoura, D.; Zavra, A.; Panoras, G.; et al. Final Report of Reusing the Reclaimed Municipal Effluent of Thessaloniki's Wastewater Treatment Plant for Irrigation Purposes in Halastra-Kalohori Agricultural Area During the Summer Time of 2008; NAGREF: Sindos, Greece, 2009; p. 54.(In Greek).
31. CMD (Common Ministerial Decision). Criteria for Reusing the Effluent from Thessaloniki's WWTP in Sindos; No. 141937. Ministry of Environment, Energy and Climate Change: Athens, Greece, 2005. (In Greek).
32. WHO (World Health Organization). Guidelines for the Safe Use of Wastewater, Excreta, and Greywater; Wastewater Use in Agriculture. WHO: Geneva, Switzerland, 2006; Volume 2.
33. Soupilas, A. Technical Report on Treated Wastewater Reuse in Thessaloniki, Greece; EYATH, A.E.: Thessaloniki, Greece, 2011. (In Greek).
34. Ayers, R.S.; Westcot, D.W. Water Quality for Agriculture; FAO Irrigation and Drainage. Paper 29, Rev. 1. Food and Agriculture Organization of the United Nations: Rome, Italy, 1985.
35. Panoras, A.; Ilias, A. Irrigation with Treated Wastewater Effluent; Giahudis: Thessaloniki, Greece, 1999; p. 190.(In Greek).
36. Angelakis, A.N.; Tsagarakis, K.P.; Kotselidou, O.N.; Vardakou, E. The Necessity for Establishment of Greek Regulations on Wastewater Reclamation and Reuse; Report for the Ministry of Public Works and Environment and Hellenic Union of Municipality Entering for Water Supply and Sewage: Larissa, Greece, 2000; p. 110.(In Greek).
37. CMD (Common Ministerial Decision). Measures, Limits and Procedures for Reuse of Treated Wastewater; No. 145116. Ministry of Environment, Energy and Climate Change: Athens, Greece, 2011. (In Greek).

There are two tables that are not available in this version of the article. To view this additional information, please use the citation on the first page of this chapter.

CHAPTER 9

Do Contaminants Originating from State-of-the-Art Treated Wastewater Impact the Ecological Quality of Surface Waters?

DANIEL STALTER, AXEL MAGDEBURG, KRISTIN QUEDNOW, ALEXANDRA BOTZAT, AND JÖRG OEHLMANN

9.1 INTRODUCTION

In urban areas, the water quality of small streams is predominantly impacted by structural degradation of stream morphology (e.g., channelization and straightening), agricultural land use in the catchment, and a high load of treated wastewater (WW). These are also the most apparent drivers of ecological conditions in anthropogenically disturbed surface waters. Together, these stressors have led to significant declines in aquatic fauna populations and biodiversity, in particular in freshwater ecosystems, which in turn has had a profound impact on the ecological integrity of many aquatic ecosystems [1]–[5].

The ecological quality of surface water is commonly assessed by analyzing benthic macroinvertebrate species assemblages [6]. The highly

Do Contaminants Originating from State-of-the-Art Treated Wastewater Impact the Ecological Quality of Surface Waters?. Stalter D, Magdeburg A, Quednow K, Botzat A, and Oehlmann J. PLoS ONE ***8**,4 (2013), doi:10.1371/journal.pone.0060616.*

diverse group of aquatic invertebrate organisms encompasses many euryoecious and stenoecious species [7]. The latter have narrow ecological requirements in terms of physico-chemical water parameters and structural characteristics of the water body. Consequently, assessing the invertebrate community can provide a picture of the 'health' of the water body and can be used to identify factors affecting its ecological quality (e.g., structural deterioration, acidification, oxygen deficiency). The technique involves comparing the composition and abundance of the present invertebrate taxa to established reference communities from anthropogenically least-disturbed reference sites.

Evaluating and improving the ecological status of surface waters is increasingly perceived to be important, which is reflected by legislation changes in this area [6], [8]. For this purpose, worldwide many different macroinvertebrate-based rating systems for ecological status evaluation with hundreds of indices, metrics, and evaluation tools have been developed [6], [8]. In the European Union, the Water Framework Directive (WFD) [9]–enacted in 2000–set an objective that all coastal and inland waters of Europe be of a 'good status' by 2015. One tool for evaluating the quality of these waters is the ASTERICS software (AQEM/STAR Ecological River Classification System), which assists the assessment of ecological status. If the outcomes of ASTERICS analysis indicate a poor ecological status, key causal factors and mitigation and remediation options must be identified and explored in order to achieve the 'good status' benchmark.

Until the early 1980s in Germany, the primary driver of the deterioration of aquatic ecosystems was the large input of nutrients via WW discharge [10]. However, as progressive WW treatment and other restoration measures increased throughout the 1980s, nutrient inputs were diminished, eutrophication reduced, and water quality impacts were ameliorated, hence allowing aquatic ecosystems to recover [11]. Generally, in high-income countries, eutrophication is largely under control at present [12], [13] because of the advances in, and the prevalence of WW treatment [14].

Despite the effective removal of nutrients and other contaminants, treated effluent often contains trace organic contaminants that have been demonstrated to have negative impacts on aquatic ecosystems [13]. These anthropogenic water pollutants and their transformation products are often present at low to very low concentrations (e.g., ng/L). Despite such minimal trace level concentrations, the large spectrum of organic contaminants occurring in surface waters [15] may pose a potential threat to aquatic wildlife, particularly with regard to mixture toxicity. Pharmaceuticals and personal care products in particular often exhibit high biological activity–some of which are also persistent in the environment [16], thus giving them the potential to affect aquatic species [17]–[23]. The presence of these contaminants in urban surface water has been primarily attributed to environmental releases of municipal and industrial WW [13]. Accordingly, we anticipate there to be a clear negative impact on macroinvertebrate assemblages and the general ecological status of surface waters in proximity to WW treatment plants.

Many studies have addressed the impact of water pollution on invertebrate communities, but frequently they consider only severely contaminated sites or individual pollutants (e.g., [24]). In this study, we have instead focused on aquatic invertebrate exposure to the 'subtle pollution' (i.e., mixtures of organic micropollutants) occurring in German lowland streams. All treatment plants currently operating in the study area are equipped with biological activated sludge treatment–the most widely used WW treatment system in high-income countries [25]. As well as the elimination of nutrients, this technology has also been shown to effectively reduce the contaminant load and toxicity of WW [26], [27]. However, a considerable fraction of toxic contaminants remains prevalent in WW treated by biological activated sludge [27].

This study aimed to enhance the understanding of factors that disturb the ecological integrity of surface waters in order to assist decision-making around remediation and mitigation of factors affecting the quality of surface waters. The primary objective was to evaluate whether lowland stream macroinvertebrate assemblages are affected by their loads of WW-associated trace organic contaminants (OC) in the water phase. Additionally, the potential impacts of heavy metals (HM) and polycyclic aromatic hydrocarbons (PAH) in sediment were investigated.

9.2 MATERIALS AND METHODS

9.2.1 STUDY AREA

The study was conducted in the Hessian Ried (HR) close to Frankfurt (Main), Germany. The region extends about 1200 km^2 and is bound by the river Main in the South, the lower mountain range Odenwald in the East, and the river Rhine in the West. The accumulation of sand and gravel, as a result of aggradational deposits, forms a large aquifer that serves as drinking water reservoir for the Frankfurt/Rhine-Main conurbation [28]. The population of the HR is approximately 800,000. Types of land use include intensive agriculture (predominantly cultivation of vegetables, fruit, wheat, and forage crops) and industry (e.g., chemical and manufacturing industries), and high-density traffic highways occur in the region [28]. Due to this high level of anthropogenic disturbance, the importance of the aquifer in the region, and the homogenous distribution of urbanization, the HR was selected as a representative urban study area to investigate the impacts of water pollution, physico-chemical factors, and structural characteristics on the ecological quality of urbanized aquatic ecosystems.

9.2.2 STUDY DESIGN

To identify key impact factors on the ecological quality, we assessed the benthic macroinvertebrate community of the streams, recorded a range of environmental and anthropogenic variables, and evaluated their potential impact on the species composition with the major focus on wastewater-associated contaminants. We conducted this study from 2005 to 2006 at 26 sampling sites in four river systems that run through the study area in an east-west direction and discharge into the Rhine River (Table 1). Most sampling sites belong to the water body type 19 (small streams in flood-plains) according to the WFD [9], [29]. Exceptions are sampling sites 1 to 6 in the Weschnitz stream (type 9: mid-sized siliceous rivers in the lower mountain ranges with fine to crude sediment) as well as Sw1 (type 6: small calcareous sediment rich streams in the lower mountain ranges) and

Mo1 (type 5: small siliceous cobble-bottom streams in the lower mountain ranges [30]). 28 sewage treatment plants are currently operating in the catchment. The location of the sampling sites relative to these plants, as well as the quantity of effluent discharged to streams, resulted in different WW loads at each of the sampling sites. This factor was defined on a scale from 0–100% (Figure S1–4). The only sampling site without WW was Wi1.

TABLE 1: River systems with respective streams, abbreviations (abbr.), and number of sampling sites (no.).

River systems	Streams	Abbr.	No.
Schwarzbach-Landgraben-system	Schwarzbach, Apfelbach	Sw	6
Schwarzbach-Landgraben-system	Landgraben, Darmbach	La	2
Modau-Sandbach-system	Sandbach	Sa	3
Modau-Sandbach-system	Modau	Mo	4
Winkelbach	Winkelbach	Wi	5
Winkelbach	Weschnitz	We	6

9.2.3 STRUCTURAL QUALITY OF THE STREAMS

To classify the streams with respect to their structural modification, we recorded the morphology or water structure quality following Zumbroich et al. [31] from class 1 (near-natural) to class 7 (completely modified). Additionally, the stream bed structure was characterized in detail (relative coverage of 10 different sediment types,) according to the macroinvertebrate sampling procedure described in Hering et al. [7].

9.2.4 BENTHIC MACROINVERTEBRATE COMMUNITY

We sampled the invertebrates in autumn and spring (September 2005 and March 2006) following the multi-habitat sampling procedure described in Hering et al. [7]. No specific permits were required for the described field studies as none of the sampling sites are privately-owned or protected in any way. The responsible Hessian State Office for Environment and Geology was informed prior to the study. Before sampling, we assessed the substrate composition of the sampling sites of a 100-m transect of each water body. The different substrate types were sampled according to their relative coverage with a total of 20 sub-samples per sampling site. We used a Surber-sampler (manufactured by precision engineers at the Goethe University of Frankfurt, Germany) with 500-μm mesh size and 20×25-cm frame size for macroinvertebrate sampling, resulting in 1-m^2 sampling plots per sampling site. Endangered or protected species were identified in the field and immediately released back to the streams. The remaining sample material was stored in 70% ethanol (denatured, Carl Roth GmbH, Darmstadt, Germany). Macroinvertebrates were separated as described in Haase & Sundermann [32] and determined to species level, if possible, but at least to the minimum taxonomic level recommended by Haase et al. [33] (Table S2–3). For separation and identification of the species we used stereo microscopes (Stemi 2000, Carl Zeiss AG, Oberkochen, Germany) and a microscope (BX 50, Olympus, Tokio, Japan) if a higher magnification was required, e.g., for the identification of *Ephemeroptera* species. We could not sample Mo4 and Sw6 during the autumn sampling campaign due to high water levels.

9.2.5 CONTAMINANTS

As WW-associated pollutants commonly correlate in their occurrence, distinguishing their individual impacts is hardly feasible. Accordingly, a number of representative pollutants were selected to account for the burden of WW-associated OCs in general. At 11 sampling campaigns, water samples from each sampling site were analyzed for 12 common organic pollutants: five organophosphates; two musk fragrances; bisphenol A; the alkylphenols, nonylphenol and octylphenol; and the insect repellent diethyltoluamide (Table 2). Pollutants were selected due to their ubiquitous occurrence in municipal WW. Unexpectedly, terbutryn–in Europe formerly used as agricultural herbicide, which is still authorized as a biocide in antifouling paints and coatings–was detected in all water samples from the beginning of the sampling campaigns, and therefore was also included in the analysis suite. For more detailed information on analytical procedures and pollutant concentrations, see Quednow & Puettmann [34]–[37]. To take into account a possible influence of sediment-bound pollutants, the sediment load of 12 HMs and metalloids as well as 16 PAHs (Table 2)–the latter selected as proposed by the US EPA [38]–was determined according to standard methods [39], [40]. Average contaminant concentrations are given in Table S4–6.

TABLE 2: Analyzed contaminants.

Phase Analyzed	Contaminant Group	Contaminants
Water	organophosphates	TBP (tributyl phosphate), TBEP (tris(2-butoxy-ethyl)phosphate), TCEP (tris(2-chloroethyl) phosphate),TCPP (tris(2-chloro-, 1-methyl-ethyl)-phosphate), TDCPP IL (tris(1 ,3-dichloro-2-propyl) phosphate)
	biphenols	BPA (bisphenol A)
	musk fragrances	HHCB (1,3,4,6,7,8-hexahydro-4,6,6,7,8,8-hexameth-ylcyclopenta-y-2-benzopyran), AHTN (6-Acetyl-1, 1,2,4,4,7- hexamethyltetraline)
	alkylphenols	octylphenol, nonylphenol
	triazines	terbutryn
	amides	DEET (diethyltoluamide)
Sediment	heavy metals	aluminium, arsenic, barium, cadmium, cobalt, chro-mium, copper, iron, manganese, nickel, lead, zinc
	polycyclic aromatic hydrocarbons	naphthalene, acenaphtylene, acenaphthene, fluorene, phenanthrene, anthracene, fluoranthene, pyrene, benzo(a)anthracene, chrysene, benzo(b)fluoranthene, benzo(k)ftuoranthene, benzo(a)pyrene, dibenz(ah)an-thracene, benzo(ghi)perylene, indeno(1,2,3-cd)pyrene

9.2.6 WATER AND SEDIMENT QUALITY

We determined the organic carbon content and average grain size of sediments according to standard methods [41], [42]. We recorded conductivity, temperature, O_2-concentration and pH (using multiparameter instrument Multi 350i, WTW, Weilheim, Germany) as well as flow velocity (using a hydrometric vane from OTT Hydromet, Kempten, Germany) during sampling events. Colorimetric on-site tests (Aquaquant®, Merckoquant®, Merck, Darmstadt, Germany) were used to determine ammonium, chloride, phosphate, total hardness, and carbonate hardness. Biological oxygen demand (BOD_5) was analyzed according to a standard method [43].

9.2.7 DATA ANALYSIS

We used the ASTERICS software [44] to calculate the ecological status (from 1–5: bad, poor, moderate, good, high) and biotic metrics (e.g., saprobic index, Shannon index, Simpson index, number of EPT taxa) for each sampling site. To reduce the number of contaminant variables, separate principal component analyses (PCAs) were performed using SPSS (version 14, SPSS Inc., Chicago, USA, 2005) with the concentrations of OCs in the water phase and concentrations of PAHs and HMs in the sediments (Table S7–9). We only considered components with an eigenvalue >1 to limit the number of resulting components. The component scores were determined by regression and were standardized with a mean of 0 and standard deviation of 1 [45]. The resulting components (OC1–4, HM1–3, and PAH1–2) with the corresponding component scores were used to replace the contaminant concentrations in all further analyses.

Using the statistic software R (R Development Core Team 2011) we applied non-metric multidimensional scaling (NMDS) using Bray-Curtis dissimilarity as implemented in R package vegan 2.0–2 [46] based on abundance data to detect differences of macroinvertebrate taxa composition among the 26 sample sites. NMDS displays dissimilarities in community composition nonlinearly onto ordination space, can cope with nonlinear species responses, and is not constrained by predictors [46]. We excluded rare species that occurred less than three times and that were present in less than three sampling sites. We used taxa scores in the biplot to illustrate differences between sampling sites. Non-correlated variables were selected (structural quality, average discharge, organic carbon content of the sediments, and following contaminant factors: OC1, OC2, HM1, HM3, PAH1, PAH2; Table S1) and were fitted post-hoc to the ordination and their significance was tested via random permutations (1000 iterations).

We used multiple linear regression models to test the impact of WW-associated contaminants and environmental variables on the following dependent variables: total abundance of macroinvertebrates, number of macroinvertebrate taxa, Simpson and Shannon diversity of macroinvertebrates, number of EPT taxa, and saprobic index. We started with models comprising the same set of non-correlated variables as used for the NMDS. We used the step function implemented in R for model selection to reach minimum adequate models. This selection process is based on minimizing information loss according to Akaike's Information Criterion values [47].

We performed Spearman rank correlation analyses using species abundances, biotic metrics, the PCA-derived contaminant factors, and all environmental variables (Table S1). To take into account the toxicity of the analyzed WW-associated contaminants, toxic units (TUs) were calculated as described in Liess & Von Der Ohe [48] and included in the correlation analysis. The TU values are based on effect concentrations (LC_{50} and 'no observed effect concentration') from acute tests with *Chirmonomus riparius* and chronic tests with *Daphnia magna*). In addition to the median pollutant concentrations, the 90th percentile of the concentrations was used for TU calculations to estimate whether the peak loads influenced the taxa composition more than the mean concentrations. TUs were summed for each sampling site and included in the correlation analysis.

9.3 RESULTS

9.3.1 SAMPLING SITES AND SPECIES DATA

The running waters of the study area were mostly structurally distorted with channel-like character, usually with trapezoidal profiles, low depth variance, large profile depth, and almost a lack of curvature erosion. Accordingly, the degree of anthropogenic deformation varied mostly between class 4 (significantly modified) to 7 (completely modified) with only La2 and Mo1 as slightly modified (class 2) and Sw1 as single near-natural stream (class 1; Figure S5). We identified a total of 141 different taxa across all sampling sites (Table S2–3). The number of taxa exceeded 50 at Wi1 and Wi2 only, while falling below 20 at La1 and Sw6 (Figure S5). The lowest Shannon diversity was determined for La1 (0.8) and the highest for Sw1 (3.0). At La1 and Sw2, the invertebrate abundance was dominated by Oligochaeta and *Gammarus roeselii*, contributing more than 50% to the total number of individuals (dominance structure in Figure S5 given as contribution of individual taxa to the total number of individuals per m^2 in decreasing order). Plecoptera were found at Sw1 and We1 only. Invasive species like *Corbicula fluminea, Potamopyrgus antipodarum,* and *Dikerogammarus villosus* were commonly found in the sampling area. At Sa2 *P. antipodarum* dominated species assemblage in terms of individuals per m^2. The ecological status was evaluated as 'good' for two sampling sites only (Wi2 and Wi4), while 50% of the sites were evaluated as insufficient or poor (Figure S5).

9.3.2 PCA-DERIVED VARIABLE REDUCTION OF CONTAMINANTS

Four components of the PCA of OCs explained ca. 88% of the total variance while component loadings (i.e., correlation coefficients between the variables and components) ranged from 0.55–0.97. The first component (OC1) encompassed the organophosphates as well as the musk fragrances and DEET (Table S7). OC1 is the contaminant factor that correlated to the highest degree with the WW load (Spearman's $\rho = 0.932$, Table S1) and is therefore regarded as predominantly WW-associated. Octylphenol and terbutryn were mainly represented by the second component (OC2), bisphenol A by the third (OC3), and nonylphenol by the fourth (OC4) component. The three components of HMs explained ca. 78% of the total variance. The component loadings ranged from 0.64–0.92. The first component (HM1) represented primarily Co, Cr, Cu, Fe, Mn, Ni, Pb, and Zn (Table S8) while the second (HM2) largely encompassed Al, As, and Ba. The third component (HM3) represented almost exclusively Cd. The two components calculated for PAHs explained more than 92% of the total variance, whereas component loadings were mainly >0.9 (except for Acyl and Fl, Table S9). The first component (PAH1) embodied 14 of the 16 PAHs while the second (PAH2) represented acenaphtylene and fluorine. PCA reduced the total number of contaminant variables from 40 variables to 9 variables, which were then used for subsequent analyses.

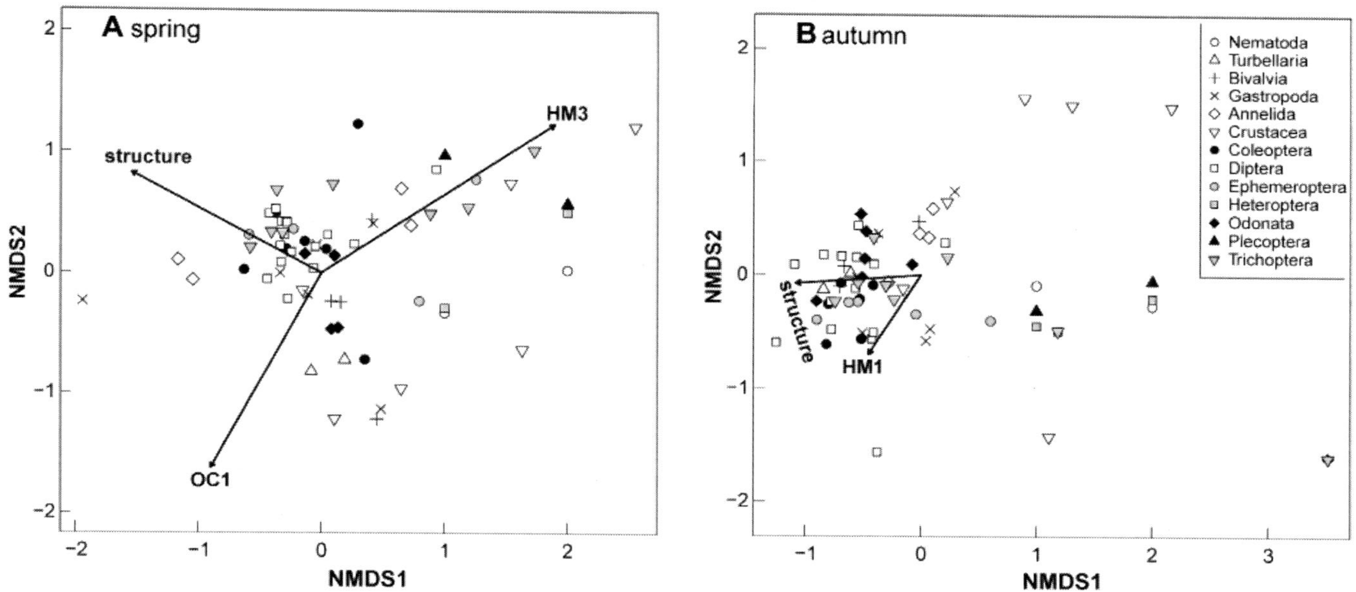

FIGURE 1: NMDS biplot of taxa and environmental variables. Displayed are variables with a significant impact ($p<0.05$) for sampling campaign in spring (A) and autumn (B). HM, components of the principal component analysis (PCA) with heavy metals; OC, components of the PCA with organic contaminants; structure, structural degradation. Spring: two convergent solutions, two dimensions, stress = 0.17; autumn: two convergent solutions, two dimensions, stress = 0.21).

9.3.3 DISSIMILARITIES IN MACROINVERTEBRATE ASSEMBLAGES RELATED TO STRUCTURAL DEGRADATION AND CONTAMINANTS

The NMDS revealed the water body structure as one of the major impact variables in both sampling campaigns ($R^2 = 0.65$ and 0.56; $p = 0.005$ and 0.002; Figure 1A, B). Other significant impact factors for the taxa composition in spring included OC1 ($R^2 = 0.43$, $p = 0.006$) and HM3 ($R^2 = 0.72$, $p = 0.005$). In the autumn sampling event, HM1 was identified as a significant impact factor ($R^2 = 0.33$, $p = 0.014$), while OC1 and HM3 showed no significant effect (OC1: $R^2 = 0.18$, $p = 0.11$; HM3: $R^2 = 0.04$, $p = 0.57$).

Structural degradation was associated with increased abundances of Diptera and Turbellaria, demonstrating the euryoecious biology of respective species (e.g., Chironomidae; Figure 1). On the other hand, abundances of more stenoecious Plecoptera species decreased with increased structural modification (Figure 1).

For the spring sampling event, the multiple linear regression model identified OC1 as the most significant impact variable on all biotic metrics apart from the number of individuals per m^2, which was more affected by PAH1 and PAH2 (Table 3). In the autumn sampling event, water body structure was the most significant impact variable on individuals per m^2 and number of taxa, while for metrics covering more sensitive species (EPT taxa and saprobic index; e.g., [49]), the WW-associated contaminant factor, OC1, was the most prominent impact variable (Table 3). This might indicate a higher susceptibility of respective taxa to corresponding organic contaminants. HM1, HM3, PAH1 and average discharge are all related to Shannon and Simpson diversity. The affecting variables identified by the regression model (water body structure, OCl, HM1 and HM3) were similar to the results of the NMDS although with the NMDS we considered the whole set of recorded species. Seasonal differences in species composition between both sampling campaigns might explain differing results between spring and autumn (e.g., [50]).

The correlation analyses–including macroinvertebrate taxa, biotic indices, and environmental variables–revealed consistent and highly significant correlations between organic pollutant concentration (OC1) and indices encompassing sensitive taxa. Biotic metrics like the saprobic index

(Figure 2), number of EPT taxa (Figure 3), and EPTCOB taxa (Ephemeroptera, Plecoptera, Trichoptera, Coleoptera, Odonata, Bivalvia) significantly correlated with OC1 ($p<0.001$) revealing correlation coefficients (Spearman's ρ) ranging from −0.63 to −0.81 ($\rho = 0.73$ and $\rho = 0.77$ in case of the saprobic index). Moreover, single genera of the orders Ephemeroptera and Coleoptera correlated significantly with OC1 (*Ephemera danica*: $\rho = -0.70$; *Elmis/Limnius/Haliplus*: ρ ranging from −0.45 to −0.53). Accordingly, numbers of Ephemeroptera and Coleoptera taxa and their abundances also highly negatively correlated with OC1 ($p<0.001$; $\rho<-0.7$).

TABLE 3: Multiple linear regression models testing the effect of environmental parameters and contaminants on biotic response variables: the total number of individuals and taxa, Simpson and Shannon diversity, number of EPT taxa and the saprobic index.

Reponse Variable	Significant Impact Variables									
	spring					autumn				
		p	df	F	R^2		p	df	F	R^2
individuals/m^2	PAH2	**	1	18.77	na	water structure	*	1	5.14	na
	PAH1	**	1	8.63	na	HM1	*	1	4.53	na
	full model	**	5,20	7.09	0.55	full model	ns	3,20	2.76	0.19
number of taxa	OC1	**	1	22.02	na	water structure	**	1	9.09	na
	PAH1	*	1	5.02	na	average discharge	**	1	8.47	na
	HM3	*	1	4.37	na	PAH1	*	1	7.39	na
	full model	**	4,21	10.62	0.61	full model	**	3,20	5.69	0.38
Simpson diversity	OC1	**	1	8.79	na	HM3	*	1	5.35	na
	OC2	*	1	6.05	na	PAH1	*	1	4.89	na
	HM1	*	1	6.19	na	full model	ns	8,15	1.88	0.23
	full model	**	5,20	10.48	0.65					
Shannon diversity	OC1	**	1	12.30	na	PAH1	**	1	12.90	na

TABLE 3: *Cont.*

	full model	**	5,20	7.78	0.58	HM3	**	1	8.97	na
						average discharge	*	1	5.93	na
						HM1	*	1	5.32	na
						full model	*	5,18	3.37	0.34
EPT taxa	OC1	**	1	10.50	na	OC1	**	1	15.82	na
	full model	**	4,21	4.99	0.39	water structure	**	1	10.15	na
						HM3	**	1	9.18	na
						full model	**	5,18	4.68	0.44
saprobic index	OC1	**	1	17.18	na	OC1	**	1	24.27	na
	HM1	*	1	4.40	na	HM3	*	1	5.70	na
	full model	**	4,21	9.07	0.56	HM1	*	1	4.86	na
						full model	**	6,17	7.48	0.63

*Given are df-, R^2-, F- and p-values for full models after stepwise deletion of non-significant terms (n.s.) and of significant model parameters. *, $p< 0.05$; **, $p< 0.01$; ***, $p< 0.001$; n.a., not available.*

9.4 DISCUSSION

9.4.1 KEY FACTORS IMPACTING SPECIES COMPOSITION

The results of NMDS and the multiple linear regressions suggest that WW-associated organic contaminants (OC1)–alongside with structural degradation–play a pivotal role in shaping macroinvertebrate species composition in anthropogenically disturbed surface waters. The correlation analyses confirmed a deleterious relationship and indicated a reduced biodiversity of EPT taxa and Coleoptera with increasing concentrations of WW-associated trace organic contaminants. The number of EPT taxa as well as the saprobic index did not correlate with other contaminant factors that were identified as significantly affecting the macroinvertebrates in NMDS and multiple linear regression models (OC2, PAH1, HM1, HM3).

This might indicate that OC1 and/or co-correlated variables were among the main negative impact factors for the macroinvertebrate communities.

Significant impact variables in the NMDS and the regression model did not necessarily have a deleterious effect. For instance, HM3 was identified as significantly affecting the species composition in the NMDS and regression analyses. HM3 represents the presence of Cd in the sediments (Table S8) suggesting a deleterious relationship due to toxic effects of Cd. However, a positive correlation of HM3 with the distance to the next WW treatment plant and a negative correlation with the WW load ($p<0.01$) indicated a non-causal relationship (Table S1). This is supported by predominantly low Cd concentrations of <0.9 mg/kg sediment (dry weight; apart from Mo1 with 2.4 mg/kg). Regarding factor PAH1, concentrations of PAHs exceed 'probable effect levels'–as proposed by Macdonald and colleagues [51]–at La1 and Mo3 by up to 20 times. At these sampling sites macroinvertebrate communities might be adversely affected. However, neither NMDS nor correlation analyses were able to support a significant impact of PAHs.

Furthermore, macroinvertebrate taxa and biotic metrics were not correlated with the water body structure. This might, however, mainly be a result of the rather homogenous sampling sites in terms of structural degradation (predominantly very highly modified, Figure S5) and does not indicate that water body structure is less relevant.

The ecological quality class determined via ASTERICS did not correlate with the contaminant factors. However, ASTERICS is designed to detect general degradation that is caused by multiple factors. The pervasive structural deterioration of the sampling sites (Figure S5) might have a predominant impact on the quality class determination masking any potential relationship between quality class and pollution.

In general, the correlation analyses revealed a principle problem with identifying environmental variables affecting species assemblages. The saprobic index and the number of EPT taxa, for example, were significantly correlated ($p<0.01$) because many EPT taxa exhibit low tolerance towards low oxygen conditions and hence have a low saprobic rate resulting in a considerable impact on the saprobic index. As the saprobic index is a measure of O_2 deficiencies due to decaying organic matter, impacts by insufficient O_2 concentrations and pollutant toxicities are hardly distinguishable. In our case, OC1 was significantly correlated with O_2 concentrations and biological oxygen demand (Table S1). In many studies low oxygen concentrations have been revealed to have a significant impact on benthic invertebrates, in particular on Ephemeroptera, Plecoptera, and Trichoptera (e.g., [52]). However, our field measurements suggested good dissolved oxygen concentrations, with the bulk of measurements falling between 9.9 and 13.2 mg/L, with one exception: La1 with an average of 7.1 mg/L. Pronounced changes in aquatic species composition are not expected at concentrations >8 mg/L [53]. However, the presence and potential impact of an oxygen deficiency may have been underestimated because we did not record oxygen concentrations at night and increased nocturnal oxygen consumption by algae or submerged aquatic macrophytes can reduce oxygen saturation considerably [54].

OC1 was also associated with increased salinity ($\rho = 0.93$, Table S1). Salinity has been suspected to affect species composition elsewhere [53] and it is possible that the relationship observed between OCl and species assemblages was more related to salinity and not the WW-associated contaminants [55]. Again, Ephemeroptera, Plecoptera, and Trichoptera have been shown to react more sensitively to salinity than other benthic invertebrate species [55]. Therefore, a negative correlation of EPT taxa with salinity might be expected. Pond [56] documented a significant loss of mayfly taxa in Appalachian mining regions when conductivity exceeded 175 μS/cm. The average conductivity at our sampling sites ranged from 419 to 1350 μS/cm indicating a potential threat towards salinity-sensitive species. However, salinity in lowland streams is generally higher compared to streams in higher mountain ranges. Hence, lowland species should be more adapted to higher salinity levels and therefore salinity effects were expected to be less pronounced.

In the present study, salinity increase was mainly caused by the WW burden as indicated by the high correlation coefficient ($\rho = 0.93$) and was therefore regarded as anthropogenic disturbance. Depending on the causal factors influencing environmental salinity levels, high salt concentrations might often be accompanied by high contaminant levels (e.g., in mining regions [49]). Accordingly, impacts by contaminants and salinity can hardly be distinguished. However, an indicator for WW-associated contaminants having an impact is the significant correlation of OC1 with Coleoptera genera (*Elmis/Limnius/Haliplus*: ρ ranging from −0.45 to −0.53) because they are much less salinity-sensitive than other groups of macroinvertebrate species [57].

In general, EPT taxa are more susceptible to toxicants than other groups of macroinvertebrates [58] and in particular to OCs [59]. Accordingly, a toxicity-related impact can be assumed. However, the contaminants represented by OC1 are apparently of minor ecotoxicological relevance to the EPT taxa at the quantified concentrations, as the lowest observed effect concentrations are 3–4 orders of magnitudes higher [60]–[67]. This indicates that organophosphates, synthetic musk fragrances, and DEET were not responsible for the relationship between OC1 and species composition. This is further supported by weak correlations of the TU-based factors (to account for toxic characteristics of the considered contaminants) with biota (data not shown). Additionally, the inclusion of peak concentrations (90% percentiles of measurements) did not enhance correlations. Nevertheless, our results demonstrate that compounds represented by OC1 are a good surrogate for contaminants associated with municipal WW and they may be a suitable set of markers for the identification of WW-contamination as their main entry pathway to surface waters is through the sewage system. Among OC1 pollutants, synthetic musk fragrances have previously been proposed as a marker for municipal WW by Buerge et al. [68]. Therefore, we assume that the basic cause for the relationship between species composition and OC1 was due to alterations of physicochemical water parameters and/or other contaminants present in sewage effluents that were not considered in our study.

The WW-associated factor responsible for the deterioration of aquatic fauna populations–oxygen concentration, salinity or contaminants–could not be unequivocally identified with the available data. It is possible that all factors contributed to the observed impact on macroinvertebrate assemblages. Long-term observations of wastewater impacted streams may help to tease apart these factors and identify the primary factor; particularly at sites upgraded with advanced WW treatment technologies for enhanced pollutant removal [27], [69]. If an upgrade with enhanced pollutant removal technologies resulted in an increase in biodiversity, it could be unequivocally determined that the pollutants were primarily responsible for the decline in invertebrate populations.

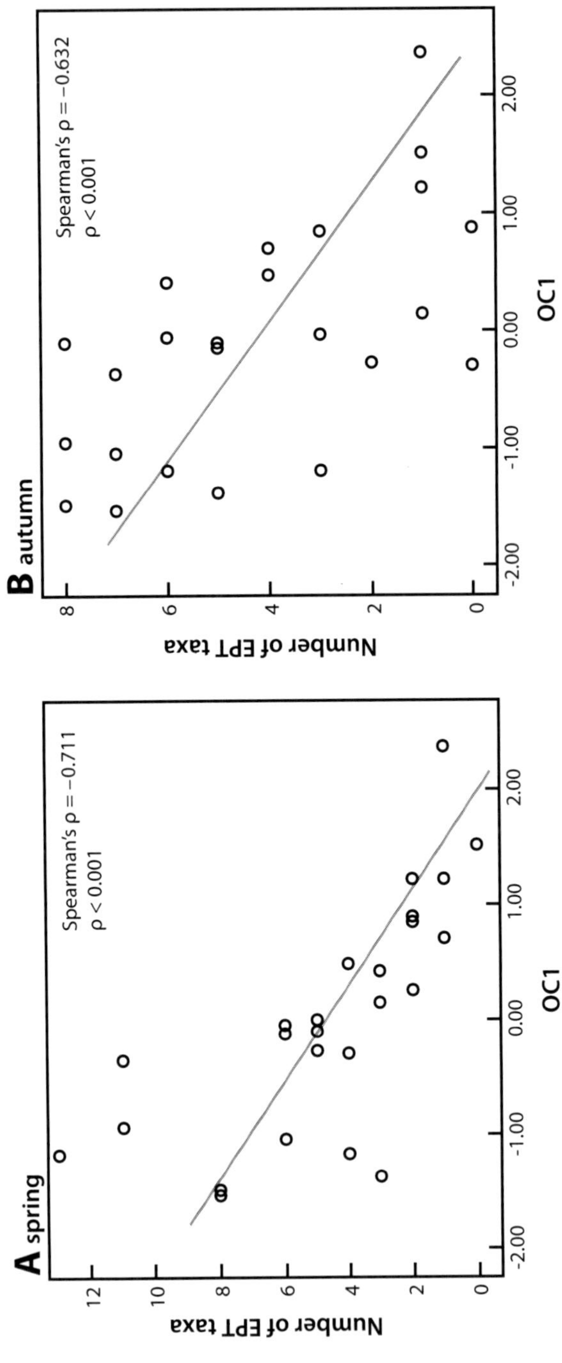

FIGURE 2: Number of EPT taxa (Ephemeroptera, Plecoptera, Trichoptera) correlating with the first component of the PCA with organic contaminants (OC1). Displayed are results for sampling campaign in spring (A) and autumn (B). Please note different scaling of y-axes in A and B.

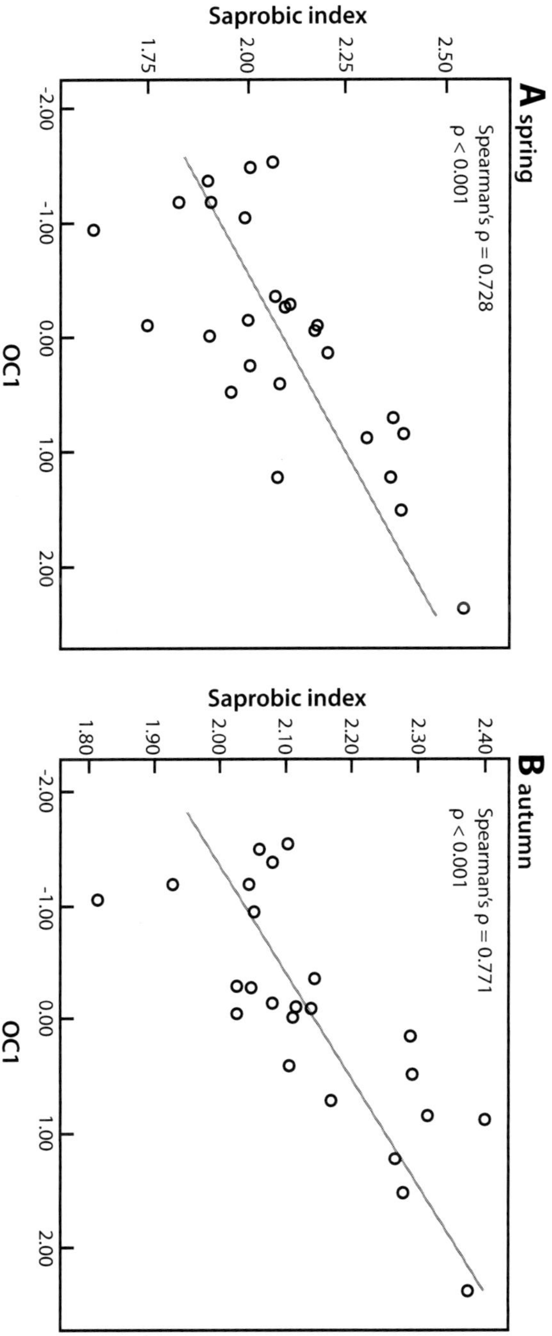

FIGURE 3: Saprobic index correlating with the first component of the PCA with organic contaminants (OC1). Displayed are results for sampling campaign in spring (A) and autumn (B). Please note different scaling of y-axes in A and B.

Advanced WW treatment is supposed to be an appropriate measure to improve the ecological status of WW-impacted surface waters via contaminant degradation/removal. However, the potential benefits and adverse impacts have to be carefully evaluated to ensure that upgrading WW treatment facilities is a worthwhile exercise. For example, a potential adverse impact associated with enhanced pollutant removal technologies, like advanced oxidation processes, is the formation of hazardous transformation products. Environmental releases of these by-products may outweigh the benefits of treatment plant upgrades to include enhanced pollutant removal [70]–[73].

9.5 CONCLUSION

Our study emphasizes a clear wastewater-associated impact on the ecological quality, e.g., decline in biodiversity, of surface waters, despite state-of-the-art wastewater treatment with biological activated sludge. As main factor threatening the ecological quality we identified wastewater-associated contaminants– despite their minute concentrations in the ng/L range– alongside with the structural modification of the streams. Consequently, we suggest that restoration measures should aim at improving both water as well as structural quality of surface waters in order to increase their ecological quality and help to conserve freshwater biodiversity.

REFERENCES

1. Ricciardi A, Rasmussen JB (1999) Extinction rates of North American freshwater fauna. Conservation Biology 13: 1220–1222. doi: 10.1046/j.1523-1739.1999.98380.x
2. Jenkins M (2003) Prospects for biodiversity. Science 302: 1175–1177. doi: 10.1126/science.1088666
3. Spanhoff B, Arle J (2007) Setting attainable goals of stream habitat restoration from a macroinvertebrate view. Restoration Ecology 15: 317–320. doi: 10.1111/j.1526-100x.2007.00216.x
4. Brack W, Apitz SE, Borchardt D, Brils J, Cardoso AC, et al. (2009) Toward a holistic and risk-based management of european river basins. Integrated Environmental Assessment and Management 5: 5–10. doi: 10.1897/ieam_2008-024.1
5. Ginebreda A, Munoz I, de Alda ML, Brix R, Lopez-Doval J, et al. (2010) Environmental risk assessment of pharmaceuticals in rivers: Relationships between hazard

indexes and aquatic macroinvertebrate diversity indexes in the Llobregat River (NE Spain). Environment International 36: 153–162. doi: 10.1016/j.envint.2009.10.003
6. Borja A, Dauer DM (2008) Assessing the environmental quality status in estuarine and coastal systems: Comparing methodologies and indices. Ecological Indicators 8: 331–337. doi: 10.1016/j.ecolind.2007.05.004
7. Hering D, Moog O, Sandin L, Verdonschot PFM (2004) Overview and application of the AQEM assessment system. Hydrobiologia 516: 1–20. doi: 10.1023/b:hydr.0000025255.70009.a5
8. Diaz RJ, Solan M, Valente RM (2004) A review of approaches for classifying benthic habitats and evaluating habitat quality. Journal of Environmental Management 73: 165–181. doi: 10.1016/j.jenvman.2004.06.004
9. Commission E (2000) Establishing a framework for community action in the field of water policy. Directive 2000/60/EC of the European parliament and of the council.
10. Karlson K, Rosenberg R, Bonsdorff E (2002) Temporal and spatial large-scale effects of eutrophication and oxygen deficiency on benthic fauna in Scandinavian and Baltic waters - A review. Oceanography and Marine Biology 40: 427–489. doi: 10.1201/9780203180594.ch8
11. Vaughan IP, Ormerod SJ (2012) Large-scale, long-term trends in British river macroinvertebrates. Global Change Biology 18: 2184–2194. doi: 10.1111/j.1365-2486.2012.02662.x
12. BMU (2007) Bundesministerium fuer Umwelt, Naturschutz und Reaktorsicherheit [German Federal Ministry for the Environment, Nature Conservation and Nuclear Safety]. Wastewater flow processed in German municipal wastewater treatment plants. Bundesministerium fuer Umwelt website. Available: http://www.bmu.de/binnengewaesser/abwasser/doc/3142.php. Accessed 2013 Mar 5.
13. Schwarzenbach RP, Escher BI, Fenner K, Hofstetter TB, Johnson CA, et al. (2006) The challenge of micropollutants in aquatic systems. Science 313: 1072–1077. doi: 10.1126/science.1127291
14. UNSD (2010) United Nations Statistic Division. Environmental Indicators: Inland Waters Resources – Wastewater (UNSD, New York).
15. Daughton CG, Ternes TA (1999) Pharmaceuticals and personal care products in the environment: Agents of subtle change? Environmental Health Perspectives 107: 907–938. doi: 10.1289/ehp.99107s6907
16. Daughton CG (2003) Cradle-to-cradle stewardship of drugs for minimizing their environmental disposition while promoting human health. I. Rationale for and avenues toward a green pharmacy. Environmental Health Perspectives 111: 757–774. doi: 10.1289/ehp.5947
17. Jobling S, Casey D, Rodgers-Gray T, Oehlmann J, Schulte-Oehlmann U, et al. (2003) Comparative responses of molluscs and fish to environmental estrogens and an estrogenic effluent. Aquatic Toxicology 65: 205–220. doi: 10.1016/s0166-445x(03)00134-6
18. Oetken M, Nentwig G, Loffler D, Ternes T, Oehlmann J (2005) Effects of pharmaceuticals on aquatic invertebrates. Part I. The antiepileptic drug carbamazepine. Archives of Environmental Contamination and Toxicology 49: 353–361. doi: 10.1007/s00244-004-0211-0

19. Parrott JL, Blunt BR (2005) Life-cycle exposure of fathead minnows (Pimephales promelas) to an ethinylestradiol concentration below 1 ng/L reduces egg fertilization success and demasculinizes males. Environmental Toxicology 20: 131–141. doi: 10.1002/tox.20087
20. Oehlmann J, Schulte-Oehlmann U, Bachmann J, Oetken M, Lutz I, et al. (2006) Bisphenol A induces superfeminization in the ramshorn snail Marisa cornuarietis (Gastropoda : Prosobranchia) at environmentally relevant concentrations. Environmental Health Perspectives 114: 127–133. doi: 10.1289/ehp.8065
21. Nentwig G (2007) Effects of pharmaceuticals on aquatic invertebrates. Part II: The antidepressant drug fluoxetine. Archives of Environmental Contamination and Toxicology 52: 163–170. doi: 10.1007/s00244-005-7190-7
22. Triebskorn R, Casper H, Scheil V, Schwaiger J (2007) Ultrastructural effects of pharmaceuticals (carbamazepine, clofibric acid, metoprolol, diclofenac) in rainbow trout (Oncorhynchus mykiss) and common carp (Cyprinus carpio). Analytical and Bioanalytical Chemistry 387: 1405–1416. doi: 10.1007/s00216-006-1033-x
23. Bundschuh M, Zubrod JP, Schulz R (2011) The functional and physiological status of Gammarus fossarum (Crustacea; Amphipoda) exposed to secondary treated wastewater. Environmental Pollution 159: 244–249. doi: 10.1016/j.envpol.2010.08.030
24. Beasley G, Kneale P (2002) Reviewing the impact of metals and PAHs on macro invertebrates in urban watercourses. Progress in Physical Geography 26: 236–270. doi: 10.1191/0309133302pp334ra
25. Liu Z, Kanjo Y, Mizutani S (2009) Removal mechanisms for endocrine disrupting compounds (EDCs) in wastewater treatment - physical means, biodegradation, and chemical advanced oxidation: A review. Science of The Total Environment 407: 731–748. doi: 10.1016/j.scitotenv.2008.08.039
26. Koh YKK, Chiu TY, Boobis AR, Scrimshaw MD, Bagnall JP, et al. (2009) Influence of operating parameters on the biodegradation of steroid estrogens and nonylphenolic compounds during biological wastewater treatment processes. Environmental Science & Technology 43: 6646–6654. doi: 10.1021/es901612v
27. Stalter D, Magdeburg A, Wagner M, Oehlmann J (2011) Ozonation and activated carbon treatment of sewage effluents: Removal of endocrine activity and cytotoxicity. Water Research 45: 1015–1024. doi: 10.1016/j.watres.2010.10.008
28. Fahrenberger G (1986) Hessisches Ried und Bergstraße. Hessen.: Staatliche Landesbildstelle.
29. Ehlert T, Hering D, Koenzen U, Pottgiesser T, Schuhmacher H, et al. (2002) Typology and type specific reference conditions for medium-sized and large rivers in North Rhine-Westphalia: Methodical and biological aspects. International Review of Hydrobiology 87: 151–163. doi: 10.1002/1522-2632(200205)87:2/3<151::aid-iroh151>3.0.co;2-a
30. Pottgiesser T, Sommerhäuser M (2003) Map of German stream types December 2003. Fließgewaesserbewertung website. Available: http://www.fliessgewaesserbewertung.de/download/typologie/. Accessed 2013 Mar 5.
31. Zumbroich T, Müller A, Friedrich G (1999) Strukturguete von Fließgewaessern - Grundlagen und Kartierung. Springer, Berlin.
32. Haase P, Sundermann A (2004) Standardisierung der Erfassungs- und Auswertungsmethoden von Makrozoobenthosuntersuchungen in Fließgewässern. Forschung-

sinstitut Senckenberg, Gelnhausen. Fließgewaesserbewertung website. Available: http://www.fliessgewaesserbewertung.de/download/probenahme-sortierung/. Accessed 2013 Mar 5.

33. Haase P, Sundermann A, Schindehütte K (2006) Informationstext zur Operationellen Taxaliste als Mindestanforderung an die Bestimmung von Makrozoobenthosproben aus Fließgewässern zur Umsetzung der EU-Wasserrahmenrichtlinie in Deutschland. Forschungsinstitut Senckenberg, Gelnhausen. Fließgewaesserbewertung website. Available: http://www.fliessgewaesserbewertung.de/download/bestimmung/. Accessed 2013 Mar 5.
34. Quednow K, Püttmann W (2007) Monitoring terbutryn pollution in small rivers of Hesse, Germany. Journal of Environmental Monitoring 9: 1337–1343. doi: 10.1039/b711854f
35. Quednow K, Puttmann W (2008) Endocrine disruptors in freshwater streams of Hesse, Germany: Changes in concentration levels in the time span from 2003 to 2005. Environmental Pollution 152: 476–483. doi: 10.1016/j.envpol.2007.05.032
36. Quednow K, Puttmann W (2008) Organophosphates and synthetic musk fragrances in freshwater streams in Hessen/Germany. Clean-Soil Air Water 36: 70–77. doi: 10.1002/clen.200700023
37. Quednow K, Puttmann W (2009) Temporal concentration changes of DEET, TCEP, terbutryn, and nonylphenols in freshwater streams of Hesse, Germany: possible influence of mandatory regulations and voluntary environmental agreements. Environmental Science and Pollution Research 16: 630–640. doi: 10.1007/s11356-009-0169-6
38. US-EPA (1982) The U.S. Environmental Protection Agency. List of Priority Pollutants. Office of Water. Water Quality Standards Database. Office of the Federal Registration (OFR) Appendix A: priority pollutants. Fed Reg. 47: 52309.
39. DIN (1996) Bestimmung von 61 Elementen durch Massenspektrometrie mit induktiv gekoppeltem Plasma (ICP-MS). Fachgruppe Wasserchemie der GDCh und Normenausschuss Wasserwesen (NAW) im DIN Deutsches Institut für Normung e.V. (Hrsg.): Deutsche Verfahren zur Wasser-, Abwasser- und Schlammuntersuchung. Band 1. VCH Weinheim, Beuth-Verlag, Berlin. ed.
40. DIN (1996) Bestimmung von 6 polyzyklischen aromatischen Kohlenwasserstoffen (PAK) mittels Hochleistungs-Flüssigkeitschromatographie.Fachgruppe Wasserchemie der GDCh und Normenausschuss Wasserwesen (NAW) im DIN Deutsches Institut für Normung e.V. (Hrsg.): Deutsche Verfahren zur Wasser-, Abwasser- und Schlammuntersuchung. Band 1. VCH Weinheim, Beuth-Verlag, Berlin. ed.
41. Hakanson L, Jansson M (1983) Principles of Lake Sedimentology. Springer, Berlin.
42. DIN (1987) Bestimmung der Partikelgrößenverteilung eines dispersen Gutes durch Siebanalyse. Fachgruppe Wasserchemie der GDCh und Normenausschuss Wasserwesen (NAW) im DIN Deutsches Institut für Normung e.V. (Hrsg.): Deutsche Verfahren zur Wasser-, Abwasser- und Schlammuntersuchung. Band 1. VCH Weinheim, Beuth-Verlag, Berlin. ed.
43. DIN (1987) Bestimmung der Sauerstoffzehrung in n Tagen. Fachgruppe Wasserchemie der GDCh und Normenausschuss Wasserwesen (NAW) im DIN Deutsches Institut für Normung e.V. (Hrsg.): Deutsche Verfahren zur Wasser-, Abwasser- und Schlammuntersuchung. Band 1. VCH Weinheim, Beuth-Verlag, Berlin. ed.

44. AQEM (2006) ASTERICS - Software manual - Version 3.0. Fließgewaesserbewertung website. Available: http://www.fliessgewaesserbewertung.de/download/berechnung/. Accessed 2013 Mar 5.
45. Backhaus K, Erichson B, Plinke W, Weiber R (2006) Multivariate Analysemethoden - Eine anwendungsorientierte Einführung. Springer, Berlin.
46. Oksanen J, Blanchet FG, Kindt R, Legendre P, Minchin PR, et al.. (2011) vegan: Community Ecology Package. R package version 2.0–2. R-project website. Available: http://CRAN.R-project.org/package=vegan. Accessed 2013 Mar 5.
47. Crawley MJ (2007) The R Book. Wiley; 1 edition (June 19, 2007).
48. Liess M, Von der Ohe PC (2005) Analyzing effects of pesticides on invertebrate communities in streams Environmental Toxicology and Chemistry. 24: 954–965. doi: 10.1897/03-652.1
49. Clements WH, Carlisle DM, Lazorchak JM, Johnson PC (2000) Heavy metals structure benthic communities in colorado mountain streams. Ecological Applications 10: 626–638. doi: 10.1890/1051-0761(2000)010[0626:hmsbci]2.0.co;2
50. Linke S, Bailey RC, Schwindt J (1999) Temporal variability of stream bioassessments using benthic macroinvertebrates. Freshwater Biology 42: 575–584. doi: 10.1046/j.1365-2427.1999.00492.x
51. Macdonald DD, Carr RS, Calder FD, Long ER, Ingersoll CG (1996) Development and evaluation of sediment quality guidelines for Florida coastal waters. Ecotoxicology 5: 253–278. doi: 10.1007/bf00118995
52. Coimbra CN, Graca MAS, Cortes RM (1996) The effects of a basic effluent on macroinvertebrate community structure in a temporary Mediterranean river. Environmental Pollution 94: 301–307. doi: 10.1016/s0269-7491(96)00091-7
53. Jacobsen D, Rostgaard S, Vasconez JJ (2003) Are macroinvertebrates in high altitude streams affected by oxygen deficiency? Freshwater Biology 48: 2025–2032. doi: 10.1046/j.1365-2427.2003.01140.x
54. Laursen AE, Seitzinger SP (2004) Diurnal patterns of denitrification, oxygen consumption and nitrous oxide production in rivers measured at the whole-reach scale. Freshwater Biology 49: 1448–1458. doi: 10.1111/j.1365-2427.2004.01280.x
55. Kefford BJ, Marchant R, Schafer RB, Metzeling L, Dunlop JE, et al. (2011) The definition of species richness used by species sensitivity distributions approximates observed effects of salinity on stream macroinvertebrates. Environmental Pollution 159: 302–310. doi: 10.1016/j.envpol.2010.08.025
56. Pond GJ (2010) Patterns of Ephemeroptera taxa loss in Appalachian headwater streams (Kentucky, USA). Hydrobiologia 641: 185–201. doi: 10.1007/s10750-009-0081-6
57. Kefford BJ, Hickey GL, Gasith A, Ben-David E, Dunlop JE, et al. (2012) Global scale variation in the salinity sensitivity of riverine macroinvertebrates: Eastern Australia, France, Israel and South Africa. Plos One 7: e35224. doi: 10.1371/journal.pone.0035224
58. Beketov MA (2004) Comparative sensitivity to the insecticides deltamethrin and esfenvalerate of some aquatic insect larvae (Ephemeroptera and Odonata) and Daphnia magna. Russian Journal of Ecology 35: 200–204. doi: 10.1023/b:ruse.0000025972.29638.46

59. Von Der Ohe PC, Liess M (2004) Relative sensitivity distribution of aquatic invertebrates to organic and metal compounds. Environmental Toxicology and Chemistry 23: 150–156. doi: 10.1897/02-577
60. WHO (1998) Flame retardants: Tris(chloropropyl) phosphate and tris(2-chloroethyl) phosphate - Environmental Health Criteria 209, Genf.
61. WHO (2000) Flame retardants: Tris(2-butoxyethyl) phosphate, tris(2 ethylhexyl) phosphate and tetrakis(hydroxymethyl) phosphonium salts - Environmental Health Criteria 218, Genf.
62. EC (2000) European Commission - IUCLID Dataset: Tributyl Phosphate. European Chemicals Bureau Luxembourg.
63. EC (2000) European Commission - IUCLID Dataset: 4-(1,1,3,3-Tetramethylbuthyl) Phenol. European Chemicals Bureau Luxembourg.
64. EC (2000) European Commission - IUCLID Dataset: Tris(2-Chloroethyl) Phosphate. European Chemicals Bureau Luxembourg.
65. EC (2002) European Commission - 4-Nonylphenol (branched) and Nonylphenol. European Union Risk Assessment Report 10. European Chemicals Bureau Luxembourg. EUR 20387 EN.
66. EC (2003) European Commission - 4,4′-Isopropylidenephenol (Bisphenol-A). European Union Risk Assessment Report 37. European Chemicals Bureau Luxembourg. EUR 20843 EN.
67. Artola-Garicano E, Sinnige TL, van Holsteijn I, Vaes WHJ, Hermens JLM (2003) Bioconcentration and acute toxicity of polycyclic musks in two benthic organisms (Chironomus riparius and Lumbriculus variegatus). Environmental Toxicology and Chemistry 22: 1086–1092. doi: 10.1002/etc.5620220516
68. Buerge IJ, Buser H-R, Mueller MD, Poiger T (2003) Behavior of the polycyclic musks HHCB and AHTN in lakes, two potential anthropogenic markers for domestic wastewater in surface waters. Environmental Science and Technology 37: 5636–5644. doi: 10.1021/es0300721
69. Hollender J, Zimmermann SG, Koepke S, Krauss M, McArdell CS, et al. (2009) Elimination of organic micropollutants in a municipal wastewater treatment plant upgraded with a full-scale post-ozonation followed by sand filtration. Environmental Science & Technology 43: 7862–7869. doi: 10.1021/es9014629
70. Magdeburg A, Stalter D, Oehlmann J (2012) Whole effluent toxicity assessment at a wastewater treatment plant upgraded with a full-scale post-ozonation using aquatic key species. Chemosphere 88: 1008–1014. doi: 10.1016/j.chemosphere.2012.04.017
71. Stalter D, Magdeburg A, Oehlmann J (2010) Comparative toxicity assessment of ozone and activated carbon treated sewage effluents using an in vivo test battery. Water Research 44: 2610–2620. doi: 10.1016/j.watres.2010.01.023
72. Stalter D, Magdeburg A, Weil M, Knacker T, Oehlmann J (2010) Toxication or detoxication? In vivo toxicity assessment of ozonation as advanced wastewater treatment with the rainbow trout. Water Research 44: 439–448. doi: 10.1016/j.watres.2009.07.025
73. Weemaes M, Fink G, Lachmund C, Magdeburg A, Stalter D, et al. (2011) Removal of micropollutants in WWTP effluent by biological assisted membrane carbon filtration (BIOMAC). Water Science and Technology 63: 72–79. doi: 10.2166/wst.2011.011

There are several supplemental files that are not available in this version of the article. To view this additional information, please use the citation on the first page of this chapter.

CHAPTER 10

Performance of a Constructed Wetland in Grand Marais, Manitoba, Canada: Removal of Nutrients, Pharmaceuticals, and Antibiotic Resistance Genes from Municipal Wastewater

JULIE C. ANDERSON, JULES C. CARLSON, JENNIFER E. LOW, JONATHAN K. CHALLIS, CHARLES S. WONG, CHARLES W. KNAPP, AND MARK L. HANSON

10.1 BACKGROUND

The environmental fate of excess nutrients and pharmaceuticals and personal care products (PPCPs) has become an area of great interest over the past decade, particularly in aquatic ecosystems [1]. In general, PPCPs are designed to be biologically active at very low doses, and the effects of exposure to these compounds, particularly under chronic exposures, are not well understood [2,3]. Micropollutants such as PPCPs are not typically targeted for removal by wastewater treatment systems [3], so these compounds are detected in surface waters globally [4-6].

In addition, antibiotic resistance genes (ARGs) have also been detected in the environment as a result of the prevalent human and veterinary use of antibacterial and antimicrobial products [7-10], which are also not eliminated by conventional wastewater treatment plants [5,11]. Genes encoding for resistance to a variety of antibiotics have been detected in surface waters, sewage, treated wastewater, and drinking water, and are ubiquitous in aquatic environments impacted by human activity [10,12-15]. Over the past decade, focus has shifted from studying antibiotic resistance primarily in a clinical context to examining the potential environmental impacts of ARGs [12]. Concern and interest are growing in regards to the role and effects of ARGs in aquatic ecosystems since there are public and environmental health implications resulting from transport and dissemination of ARGs into water bodies [7,10,14,16,17]. Primarily, ARGs are a concern due to the potential for persistence of antibiotic resistance and future outbreaks via antibiotic-resistant pathogens [5,12]. The World Health Organization has identified antibiotic resistance as a major health concern [17] and it has been reported that diseases that were previously eradicated (e.g. tuberculosis) may soon pose a severe global risk to human health due to the prevalence of ARGs and resistant pathogens [18].

Treatment wetlands offer a potential option for cost-effective removal of PPCPs and ARGs from municipal wastewater. Wetlands can be used as a secondary or tertiary treatment step, following chemical and/or biological treatments, and rely upon natural processes in shallow water or temporarily flooded land that is able to support aquatic life [18]. These systems tend to be less resource-intensive than conventional wastewater treatment plants [5,18], and have been used successfully for treatment of municipal sewage in small communities, as well as for some industrial wastewaters [19]. While most research has focused on the use of wetlands for reduction of nutrients and biochemical oxygen demand (BOD) in water bodies receiving runoff from agricultural or urban sources [6,20], recent studies have shown that these systems might remove PPCPs as well [1,6,18,21]. Specifically, wetlands have shown potential for removal of antibiotics via sorption, uptake by plants, and partial or complete physico-chemical and/or biological degradation [5]. However, removal efficiency in wetlands is affected by a number of factors, including age of the wetland, seasonality, and presence or absence of plants [19-21]. Effects of climate and seasonality are particularly important considerations for wetlands in the Canadian Prairies [4,22] as many studies of treatment wetlands have been conducted in the southern United States (e.g. [1]) and Europe (e.g. [5,18]). These climates are quite different from Canada, and the published results may not be applicable to this geographical region as wetlands rely heavily on climatic and biological factors. To optimize these systems for removal of PPCPs and ARGs in the Canadian prairie climate, a better understanding of the numerous interacting parameters is required, as well as some sense of how current systems are functioning, if at all, in this regard.

Within the province of Manitoba, Canada, there are many small communities (populations $\leq$ 10,000) where full-scale conventional wastewater treatment plants are not financially or operationally feasible. It has been estimated that upwards of 350 communities in Manitoba rely on lagoons for the treatment of their waste prior to direct release into surface waters [23]. With the implementation of stricter provincial and federal guidelines around municipal wastewater release [24], alternative treatment systems, such as wetlands, need to be characterized for their efficacy at removing nutrients, PPCPs, and ARGs in a rural, prairie context. Preliminary work has been done in other communities in Manitoba to quantify the concentrations of pharmaceuticals in wastewater lagoon effluent [4], but the effectiveness of wetland treatment in this region is currently unknown. The community of Grand Marais uses one of the few operating sewage lagoon/constructed wetland treatment systems in the province and was selected as a model system for this study. The overall objectives of this study were to characterize the presence of nutrients and emerging wastewater contaminants (i.e., PPCPs and ARGs) in the Grand Marais system and to evaluate the effectiveness of treatment wetlands in removal of these contaminants. It was hypothesized that the use of a treatment wetland would enhance degradation and elimination of these target compounds, and therefore, could be an option to complement the current lagoon wastewater treatment system in communities that rely on lagoon treatment alone.

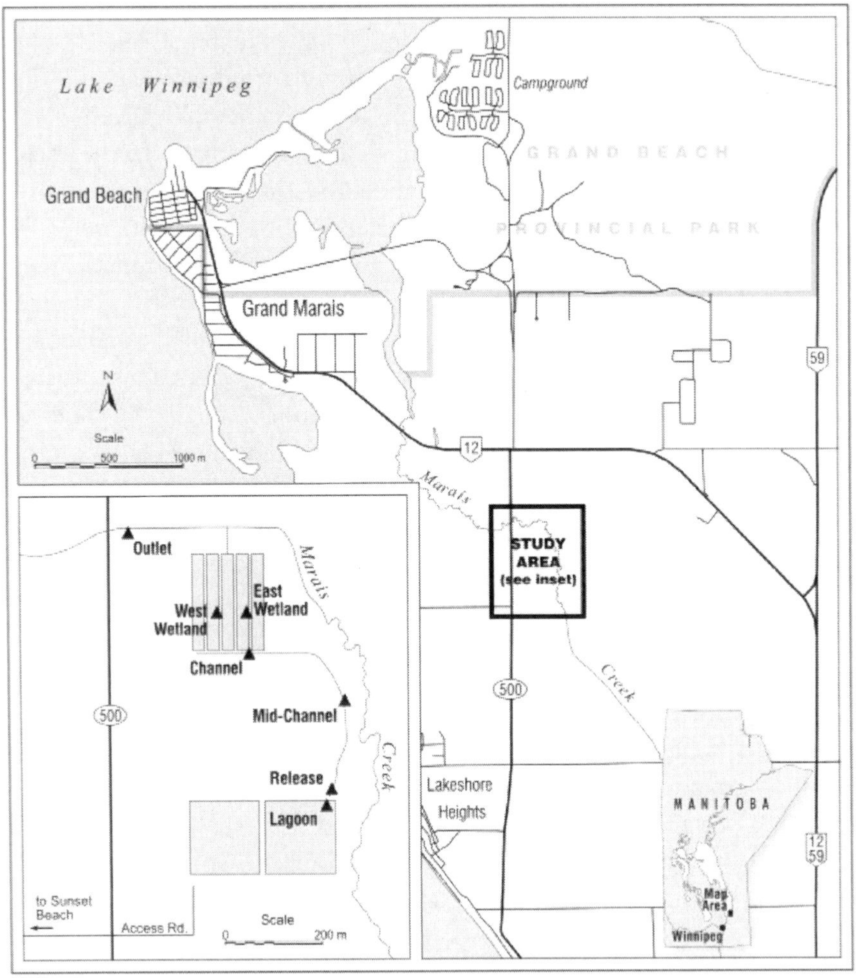

FIGURE 1: Map showing the seven sampling site locations in the Grand Marais treatment system in Manitoba, Canada. Sites were Lagoon, Release, Mid-Channel, Channel, East Wetland, West Wetland, and Outlet.

TABLE 1: Water quality parameters measured in sampling sites near the Grand Marais treatment wetland during 2012

Date	Site	Total suspended Solids (mg/L)	Nitrite + nitrate (mg/L)	Total ammonia + ammonium (mg/L)	Total phosphorus (mg/L)	Chl A (μg/L)	DO (mg/L)	T (°C)	Conductivity (mS/cm)	pH
May 22	Lagoon	29 ±3	0.14 ± 0.01	1.7 ± 0.1	3.1 ± 0.10	1.5 × 102	1.5	15.3	1.1	NA
	Outlet	8.2 ± 0.8	<LOD	0.010	0.030 ± 0.002	12	7.4	14.8	0.42	NA
June 15	Lagoon	34 ± 6	<LOD	0.060 ± 0.01	1.5 ± 0.04	1.6 × 102	1.5	18.1	1.1	9.25
	Outlet	8.6 ± 1	<LOD	0.020 ± 0.01	0.040 ± 0.01	15	7.6	17.5	0.38	7.56
July 16	Release	12 ± 0.8	<LOD	0.17 ± 0.01	0.68 ± 0.01	11	5.6	20.1	0.99	9.68
	Channel	12 ± 1	<LOD	0.22 ± 0.02	0.46 ± 0.01	22	0.60	19.4	0.89	9.26
	East Wetland	7.4 ± 2	<LOD	0.18 ± 0.04	0.40 ± 0.04	80	0.90	18.6	0.89	7.85
	West Wetland	12 ± 2	<LOD	0.030 ± 0.02	0.10 ± 0.04	1.7 × 102	0.90	17.2	0.54	7.10
	Outlet	7.0 ± 1	<LOD	0.020 ± 0.001	0.010 ± 0.007	16	4.4	19.3	0.41	7.44
July 23	Release	9.7 ± 0.9	<LOD	0.25	0.39 ± 0.03	24	6.0	24.2	1.1	9.95
	Mid-Channel	31 ± 8	<LOD	0.14	0.61 ± 0.02	76	0.20	21.7	1.1	8.89
	Channel	13 ± 2	<LOD	0.040	0.51 ± 0.03	24	0.50	20.2	1.1	8.33
	East Wetland	5.3 ± 0.5	<LOD	0.060	0.10 ± 0.04	1.2 × 102	0.30	19.5	1.1	7.31
	West Wetland	15 ± 5	<LOD	0.030	0.040 ± 0.01	1.3 × 102	0.30	18.9	0.72	6.92
	Outlet	4.2 ± 1	<LOD	0.020	<LOQ	15	4.3	23.3	0.38	7.46

Measurements are presented as mean value ± SD; <LOD = below the limit of detection; <LOQ = below the limit of quantification; NA= not available; Chl A = chlorophyll-a; DO = dissolved oxygen. LODs: Nitrite + nitrate – 0.05 mg/L, Total ammonia + ammonium – 0.010 mg/L; LOQs: Phosphorus – 0.010 mg/L.

10.2 RESULTS

10.2.1 GENERAL WATER QUALITY PARAMETERS

Samples were collected from the lagoon and from six sites within the treatment wetland between the influent entry point and the outlet into receiving surface waters. Upstream to downstream (direction of lagoon effluent flow), the sites were as follows: Lagoon, Release, Mid-Channel, Channel, East Wetland, West Wetland, and Outlet (Figure 1). Results of water quality monitoring at the seven sites in 2012 are reported in Table 1. The measured temperatures varied over the course of the sampling season, as expected, and among sites by as much as 5.3°C on the same sampling day. Conductivity was generally least at the Outlet site and greatest at the Lagoon or Release sites. Concentrations of chlorophyll-a (measured at ~ 30 cm below the surface) were quite variable among sites, with the greatest concentrations measured at the East Wetland, West Wetland, and Lagoon sites. In general, the concentrations of DO (dissolved oxygen) were quite low in the lagoon and wetland, with several measurements below 1 mg/L. The greatest concentration of DO was measured at the Release and Outlet sites, and the least concentration of DO was measured in the channel and lagoon. Measured pH ranged from 6.9 to 10.0 with the greatest pH values observed at the Lagoon, Release, and Channel. The Outlet and East Wetland sites typically had the lowest values of total suspended solids (TSS), and the Lagoon had the greatest values of TSS.

An approximate discharge rate was calculated using the distance from lagoon release to the Channel site. Assuming a discharge volume of 23,200 m^3, discharge rate was ~0.02 m^3/s, averaged over the course of the entire lagoon release period (July 11 to 24), and residence time within the length of the channel was approximately 20 hours. The channel itself is a ditch with wetland plants lining the sides. Residence time in the wetland was not determined due to the complexity of the flow patterns and the altered channels, which no longer followed the engineered 'snaking' flow pattern through winding rows. When the wetland was constructed in 1996, it was recommended that it receive inputs from the secondary lagoon in the fall (September 1 to October 31) with anticipated retention times of at least five to ten days.

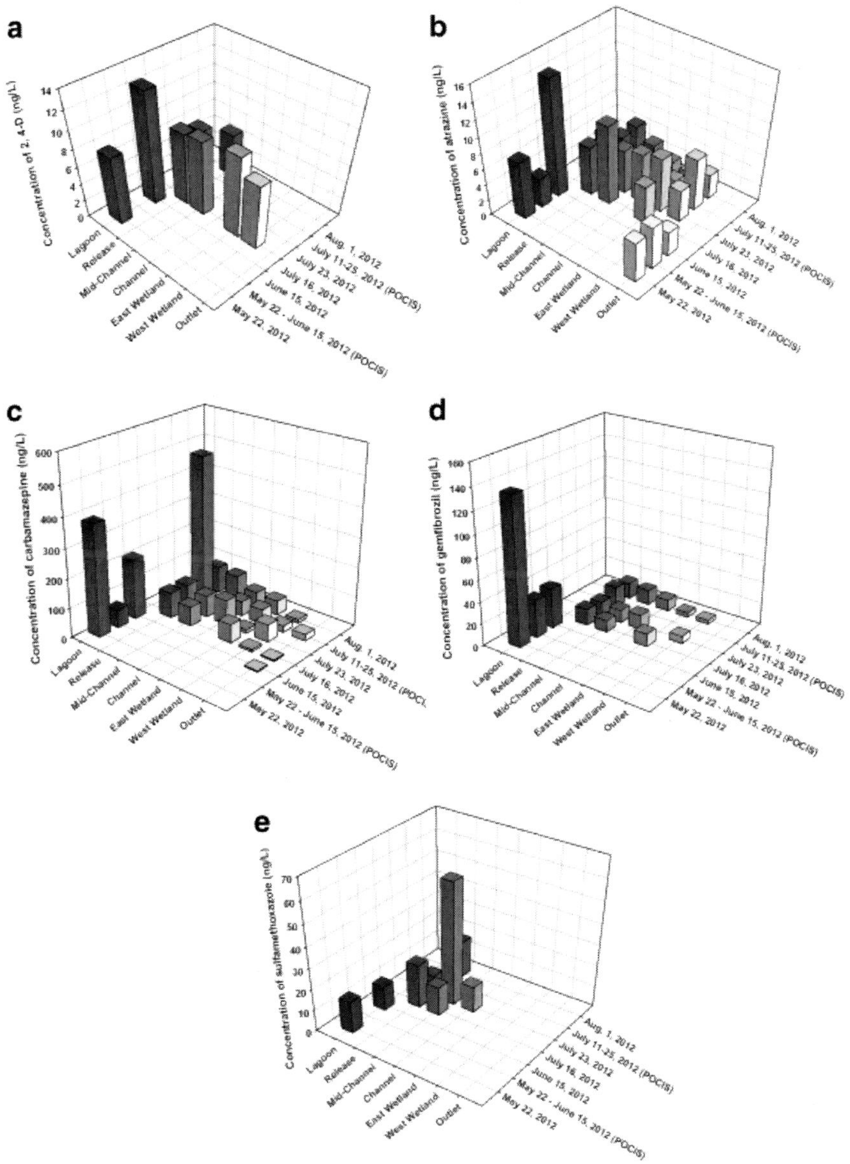

FIGURE 2: Mean concentrations of a) 2,4-D, b) atrazine, c) carbamazepine, and d) gemfibrozil and e) sulfamethoxazole measured at locations in the Grand Marais treatment wetland in summer 2012 by POCIS or SPE sampling.

TABLE 2: Calculated hazard quotients for pesticides and PPCPs detecteda in the Grand Marais treatment wetland and surrounding sampling sites in 2012 (adapted from Carlson et al., 2013)[4]

Compound	Species	Toxicity endpoint	Toxicity value (mg/L)	MECb (mg/L)	HQc	Reference for toxicity value
2,4-D	*Ranunculus aquatilis* (Water buttercup)	EC50 – 4 week relative growth	0.2	1.3×10^{-5}	6.4×10^{-2}	Belgers et al., 2007 [26]
	Daphnia magna	EC50 – 48 h immobilization	25	1.3×10^{-5}	5.1×10^{-4}	Martins et al., 2007 [27]
	Oncorhynchus mykiss (Rainbow trout)	LC50 – 96 h exposure	100	1.3×10^{-5}	1.3×10^{-4}	Little et al., 1990 [28]
Atrazine	*Lemna minor*	IC50 – 7 day growth inhibition	61.7	1.5×10^{-5}	2.4×10^{-4}	Teodorovic et al., 2012 [29]
	Daphnia magna	EC50 – 48 h immobilization	25.3	1.5×10^{-5}	1.7×10^{-2}	Phyu et al., 2004 [30]
	Oncorhynchus mykiss	LC50 – 28 day exposure	0.87	1.5×10^{-5}	5.8×10^{-4}	Giddingset al., 2005 [31]
Carbamazepine	*Lemna minor*	EC50 – 7 day growth inhibition	22.5	5.0×10^{-4}	2.2×10^{-2}	Cleuvers, 2003 [32]
	Daphnia magna	EC50 – 48 h immobilization	>100	5.0×10^{-4}	5.0×10^{-3}	Cleuvers, 2003 [32]
	Oryzias latipes (Japanese medaka)	LC50 – 48 h exposure	35.4	5.0×10^{-4}	1.4×10^{-2}	Kim et al., 2007 [33]
Gemfibrozil	*Chlorella vulgaris*	LC50 – 24 h exposure	60	1.4×10^{-4}	2.3×10^{-3}	El-Bassat et al., 2011 [34]
	Daphnia spp.	ECOSAR EC50 (acute)	6	1.4×10^{-4}	2.3×10^{-2}	Sanderson et al., 2003 [35]
	Fish spp.	ECOSAR EC50 (acute)	0.9	1.4×10^{-4}	1.5×10^{-1}	Sanderson et al., 2003 [35]
Sulfamethoxazole	*Pseudokirchneriella subcapita*	EC50 – 72 h growth inhibition	0.52	5.8×10^{-5}	1.1×10^{-1}	Isidori et al., 2005 [36]
	Daphnia magna	EC50 – 24 h immobilization	25.2	5.8×10^{-5}	2.3×10^{-3}	Isidori et al., 2005 [36]
	Oryzias latipes	LC50 – 96 h exposure	562.5	5.8×10^{-5}	1.0×10^{-4}	Kim et al., 2007 [33]

TABLE 2: *Cont.*

Compound	Species	Toxicity endpoint	Toxicity value (mg/L)	MECb (mg/L)	HQc	Reference for toxicity value
Sulfapyridine	*Pseudokirchneriella subcapitata*	IC50 – 72 h growth inhibition	10.2	7.9×10^{-6}	7.7×10^{-4}	Blaise et al., 2006 [37]
	Thamnocephalus platyurus (Beavertail fairy shrimp)	LC50 – 24 h exposure	144.4	7.9×10^{-6}	5.5×10^{-5}	Blaise et al., 2006 [37]
	Oncorhynchus mykiss	48 h TEC – primary hepato-cyte exposure	>249	7.9×10^{-6}	3.2×10^{-5}	Blaise et al., 2006 [37]

a A full list of screened compounds and their limits of detection can be found in Additional file 1: Table S1.
b MEC = Maximum environmental concentration measured in the current study.
c HQ = Hazard quotient.

10.2.2 CONCENTRATIONS OF NUTRIENTS

The concentrations of nitrate + nitrite, total ammonia + ammonium, and total phosphorus are also reported in Table 1. Only one sample, from the Lagoon site, had a detectable and quantifiable concentration of nitrate + nitrite of 0.14 mg/L. Measurements of total ammonia + ammonium ranged from 0.02 to 1.7 mg/L. These measured concentrations were generally greatest at the Lagoon, Release, and Channel sites and least at the Outlet site. Finally, total phosphorus was measured between 0.01 and 3.1 mg/L, with the greatest concentrations occurring at the Lagoon site and the least concentrations at the Outlet site.

10.2.3 CONCENTRATIONS OF PHARMACEUTICALS AND PESTICIDES

Only six of the thirty-nine target pharmaceuticals and pesticides were detected in samples from the Grand Marais study area: the herbicides 2,4-D and atrazine, the anticonvulsant carbamazepine, the lipid regulator gemfibrozil, and the antibiotics sulfamethoxazole and sulfapyridine (Additional file 1: Table S1 for full list of compounds and LODs and Additional file 1: Table S2 for full list of concentrations observed). Attempts were made to determine dissipation rate constants for these compounds based upon collected field data. However, constants could not be calculated since consistent dissipation was not observed between sites along the channel, possibly due to insufficient retention time in the wetland. The range of concentrations measured for each compound and the differences among sites are discussed below. There were only two sampling events (June 15 and July 23/25) for which Polar Organic Chemical Integrative Sampler (POCIS) and solid phase extraction (SPE) samples could be compared quantitatively. The concentrations measured from POCIS samples were quite consistent with those measured by SPE, which is in agreement with previous comparisons of these techniques at similar sites in Manitoba [4]. This agreement suggests that the time-weighted-average concentrations, observed by POCIS, may likely be in line with the day-to-day fluctuations expected in a dynamic system, and thus are an integrator of changing temporal levels of chemicals with time [25]. It is important to note, however, that such agreement does not necessarily prove that time-weighted-average concentrations must be at the same concentration ranges as that of grab measurements, which could fortuitously measure chemicals at abnormally high or low concentrations.

In the majority of the water samples analyzed, 2,4-D was either not detected or below the limit of quantification (LOQ) (Figure 2a), similar to results observed elsewhere in rural Manitoba [4]. Most of the detections occurred on July 16, 2012, with very similar concentrations measured across the sites, in the range of 7 to 9 ng/L. The greatest concentration of 2,4-D measured was 13 ng/L at the Lagoon site using SPE. The Lagoon site had significantly more 2,4-D present than the Channel, West Wetland, or Outlet sites ($p<0.05$). There were no significant differences between concentrations of 2,4-D in the Channel and the Outlet ($p>0.05$), so elimination of 2,4-D was not significant within the wetland.

Concentrations of atrazine varied from non-detectable to 15 ng/L, with at least one detection in all sampling locations (Figure 2b). Atrazine was detected in the Lagoon and Outlet sites in the spring sampling and consistently in the wetland and channel during the summer months. There was a significant difference between the Channel site upstream and the Outlet site downstream of the wetland ($p<0.05$), suggesting that elimination processes occurred in the wetland.

The greatest concentrations of carbamazepine in individual samples were measured by POCIS at the Release site (500 ng/L) and by SPE at the Lagoon (380 ng/L) (Figure 2c). Generally, concentrations of carbamazepine were below 100 ng/L and detections were recorded for all sampling sites over the course of the study period. There was a significant reduction observed between entry and release points at the wetland (i.e. Channel and Outlet, respectively) ($p<0.05$), but there were no significant differences in concentrations of carbamazepine among any of the other sites. These results suggest processes within the wetland may significantly reduce concentrations of carbamazepine.

Gemfibrozil was detected at all sites except for the Outlet and the greatest concentration of 140 ng/L was measured by SPE at the Lagoon (Figure 2d). Concentrations of gemfibrozil were generally greater at the Release site than at the Wetland or Channel sites. The Lagoon site had significantly greater concentrations of gemfibrozil than any other sampling location ($p<0.05$), but there was no significant reduction in concentration observed as a result of passage through the treatment wetland ($p>0.05$).

Sulfamethoxazole was detected on five sampling days and only at four of the sampling sites: Lagoon, Release, Mid-Channel, and Channel (Figure 2e). The greatest concentration measured in an individual sample was 58.1 ng/L, which was measured at Mid-Channel by SPE. Statistical analyses found no differences among any of the sampling sites in terms of concentrations of sulfamethoxazole or between locations upstream and downstream of the wetland ($p>0.05$), indicating that elimination of sulfamethoxazole was not occurring within the Grand Marais treatment system.

Finally, sulfapyridine was only measured once at a quantifiable concentration (7.9 ng/L) and this was at the Outlet site. It was detected a few other times below LOQ, and the majority of samples had non-detection of sulfapyridine. There were no trends observed among sites for concentrations of sulfapyridine since it did not persist in the environment and was therefore not detected regularly in samples.

The hazard quotients (HQs) ranged from 3.2×10^{-5} to 1.5×10^{-1} (Table 2) so none of the pesticides or PPCPs quantified were deemed to pose a significant hazard (HQ> 1) to aquatic plants, invertebrates, or fish. The greatest HQ values were for gemfibrozil and sulfamethoxazole, calculated for fish and primary producers, respectively. Sulfapyridine, atrazine, and 2,4-D were expected to pose the least hazard to primary producers, invertebrates, and fish based upon the calculated HQs.

10.2.4 PRESENCE OF ARGS

Abundances of 16S rRNA genes (a surrogate measure of total bacteria) were fairly consistent over time at each site, with values ranging between 10^5 and 10^7 genes per mL of water sampled (Additional file 1: Table S3). Abundances of ARGs were standardized to the abundance of 16S in each sample to provide an indication of the proportion of the bacterial genes that could impart microbial resistance (Figure 3a and 3b). All of the

ARGs of interest were measured at each site and during every sampling event, except for *tet*(W) at the Release and Channel sites on August 1 and bla_{SHV} at the Outlet site on June 19. The *tet* gene series confers resistance to tetracycline, which includes ribosomal protection proteins and efflux pumps. The *bla* genes are for enzymes that provide beta-lactam resistance, with bla_{TEM} being most commonly found. *Sul* are genes for sulfonamide resistance.

Of the ten ARGs investigated in this study, the third multi-plex tet-gene series, (*tet*(K, L, M, O, S)) and bla_{TEM} generally had the greatest abundances in the samples from the Grand Marais treatment system. There was no obvious pattern of abundances of ARGs with movement upstream to downstream in the system, which did not warrant investigating individual determinants, but often the least measured abundance of ARGs was in the channel (Figure 3a and 3b). Concentrations of sulfonamide compounds were compared to abundances of *sul*-I, *sul*-II, and *sul*-III, but there was no significant linear relationship between abundances of these ARGs and measured concentrations of sulfonamides in the Grand Marais system (Figure 4). This is not surprising, as drug concentrations are below the Minimum Inhibitory Concentration (MIC) for most bacteria [38], and residence times are too short to monitor any effects at sub-inhibitory concentrations [39]; Most importantly, antibiotic resistance develops in the guts of treated organisms and therefore has different fates than the chemical antibiotic once released into the environment. Due to analytical issues, it was not possible to measure the concentrations of beta-lactam or tetracycline antibiotics in the system, so comparisons between those compounds and abundances of corresponding ARGs were not possible.

There was significant removal of bla_{SHV} between West Wetland and Outlet ($p<0.05$), but none of the other antibiotic resistant bacteria were significantly removed by the wetland. Overall, the abundance of each of the ARGs was less than 1% of the abundance of 16S genes, suggesting less than 1% of the bacterial population had the potential for resistance via one particular gene, which is typical for many lagoon systems, but the presence of multiple ARGs within a bacterium is also possible [40].

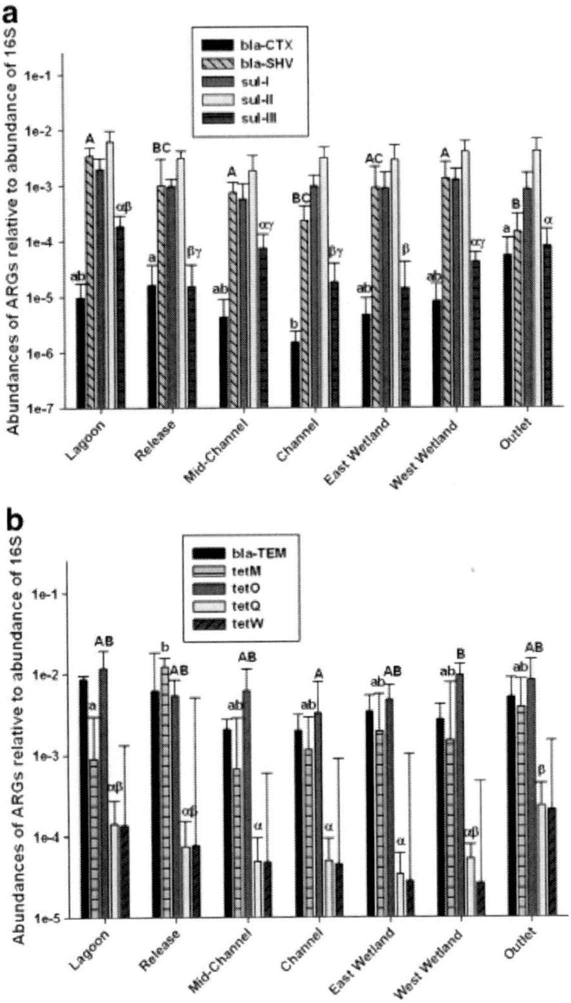

FIGURE 3: a) Mean (±SD) abundances of blaCTX, blaSHV, sul-I, sul-II, and sul-III antibiotic resistance genes standardized to abundances of 16S-rRNA from samples collected at locations in the Grand Marais treatment wetland in summer 2012 and analyzed using qPCR. Statistically significant differences ($p<0.05$) in abundances of individual genes are indicated using different lower case, upper case, and Greek letters. b) Mean (±SD) abundances of blaTEM and tetr antibiotic resistance genes standardized to abundances of 16S r-RNA from samples collected at locations in the Grand Marais treatment wetland in summer 2012 and analyzed using qPCR. Statistically significant differences ($p<0.05$) in abundances of individual genes are indicated using different lower case, upper case, and Greek letters.

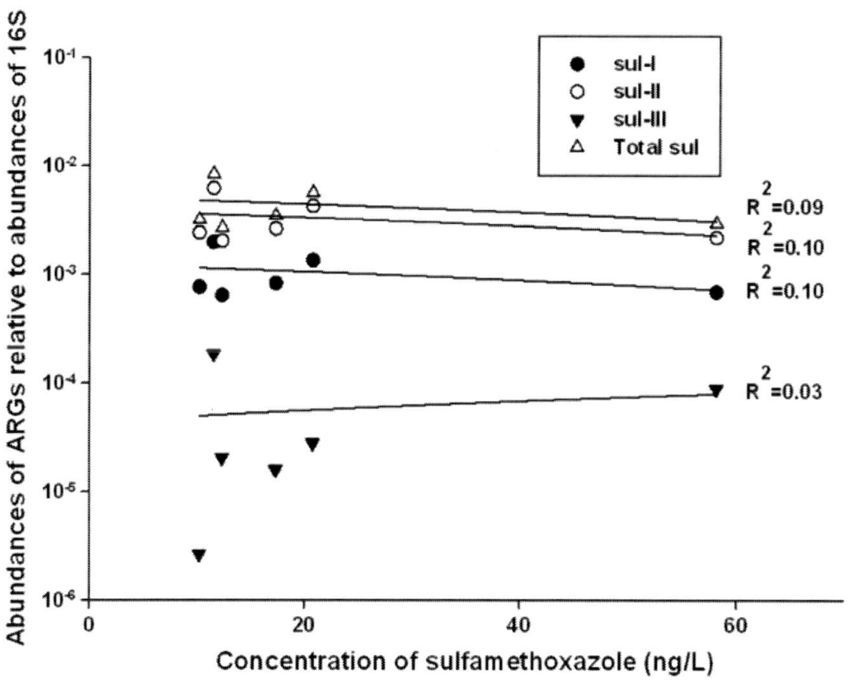

FIGURE 4: Abundances of sulfonamide resistance genes (sul-I, sul-II, sul-III, and sum of all three) compared to concentrations of sulfamethoxazole measured in the Grand Marais treatment wetland in summer 2012. There were no significant correlations between the abundances of ARGs and the concentration of antibiotics in the water ($p>0.05$)

10.3 DISCUSSION

10.3.1 PRESENCE AND REMOVAL OF NUTRIENTS

Performance of the Grand Marais treatment wetland system was comparable to other wetlands, particularly in Europe, where some removal of nutrients (typically 30 to 50% of N and P) is expected, assuming loadings are not excessive [41]. Concentrations of phosphorus measured in the wetland were consistent with previous studies of other wetlands from the Interlake region of south-central Manitoba [42] and were below trigger levels for all lake types. Therefore, phosphorus was not considered a hazard for aquatic organisms. Nitrate and nitrite were not a concern at any of the sites sampled as they were only detected in one sample during the entire study duration. There were several instances where concentrations of total ammonia + ammonium surpassed the Canadian regulations to protect aquatic life, as specified by the particular pH and temperature conditions during the time of sampling [43]. Excess total ammonia + ammonium was measured in both the channel and in the lagoon and may be a result of processes within the treatment system whereby anoxic conditions in stagnate water can produce ammonia [44]. The elevated ammonia in the lagoon appeared to be more transient than that in the channel since several consecutive samples from the Channel and Mid-Channel sites had excess ammonia. However, concentrations of total ammonia, calculated according to Canadian wastewater regulations [45], did not surpass the requirements for wastewater.

As mentioned above, pH played a role in the allowable concentration of ammonia, and there were several instances where pH was measured above levels that are recommended for fresh water organisms (i.e. > 9.0) [43]. However, measured pH and conductivity in the Grand Marais wetland were very similar to other wetland areas in the Rural Municipality of St. Clements [46]. The DO levels were quite low in both the lagoon and the wetland system (except for the Outlet site) and generally were well below the recommended guidelines for freshwater habitats (i.e. <5.5 mg/L) [43]. The East Wetland and West Wetland sites had concentrations of DO that were below those typically measured in other Manitoba wetlands. However, DO at the Outlet site was consistent with concentrations measured in other local wetlands [42,47]. The Canadian wastewater regulations for TSS require concentrations no greater than 25 mg/L for a short-term duration, and all measured values were at or below that level so TSS was not a concern in this system [45].

In general, concentrations of nutrients decreased from upstream of the wetland to downstream, indicating that the treatment wetland system was attenuating concentrations of nutrients in wastewater. In addition, many of the measured water quality parameters improved with movement from the lagoon to the outlet, so the wetland represented a fairly effective means of secondary treatment for municipal wastewater produced by small communities. The East Wetland had consistently greater concentrations of nutrients and conductivity than the West Wetland. This result was consistent with the longer travel time to the West Wetland than the East Wetland and greater opportunity for removal of excess nutrients. Although the retention time within the wetland was shorter than originally intended, as discussed further in the site description, a large-scale reconfiguration would not be necessary to meet standards for nitrate, nitrite, or TSS. However, modifications to the current operation and configuration should be considered in order to improve the pH, DO, and ammonia in the system. These parameters should continue to be monitored since they were measured at levels of concern over the course of the study.

10.3.2 PRESENCE AND REMOVAL OF PESTICIDES AND PPCPS

The presence or absence of specific micropollutants is partly attributable to the residence time within septic tanks prior to entry into the sewage lagoons. While photodegradation cannot occur in septic tanks, other degradative processes such as anaerobic microbially-mediated biotransformation likely do occur. Consequently, relatively labile compounds such as naproxen and diclofenac [48] were probably degraded to some extent, and possibly below detection limits prior to arriving at the lagoons. Sorption of analytes to septic tank particulates is also likely. The two sulfonamides are photolabile, so photolysis within the sewage lagoon could also have contributed to the resulting non-detection in the majority of samples. On those occasions where sulfamethoxazole or sulfapyridine were detected, it may have been a result of light attenuation and shielding due to turbidity and dissolved organic carbon (DOC) within the lagoon cells [49]. Atrazine and 2,4-D were measured at very low concentrations (typically <10 ng/L). Since only 2,4-D has been reported as applied agriculturally (at very low total loadings) in the municipality [50], the presence of atrazine was possibly due to use on private residential properties.

All detected and quantifiable micropollutants were measured in the ng/L range in samples from the Grand Marais lagoon and treatment wetland. There was a distinct spike in concentrations of micropollutants downstream of the lagoon during discharge and a subsequent reduction in concentrations with time. However, concentrations for some of the compounds, including carbamazepine and gemfibrozil, remained well above pre-discharge levels as of August 1 (nearly a week post-release), indicating that with the cease in flow from the lagoon, there is likely residual wastewater stagnating within the wetland. It is possible that some changes in concentration may be the result of water evaporation or addition; however, concentration and dilution effects would affect all analytes equally, which was not observed. While no measurements of pesticides and PPCPs occurred in winter, we note that these shallow wetland and stream systems are predominantly or completely frozen over the winter. This would presumably result in no removal of analytes by either microbial activity or photodegradation (i.e., light penetration would be prevented almost completely by ice cover and would be of low intensity in any event) until spring melt.

A hazard assessment was conducted using the maximum concentration of each compound measured in the environment and comparing it to toxicity thresholds for aquatic plants, invertebrates, and fish (Table 2). While none of the calculated HQs surpassed a value of unity, those for gemfibrozil and sulfamethoxazole approached the threshold of concern for fish and aquatic plants, respectively, so these compounds might warrant more regular monitoring.

There was significant removal of atrazine and carbamazepine by the treatment wetland, but the wetland did not significantly or consistently attenuate concentrations of 2,4-D, gemfibrozil, or sulfamethoxazole. Due to the very infrequent measurement of sulfapyridine above the limits of detection or quantification, it was not possible to determine the impact of the wetland on this compound. In general, concentrations of these compounds decreased from upstream to downstream, lagoon to outlet, but there was no evidence for significant elimination within the wetland itself.

In previous studies, removal of atrazine within wetlands was dependent upon retention time [51]. Kadlec and Hey [51] reported between 25 and 95% removal of atrazine in different wetland cells after 3 to 4 weeks of retention time. Similarly, Kao et al. [52] observed up to 99% removal of atrazine within 15 days in anaerobic cells spiked with sucrose media, but less than 9% removal in control wetlands that were not inoculated with media or a nitrogen source. While removal of atrazine from wastewater can be quite variable and very dependent upon the specific substrates and characteristics of the wetland, the results from the current study suggest that the Grand Marais wetland conditions are conducive to removal of atrazine.

Previous studies have reported relatively effective removal of carbamazepine, with 51% removal of carbamazepine via treatment in a forested wetland for 27 days, and up to 80% removal of carbamazepine in *Typha*-inhabited freshwater wetlands over the course of 6 days [1,21]. These results agree with those of the current study where lagoon discharge was treated in a *Typha*-dominated wetland with a residence time of approximately 20 hours. While carbamazepine is relatively persistent, it may be removed to some extent by sorption to suspended particles and uptake by plants, including biotransformation by *Typha* spp. [53,54]. That having been said, sorption is unlikely to be a significant removal process for the analytes that were consistently detected. For example, over 99% of carbamazepine is expected to be in the dissolved phase, given the octanol-water partition coefficient of $10^{2.45}$[55] of the neutral species (predominant at our observed pH values, Table 1) and the maximum observed suspended matter in the lagoon/wetland system (TSS of 29 mg/L, Table 1), assuming all of this matter is organic carbon. While polar organic chemicals can also sorb by other mechanisms, e.g., ion exchange, the low particulate levels observed preclude the likelihood that sorption to such is a major attenuation process, at least in this system.

Unlike the current study, Conkle et al. [1] noted >90% removal of sulfonamides and 95% removal of gemfibrozil, however, the differences may have been a result of the 27 day retention period. In a comparative study, treatment wetlands were found to be ineffective for removal of sulfamethoxazole [18], so removal of this class of PPCPs may be site-specific. Microbial degradation of gemfibrozil has been observed to be relatively rapid in groundwater conditions [56], and in the current study, there was a general pattern of reduction in concentration with passage through the wetland. The lack of statistical significance might be due to the small number of samples collected at the Channel site or the relatively low concentrations found following release from the lagoon.

10.3.3 PRESENCE AND REMOVAL OF ARGS

Similar abundances of sulfonamide resistance genes were measured in a previous study of a river impacted by both urban and agricultural impacts [57]. Sulfonamides are used in both human and veterinary medicine and target the enzyme dihydropteroate synthase (DHPS), which is part of the folic acid pathway [57]. A previous study reported normalized abundances of sulfonamide resistance genes between 0.02 and 7.7% [12], which agrees with the findings in the Grand Marais system (~0.5%). The sulfonamide resistance genes assessed in the current study (*sul*-I, *sul*-II, and *sul*-III) were measured at relatively high concentrations compared to other ARGs. However, since there was no significant relationship between concentrations of sulfonamides and abundances of sulfonamide resistance genes (Figure 4), presence of these genes within the wetland are probably well established as a result of repeated inputs into the system, both past and present. Concentrations of other types of antibiotics (e.g. tetracyclines, beta-lactams) were not measured, so it is unclear whether there was any cross-resistance within the system as a result of the presence of those specific compounds.

Tetracyline resistance genes (e.g. *tet*(M), *tet*(O), *tet*(Q), and *tet*(W)) have been investigated in other studies due to their common transmission in the environment [58] and these ARGs had relatively great abundances in the current study. Smith et al. [58] measured abundances of ARGs in cattle feedlot lagoons and reported concentrations of tetracycline resistance genes ranging from approximately 10^4 to 10^6 copies per mL, which is within 1 or 2 orders of magnitude of the concentrations measured in the current study. The abundances of tetracycline resistance genes measured by Pei et al. [57] were 2 to 3 orders of magnitude less than those measured in the current study. Some of the differences in abundances may be due to sampling in sediments by Pei et al. [57] rather than in water, as in the current study.

There were no obvious trends when upstream (i.e. lagoon) and downstream (i.e. output) abundances of ARGs were compared. The only ARG for which the relative abundance was significantly less at the output than in the treatment wetland was bla_{SHV}. There may have been some removal of microbes bearing this gene in the wetland, but none of the other ARGs were significantly reduced by treatment with the wetland. Previous studies with full-scale and bench-scale wetlands have demonstrated significant removal of bacteria from wastewater, resulting in an approximate reduction of two orders of magnitude or up to 99% of bacteria [40,59]. However, Vacca et al. [59] noted that removal efficiency was highly dependent upon the operation conditions of the wetland, as well as the presence of plants. Removal of bacteria from the Grand Marais treatment wetland likely occurred via a combination of filtering by those plants that were present and sedimentation since DO levels were insufficient in many sites to promote predation by micro-invertebrates [41,59].

With the qPCR method of quantifying abundances of genes within samples from a system, genes from both living and dead bacteria are included so the results may not necessarily represent the true proportion of living bacteria that might be resistant to antibiotics [57]. This should be taken into consideration when quantificatiying of abundances of ARGs within the system. While the Grand Marais treatment wetland appeared to remove bacteria in general, there was no indication that there is any targeted removal of ARGs in the wetland. As a result, the Grand Marais treatment wetland does not appear to be an optimal system for removal of ARGs in its current operational state.

10.4 MATERIALS AND METHODS

10.4.1 STUDY AREA

The Grand Marais treatment wetland (50° 31' N and 96° 35' W) is located in the Rural Municipality of St. Clements, near Grand Marais, MB, and Lake Winnipeg (Figure 1). The wetland receives rural wastewater from the secondary lagoon of a two-lagoon system located directly to the south. Each lagoon is approximately 134 m by 134 m and 2.3 m in depth, with a total storage volume of 29,400 m^3 and licensing to allow up to 1.5 m of liquid within the lagoon cells [60]. There are no direct sewage lines into the lagoon facility, so sewage is aged for an unknown length of time in septic tanks before hauling by septic trucks to the lagoon. Consequently, retention time within the primary lagoon cell is also not well defined. While time within the secondary cell is better known, understanding the residence times in the lagoons was not central to this study since the wetland performance was the main area of focus, though determining this would help to better understand inter-year variability. Prior to the 2012 release, the last release event was July 2011, meaning some waste had aged a maximum of approximately one year in the secondary lagoon.

The treatment wetland is composed of a 0.7 km long wetland channel from the lagoon to the five channel "rows"; the rows collect discharged lagoon water from the channel and direct it through the wetland. The five rows were intended to achieve a 'snaking' configuration whereby water would enter the wetland at a single point and exit after passing through all of the rows. The wetland was designed to retain water at a depth of 15 to 30 cm throughout the year. Prior to release, the wetland contained water, which would have been inputted from snow melt, precipitation, and remaining effluent from the previous year. In reality, the residence time in the wetland is likely much shorter than originally anticipated (five to ten days). This is due to water entering the wetland via all of the rows and flowing directly through to Marais Creek as a result of the loss of the discrete rows since construction in 1996, and a lack of sufficient hydraulic head to maintain flow at the designed hydraulic residence time. Treated wastewater from the wetland ultimately flows into Lake Winnipeg. Lagoon water is released into the treatment wetland one or two times per year (i.e., summer, normally June or July, and fall, normally October) depending on lagoon capacity. This summer and possible fall release is typical of most lagoon systems in Manitoba [4]. The volume, frequency, and timing of releases have varied over recent years because the size of the primary lagoon cell has increased.

10.4.2 STUDY SITES

Sampling was performed both before and after lagoon release in 2012. There were a total of six sampling sites in the wetland, as well as one site in the secondary treatment lagoon (Figure 1). The six sites were selected at different locations within the treatment wetland between the influent entry point and the outlet into the surrounding water. The site names from upstream to downstream were as follows: Lagoon, Release, Mid-Channel, Channel, East Wetland, West Wetland, and Outlet.

The Release and Mid-Channel sites were dominated by submergent plants, as well as *Lemna* spp., and had water depths of ~1 m. Emergent species, particularly *Typha* spp., and some small bushes dominated the East Wetland and West Wetland sites. West Wetland had a water depth of about 40 cm while East Wetland was about 60 cm deep. In the deeper areas of both wetland sites, *Lemna* spp. and several submergent species were present where the wetland water levels are sustained during dry years [60]. The Outlet site was relatively deep (~1-1.5 m deep, depending upon precipitation and evaporation) and wide (2 m wide at culvert) compared to the other sites thus resulting in greater flow. No submergent or emergent wetland plant species were present at the Outlet, but there were grasses and other terrestrial vegetation growing along the creek bank. The hydrology of Marais Creek (which receives flow from the Outlet) is not defined due to a lack of gauging stations, but it is ~3 m wide and discharge of the creek has been measured at 0.06 m^3/s [60].

10.4.3 GENERAL WATER QUALITY PARAMETERS

General water quality and physico-chemical parameters (dissolved oxygen (DO), conductivity, chlorophyll-a, pH, and water temperature) were measured during each sampling event using a YSI 6600 Multi Parameter Water Quality Meter sonde (YSI Inc., Yellow Springs, OH).

10.4.4 SAMPLE COLLECTION

Grab samples for nutrient analyses, total suspended solids (TSS), ARGs, and PPCPs were collected on May 22, June 15, July 16, July 23, and August 1. All sample types were collected on each sample day with the exception of: August 1, where samples were only taken for PPCP analysis and ARGs, and May 22, where no antibiotic resistance genes samples were taken. Summer release from the lagoons into the treatment wetland occurred from July 11 to 24, 2012. Prior to release (May 22 and June 15), samples were taken in the lagoon and at the Outlet site, and during and after release (July 16 and 23, and August 1) samples were taken in the treatment wetland.

Samples were collected using sterile 500 mL polyethylene bottles and 4 L amber glass bottles, as required for the analytical procedures. Each bottle and cap was rinsed three times with sample water and the rinsate was discarded downstream from the sampling location. The bottle was then lowered into the water to a depth of approximately 30 cm below the water surface, filled, and capped underwater with care taken to ensure no headspace was left in the bottle. Extra sample bottles filled with nanopure (18 MΩ cm) Milli-Q water (Millipore Corporation, Billerica, MA) were opened at the sampling sites to serve as field blanks. During the wastewater release event, all equipment and the exteriors of sample bottles were disinfected after contact with wetland water using either isopropanol or bleach. Following collection, samples were stored at 4°C for up to 24 h for ARG samples and for 24-48 h prior to analysis of nutrients or extraction by solid phase extraction (SPE) prior to further analytical analysis for PPCPs. Extracted samples were stored at -20°C for no more than 6 weeks prior to analysis by LC/MS [61].

In addition to grab samples, Polar Organic Chemical Integrative Samplers (POCIS) (Environmental Sampling Technologies, St. Joseph, MO) were used for continuous time-weighted-average passive sampling of pharmaceuticals, as described in detail previously (refer to [4]). POCIS samplers were deployed at the lagoon and wetland outlet sites prior to release in 2012 (from May 22 to June 15), and at five wetland sites during release in 2012 (from July 11 to July 25). Samplers were prepared prior to deployment as described by Carlson et al. [4] and transported to each site in pre-cleaned containers filled with Milli-Q water. They were then suspended near the bottom of the river, wetland, or lagoon using aircraft cable tethered to rebar stakes. A triplicate set of POCIS samplers was deployed in each cage per sampling location for a 2-4 week period. After collection, samplers were rinsed with Milli-Q water, wrapped in foil that had been pre-ashed at 450°C, transported on ice, and frozen at -20°C for up to 2 months prior to extraction.

For extracted SPE samples and collected POCIS samples, minimal losses have been previously observed for the compounds of interest during frozen storage for 2-3 months (<7%) and up to 20 months (<20%) [61]. Therefore, any losses incurred during the storage period were deemed to be negligible and thus, no corrections were required to account for sample losses between collection and analysis.

10.4.5 NUTRIENT AND TSS ANALYSES

Concentrations of nitrate + nitrite, total ammonia + ammonium, and total phosphorus were measured in the water samples. All nutrient analyses were performed by ALS Laboratory Group Analytical Chemistry and Testing Services (Winnipeg, MB), or in-house. Concentrations of nitrogen species were determined at ALS by flow injection analysis (Lachat Instruments, Loveland, CO), as per the manufacturer's standard methods. The limits of detection (LOD) for ammonia and nitrate + nitrite were 0.050 mg/L and 0.010 mg/L, respectively. Total reactive phosphorus was measured in-house with a limit of quantification (LOQ) of 0.010 mg-PO_4^{3-}-P/L. Concentrations of phosphorus species were measured according to standard methods [62]. Total suspended solids (TSS) were quantified according to a modified procedure based on *Standard Methods for the Examination of Water and Wastewater* [62].

10.4.6 PESTICIDES AND PPCP ANALYSES

10.4.6.1 ANALYTICAL STANDARDS

A number of pharmaceutical classes were monitored, including estrogenic compounds, beta-blockers, antibacterial agents, antidepressants, NSAIDs, antibiotics, and lipid regulators. The specific compounds were selected due to their prevalence and/or persistence in the environment, based on published literature [63]. Analyses were conducting using analytical standards for thirty-nine pharmaceutical compounds and pesticides, with compounds and sources described in detail by Carlson et al. [4]. Tylosin and erythromycin standards were 97% and 95% pure, respectively, and all other chemicals were >98% purity. Stable isotope standards were >99% isotopically pure. Isotope sources are found in Carlson et al. [4]. A full list of the compounds and their LOQs can be found in Additional file 1: Table S1 of the online Supplemental Information.

10.4.6.2 SAMPLE EXTRACTION

Grab samples from the lagoon and wetland were processed by solid phase extraction (SPE). Samples were sub-sampled into triplicate 500 mL samples (May 22, 2012) or 250 mL samples (all other dates), prior to filtration through 0.45 μm Metricel membrane filters (Pall Life Sciences, Mississauga, ON). A 25 ng aliquot of internal standard was added to each sample prior to extraction by 3 cc/60 mg OASIS™ HLB cartridges (Waters Corporation, Milford, MA). Samples were pre-conditioned with 2 mL of methanol, then 2 mL of water, and drawn through the cartridges at <5 mL/min. Cartridges were eluted with 3 mL of methanol at 0.5 mL/min. Extracts were evaporated under a stream of nitrogen at 40°C, reconstituted in 0.5 mL of 10:90 methanol:water, and filtered using a 0.22 μm polytetrafluoroethylene syringe filter (Restek Corporation, Bellefonte, PA). The final extracted volume was stored in darkness at 4°C for no longer than one week prior to analysis. One laboratory blank containing only Milli-Q water and internal standards and one field blank were extracted for each set of samples extracted by SPE.

POCIS samples were extracted by a similar method. Samplers were placed in Milli-Q water for 15 min to wet the HLB phase then were extracted in a 60 mL glass clean-up column containing 3-5 g of anhydrous sodium sulfate (Sigma, pre-dried at 450°C). Using 25-35 mL of methanol, individual POCIS sorbent was washed into the column and 50 ng of each internal standard was added to the solution. The extract was gravity-drained into a round bottom flask, and rotary-evaporated at 47-52°C to ca. 5 mL, then dried under a slow stream of nitrogen at 40°C. Samples were reconstituted in 0.5 mL of 10:90 methanol:water and filtered through a 0.22 μm syringe filter, then stored at 4°C for a maximum of one week before analysis. One laboratory blank POCIS, containing only the internal standards, and one field blank were extracted for each set of POCIS samplers.

10.4.6.3 INSTRUMENTAL ANALYSIS

Concentrations of organic micropollutants were measured by liquid chromatography coupled with tandem mass spectrometry (LC/MS/MS). The standards and HPLC mobile phases were prepared using Milli-Q water and HPLC grade methanol (Fisher Scientific, Ottawa, ON) and buffered with 10 mM ammonium acetate (Sigma Aldrich, St. Louis, MO) or 90% formic acid (Fisher Scientific). Stock solutions of all micropollutants were prepared in HPLC grade methanol (Fisher Scientific). Details of the LC/MS/MS systems and their specifications have been described in detail previously [4].

External calibrations were performed using standards over a concentration range of 2-500 μg/L. Analytes were quantified using isotope dilution when possible, or via internal standardization [4]. Extraction efficiencies from SPE and POCIS extracts were 40-100%, but after correction with internal standards, based on spike-and-recovery experiments, efficiencies were 90-110% (data not shown). Relative standard deviations (RSDs) were <20% for triplicates from POCIS extractions and <8% for triplicates from SPE extractions. Concentrations of individual compounds were calculated using literature values for standard POCIS sampling rates [4]. In cases where these were unavailable, such as for diazinon, an average sampling rate for a suite of twenty-nine other pesticides and pharmaceuticals was used [63].

10.4.7 ANTIBIOTIC RESISTANCE GENES

10.4.7.1 SAMPLE PREPARATION

Prior to sampling, 500 mL polypropylene bottles (Chromatographic Specialties Inc., Brockville, ON) were autoclaved at 121°C for 2 h and capped until time of sampling. Samples for ARGs were collected as described above and stored for no more than 24 h at 4°C before extraction. Each ARG sample was filtered using a sterile, disposable Nalgene cup with a pre-installed 0.2 μm filter (Thermo Fisher Scientific Inc., Waltham, MA). The filter was removed using flame-sterilized forceps, folded, and placed into a 1.5 mL polypropylene centrifuge tube. The centrifuge tube was stored frozen at -20°C, and shipped on ice to the University of Strathclyde (Glasgow, UK) for analysis.

10.4.7.2 DNA EXTRACTION

A PowerSoil DNA Isolation Kit (MoBio Laboratories Inc., Carlsbad, CA) was used for DNA extraction. Filters were digested in a buffered solution with sodium dodecyl sulfate (SDS), which was provided by the kit. Cell disruption was achieved by a FastPrep24 instrument run twice for 20 s at a setting of 6.0. The remaining chemical precipitations and centrifugation procedures followed the manufacturer's protocols. The DNA was eluted with molecular-grade DNase- and RNase-free water and stored at -80°C until further analysis.

10.4.8 QUANTITATIVE PCR

Abundances of 16S rRNA and ten ARGs were quantified by quantitative PCR (qPCR) using the Bio-Rad SsoFast™ EvaGreen® reagent system (Bio-Rad Laboratories Ltd., Mississauga, ON). The genes of interest were: *sul*-I, *sul*-II, *sul*-III (sulfonamide resistance genes), a series of multiplex primers for tetracycline resistance ([64], Additional file 1: Table S3), bla_{CTX}, bla_{TEM}, bla_{SHV} (beta-lactam resistance genes), and 16S-rRNA (a surrogate measure of total bacteria). A reaction with total volume of 10 μL was set up by adding 1 μL of DNA to 5 μL of SsoFast reagent and appropriate primers (from [65]) at 500 nM concentrations, and topping up with molecular-grade water. The Bio-Rad iQ5 was run for 2 min at 95°C for DNA denaturation, followed by 40 cycles at 95°C for 5 s, annealing temperature for 10 s (Additional file 1: Table S3), and 72°C for 10 s for DNA elongation. Reactions were monitored continuously by tracking the intensity of fluorescence.

Serially diluted plasmid DNA of known quantity was used for reaction standards and run in all reactions. Molecular-grade water was used as a reaction negative control. All standards and blanks were run according to the same procedures as the samples. For quality control purposes, a portion of the samples were selected at random and spiked with standards to assess reaction efficiencies. In addition, post-analytical melt curves from 55°C to 95°C were used to verify reaction quality. Abundances of genes are presented as log-transformed values, and were normalized to 16S-rRNA values to represent resistance per total bacteria.

10.4.9 HAZARD ASSESSMENT

Hazard quotients (HQs) were calculated for each micropollutant of interest using standard tests and endpoints for aquatic toxicity assays, specifically those for primary producers, invertebrates, and fish. Briefly, estimates of effective concentrations (EC50) or lethal concentrations (LC50) were obtained from the appropriate literature. A predicted 'no effect concentration' (PNEC) was estimated for each target compound by dividing the lowest EC50 or LC50 by an uncertainty factor of 1000 [66]. The greatest measured environmental concentration (MEC) was then divided by the PNEC to obtain the HQ. Quotients less than 1 were considered unlikely to pose a concern, while those greater than 1 were considered to be of possible concern [67].

10.4.10 STATISTICAL METHODS

The experimental unit used was the individual sample or subsample and data is presented as mean ± standard deviation (SD) unless otherwise indicated. All analyses were conducted using SigmaStat (version 3.5, Systat Software, Inc.). Statistical differences between concentrations of pharmaceuticals at each sampling location, as measured by SPE and POCIS, were determined by two-way ANOVA tests followed by Holm-Sidak post-hoc tests where either raw or transformed data met the assumptions of normality and equality of variance. Concentrations of pharmaceuticals upstream (Channel) and downstream (Outlet) of the treatment wetland were compared using Student's t-tests or Mann-Whitney tests.

Abundances of ARGs were standardized relative to abundance of 16S, whereby relative abundance of a particular ARG was equal to 'log (ARG/16S)'. The relative abundances were then compared by two-way ANOVA tests followed by Holm-Sidak post-hoc tests where log-transformed data met the assumptions of normality and equality of variance. Where data did not meet the assumption of normality, Kruskal-Wallis ANOVA by Ranks tests were used and followed by Dunn's post-hoc tests. Differences were considered significant at $p<0.05$.

10.5 CONCLUSIONS

In the current study, there was a clear nutrient and micropollutant pulse into the treatment wetland as a result of lagoon release. The Grand Marais treatment wetland removed nutrients, suspended solids, and several pharmaceutical compounds. However, in its current configuration, it was not an effective treatment for most of the micropollutants that were quantifiable within the system or for removal of ARGs. Micropollutants were degraded with time and movement through the system and there was some reduction in bacterial counts from upstream to downstream. However, our results suggest that treatment wetlands operating in a manner similar to that of Grand Marais, and found in conditions akin to the Canadian Prairies, may not be optimal approaches for treating wastewater with detectable concentrations of micropollutants. The retention time within the current configuration of the Grand Marais wetland is shorter than originally designed. Therefore, upgrading the system to extend the retention time (e.g. fixing and cleaning out the channels to promote 'snaking') may be required to specifically target micropollutants and ARGs using these types of treatment systems.

REFERENCES

1. Conkle JL, White JR, Metcalfe CD: Reduction of pharmaceutically active compounds by a lagoon wetland wastewater treatment system in southeast Louisiana. Chemosphere 2008, 73:1741-1748.
2. Comeau F, Surette C, Brun GL, Losler R: The occurrence of acidic drugs and caffeine in sewage effluents and receiving waters from three coastal watersheds in Atlantic Canada. Sci Tot Environ 2008, 396:132-46.
3. Fent K, Weston AA, Caminoada D: Ecotoxicology of pharmaceuticals. Aquat Toxicol 2006, 76:122-59.
4. Carlson JC, Anderson JC, Low JE, Cardinal P, MacKenzie SD, Beattie SA, Challis JK, Bennett RJ, Meronek SS, Wilks RPA, Buhay WM, Wong CS, Hanson ML: Presence and hazards of nutrients and emerging organic micropollutants from sewage lagoon discharges into Dead Horse Creek, Manitoba, Canada. Sci Tot Environ 2013, 445-446:64-78.
5. Hijosa-Valsero M, Fink G, Schlüsener MP, Sidrach-Cardona R, Martín-Villacorta J, Ternes T, Bécares E: Removal of antibiotics from urban wastewater by constructed wetland optimization. Chemosphere 2011, 83:713-719.
6. Matamoros V, Arias CA, Nguyen LX, Salvadó V, Brix H: Occurrence and behavior of emerging contaminants in surface water and a restored wetland. Chemosphere 2012, 88:1083-1089.
7. Kummerer K: Antibiotics in the aquatic environment. A review – part II. Chemosphere 2009, 75:435-441.
8. Munir M, Wong K, Xagoraraki I: Release of antibiotic resistant bacteria and genes in the effluent and biosolids of five wastewater utilities in Michigan. Water Res 2010, 45:681-693.

9. Negreanu Y, Pasternak Z, Jurkevitch E, Cytryn E: Impact of treated wastewater irrigation on antibiotic resistance in agricultural soils. Environ Sci Technol 2012, 46:4800-4808.
10. Pruden A, Pei R, Storteboom H, Carlson KH: Antibiotic resistance genes as emerging contaminants: studies in northern Colorado. Environ Sci Technol 2006, 40:7445-7450.
11. Rowan NJ: Defining established and emerging microbial risks in the aquatic environment: current knowledge, implications, and outlooks. Int J Microbiol 2010, 2011:462832.
12. Czekalski N, Berthold T, Caucci S, Egli A, Burgmann H: Increased levels of multiresistant bacteria and resistance genes after wastewater treatment and their dissemination into Lake Geneva. Front Microbiol 2012, 3:106.
13. Graham DW, Olivares-Rieumont S, Knapp CW, Lima L, Werner D, Bowen E: Antibiotic resistance gene abundances associated with waste discharges to the Almendares River near Havana, Cuba. Environ Sci Technol 2011, 45:418-424.
14. Dodd MC: Potential impacts of disinfection processes on elimination and deactivation of antibiotic resistance genes during water and wastewater treatment. J Environ Monitor 2012, 14:1754-1771.
15. Zhang XX, Zhang T, Fang HHP: Antibiotic resistance genes in water environment. Appl Microbiol Biot 2009, 82:397-414.
16. Knapp CW, Lima L, Olivares-Rieumont S, Bowen E, Werner D, Graham DW: Seasonal variations in antibiotic resistance gene transport in the Almendares River, Havana, Cuba. Front Microbiol 2012, 3:396.
17. World Health Organization (WHO): WHO annual report on infectious disease: overcoming antimicrobial resistance. 2000. http://www.who.int/infectious-disease-report/2000/] (Accessed 31/10/12)
18. Breitholtz M, Näslund M, Stråe D, Borg H, Grabic R, Fink J: An evaluation of free water surface wetlands as tertiary sewage water treatment of micropollutants. Ecotox Environ Safe 2012, 78:63-71.
19. Hijosa-Valsero M, Sidrach-Cardona R, Bécares E: Comparison of interannual removal variation of various constructed wetland types. Sci Tot Environ 2012, 430:174-183.
20. Dordio AV, Estêvão Candeias AJ, Pinto AP, Teixeira da Costa C, Palace Carvalho AJ: Preliminary media screening for application in the removal of clofibric acid, carbamazepine, and ibuprofen by SSF-constructed wetlands. Ecol Eng 2009, 35:290-302.
21. Reyes-Contreras C, Hijosa-Valsero M, Sidrach-Cardona R, Bayona JM, Bécares E: Temporal evolution in PPCP removal from urban wastewater by constructed wetlands of different configuration: a medium term study. Chemosphere 2012, 88:161-167.
22. Werker AG, Dougherty JM, McHenry JL, Van Loon WA: Treatment variability for wetland wastewater treatment in cold climates. Ecol Eng 2002, 19:1-11.
23. Manitoba Land Initiative (MLI): Manitoba land initiative – wastewater layer. 2004. http://mli2.gov.mb.ca] (Accessed 27/11/12)
24. Manitoba Water Stewardship: Manitoba water quality standards, objectives, and guidelines. Manitoba: Water Stewardship Report; 2011:72.

25. Ort C, Lawrence MG, Rieckermann J, Joss A: Sampling for pharmaceuticals and personal care products (PPCPs) and illicit drugs in wastewater systems: are your conclusions valid? A critical review. Environ Sci Technol 2010, 44:6024-6035.
26. Belgers JDM, Van Lieverloo RJ, Van der Pas LJT, Van den Brink PJ: Effects of the herbicide 2,4-D on the growth of nine aquatic macrophytes. Aquat Bot 2007, 86:260-268.
27. Martins J, Oliva Teles L, Vasconcelos V: Assays with Daphnia magna and Danio rerio as alert systems in aquatic toxicology. Environ Intern 2007, 33:414-425.
28. Little EE, Archeski RD, Flerov BA, Kozlovskaya VI: Behavioral indicators of sublethal toxicity in rainbow trout. Arch Environ Contam Toxicol 1990, 19:380-385.
29. Teodorović I, Knežević V, Tunić T, Čučak M, Lečić JN, Leovac A, Tumbas II: Myriophyllum aquaticum versus lemna minor: sensitivity and recovery potential after exposure to atrazine. Environ Toxicol Chem 2012, 31:417-426.
30. Phyu YL, Warne M, St J, Lim RP: Toxicity of atrazine and molinate to the cladoceran Daphnia carinata and the effects of river water and bottom sediment on their bioavailability. Arch Environ Con Tox 2004, 46:308-315.
31. Giddings JM, Anderson TA, Hall LW Jr, Hosmer AJ, Kendall RJ, Richards RP, Solomon KR, Williams WM: Atrazine in north American surface waters: a probabilistic aquatic ecological risk assessment. Pensacola, FL: Society of Environmental Toxicology and Chemistry (SETAC); 2005.
32. Cleuvers M: Aquatic ecotoxicity of pharmaceuticals including the assessment of combination effects. Toxicol Lett 2003, 142:185-94.
33. Kim Y, Choi K, Jung J, Park S, Kim P-G, Park J: Aquatic toxicity of acetaminophen, carbamazepine, cimetidine, diltiazem, and six major sulfonamides, and their potential ecological risks in Korea. Environ Int 2007, 33:370-375.
34. El-Bassat RA, Touliabah HE, Harisa GI, Sayegh FAQ: Aquatic toxicity of various pharmaceuticals on some isolated plankton species. Int J Med Sci 2011, 3:170-80.
35. Sanderson H, Johnson DJ, Wilson CJ, Brain RA, Solomon KR: Probabilistic hazard assessment of environmentally occurring pharmaceuticals toxicity to fish, daphnids, and algae by ECOSAR screening. Toxicol Lett 2003, 144:383-395.
36. Isidori M, Lavorgna M, Nardelli A, Pascarella L, Parrella A: Toxic and genotoxic evaluation of six antibiotics on non-target organisms. Sci Tot Environ 2005, 346:87-98.
37. Blaise C, Gagné F, Eullafrroy P, Férard JF: Ecotoxicity of selected pharmaceuticals of urban origin discharged to the Saint-Lawrence River (Quebec, Canada): A review. Braz J Aquat Sci Technol 2006, 10:29-51.
38. EUCAST (European Committee for Antimicrobial Susceptibility Testing): Determination of minimum inhibitory concentrations (MIC) of antibacterial agents by agar dilution. EUCAST Definitive Document, E. Def 3.1. Basel, Switzerland: European Society of Clinical Microbiology and Infectious Diseases; 2000.
39. Knapp CW, Engemann CA, Hanson ML, Keen PL, Hall KJ, Graham DW: Indirect evidence of transposon-mediated selection of antibiotic resistance genes in aquatic systems at low-level oxytetracycline exposures. Environ Sci Technol 2008, 42:5348-5353.
40. Peak N, Knapp CW, Yang RK, Hanfelt MM, Smith MS, Aga DS, Graham DW: Abundance of six tetracycline resistance genes in wastewater lagoons at cattle feedlots with different antibiotic use strategies. Environ Microbiol 2007, 9:143-151.

41. Verhoeven JTA, Meuleman AFM: Wetlands for wastewater treatment: opportunities and limitations. Ecol Eng 1999, 12:5-12.
42. Murkin HR, Stainton MP, Boughen JA, Pollard JB, Titman RD: Nutrient status of wetlands in the Interlake region of Manitoba, Canada. Wetlands 1991, 11:105-122.
43. Canadian Council of Ministers of the Environment: Canadian environmental quality guidelines. 2010. http://ceqg-rcqe.ccme.ca/] (Accessed 10/10/12)
44. Koike I, Hattori A: Denitrification and ammonia formation in anaerobic coastal sediments. Appl Environ Microbiol 1978, 35:278-282.
45. Canadian Minister of Justice: Wastewater systems effluent regulations (SOR/2012-1239. http://laws-lois.justice.gc.ca/PDF/SOR-2012-139.pdf] (Accessed 07/12/12
46. Jones GBJ, Greenall J, Punter E: A preliminary vegetation survey of the Gull Lake wetland areas. Winnipeg MB: Terrestrial Quality Management Section, Manitoba Environment; 1999. Report No. 99-04
47. Gabor TS, Murkin HR, Stainton MP, Boughen JA, Titman RD: Nutrient additions to wetlands in the Interlake region of Manitoba, Canada: effects of a single pulse addition in spring. Hydrobiol 1994, 279–280:497-510.
48. Carballa M, Omil F, Ternes T, Lema JM: Fate of pharmaceutical and personal care products (PPCPs) during anaerobic digestion of sewage sludge. Wat Res 2007, 41:2139-2150.
49. Challis JK, Carlson JC, Friesen KJ, Hanson ML, Wong CS: Aquatic photochemistry of the sulfonamide antibiotic sulfapyridine. Photochem Photobiol A 2013. Submitted
50. Manitoba Agricultural Services Corporation: Manitoba management plus program: pesticide data browser. 2012. http://www.mmpp.com/mmpp.nsf/mmpp_browser_pesticide.html] (Accessed 22/03/12)
51. Kadlec RH, Hey DL: Constructed wetlands for river water quality management. Wat Sci Technol 1994, 29(4):159-168.
52. Kao CM, Wang JY, Wu MJ: Evaluation of atrazine removal processes in a wetland. Wat Sci Technol 2001, 44:539-544.
53. Dordio AV, Belo M, Martins Teixeira D, Palace Carlvalho AJ, Dias CMB, Picó Y, Pinto AP: Evaluation of carbamazepine uptake and metabolism by Typha spp., a plant with potential use in phytoremediation. Biores Technol 2011, 102:7827-7834.
54. Tanoue R, Sato Y, Motoyama M, Nakagawa S, Shinohara R, Nomiyama K: Plant uptake of pharmaceutical chemicals detected in recycled organic manure and reclaimed wastewater. J Agri Food Chem 2012, 60:10203-10211.
55. Barnes KK, Kolpin DW, Furlong ET, Zaugg SD, Meyer MT, Barber LB: A national reconnaissance of pharmaceuticals and other organic wastewater contaminants in the United States - Groundwater. Sci Tot Environ 2008, 402:192-200.
56. Fang Y, Karnjanapiboonwong A, Chase DA, Wang J, Morse AN, Anderson TA: Occurrence, fate, and persistence of gemfibrozil in water and soil. Environ Toxicol Chem 2012, 31:550-555.
57. Pei R, Kim S-C, Carlson KH, Pruden A: Effect of river landscape on the sediment concentrations of antibiotics and corresponding antibiotic resistance genes (ARG). Water Res 2006, 40:2427-2435.
58. Smith MS, Yang RK, Knapp CW, Niu Y, Peak N, Hanfelt MM, Galland JC, Graham DW: Quantification of tetracycline resistance genes in feedlot lagoons by real-time PCR. Appl Environ Microbiol 2004, 70:7372-7377.

59. Vacca G, Wand H, Nikolausz M, Kuschk P, Kästner M: Effects of plants and filter materials on bacteria removal in pilot scale constructed wetlands. Water Res 2005, 39:1361-1373.
60. I.D. Group (Manitoba) Inc: Environment act proposal for: Sewage lagoon R.M. of St. Clements. 1995. File no. 12-037
61. Carlson JC, Challis JK, Hanson ML, Wong CS: Stability of pharmaceuticals and other polar organic chemicals stored on polar organic chemical integrative samplers and solid phase extraction cartridges. Environ Toxicol Chem 2013, 32:337-344.
62. Eaton AD, Franson MAH: Standard methods for the examination of water and wastewater. 21st edition. Washington, DC: American Public Health Association (APHA); 2005.
63. Li HX, Helm PA, Paterson G, Metcalfe CD: The effects of dissolved organic matter and pH on sampling rates for polar organic chemical integrative samplers (POCIS). Chemosphere 2011, 83:271-280.
64. Ng LK, Martin I, Alfa M, Mulvey M: Multiplex PCR for the detection of tetracycline resistant genes. Mol Cell Probes 2001, 15:209-215.
65. Knapp CW, Dolfing J, Ehlert PAI, Graham DW: Evidence of increasing antibiotic resistance gene abundances in archived soils since 1940. Environ Sci Technol 2010, 44:580-587.
66. Waiser MJ, Humphries D, Tumber V, Holm J: Effluent-dominated streams, part II: presence and possible effects of pharmaceuticals and personal care products in wascana creek, Saskatchewan, Canada. Environ Toxicol Chem 2012, 30:508-19.
67. Hanson ML, Solomon KR: Haloacetic acids in the aquatic environment. Part II: ecological risk assessment. Environ Pollut 2004, 130:385-01.

There are several supplemental files that are not available in this version of the article. To view this additional information, please use the citation on the first page of this chapter.

CHAPTER 11

Irrigation with Treated Wastewater: Quantification of Changes in Soil Physical and Chemical Properties

PRADIP ADHIKARI, MANOJ K. SHUKLA, JOHN G. MEXAL, AND DAVID DANIEL

11.1 INTRODUCTION

Southern New Mexico is characterized as semi-arid region where wastewater reclamation and reuse for irrigation has become important part of water resources planning. This has occurred as a result of the increasing fresh water scarcity, high nutrients in wastewater, and the high cost of advanced treatment required for other wastewater uses. United Nations Millennium Development Goal also targets the use of wastewater as irrigation to reduce the water deficit [1]. Certain quality criteria should be met prior to using wastewater for irrigation. Some of the parameters requiring close attention are electrical conductivity (EC), total dissolved solids (TDS), sodium adsorption ratio (SAR), suspended heavy metals and organic matter (OM). Without proper management, wastewater application

Irrigation with Treated Wastewater: Quantification of Changes in Soil Physical and Chemical Properties. Irrigation & Drainage Systems Engineering ***3**,117 (2014), doi: 10.4172/2168-9768.1000117.*

can pose serious risks to human health and the environment [1]. Treatment of urban and industrial wastewater is complex, expensive, and requires energy and technology. The safe disposal of the treated wastewater is also a challenge because the effect of wastewater on the soil and plant environment is complex and depends upon the amount of various elements present in the wastewater. Reuse of effluent could be beneficial especially in areas where water stress is a major concern primarily due to limited water resources, higher water demands and limited economic resources. Wastewater can add nutrients to the soil system stimulating plant growth, increasing plant NO_3^- uptake, and the turnover of soil NO_3^- and denitrification. A major objective of land application systems is to allow the physical, chemical, and biological properties of the soil-plant environment to assimilate wastewater constituents without adversely affecting beneficial soil properties [2]. However, when wastewater is irrigated beyond the assimilation capacity of the soil-plant system, it can provide a source of readily leachable nutrient or contaminant [3]. Waste water can also affect soil physical properties, including bulk density (BD), drainable porosity (d), soil moisture retention and hydraulic conductivity (K_s). Recent study on the same location reported lower K_s and macroporosity in the wastewater irrigated areas than in the unirrigated areas [4]. The levels of dissolved OM and suspended solids in effluent depend on the quality of the raw sewage water and the degree of treatment [5,6] Suspended solids present in effluents accumulate in soil voids and physically block water-conducting pores leading to a sharp decline in soil hydraulic properties [4,5]. The reduction in K_s could be due to the retention of OM during infiltration and the change of pore size distribution as a result of expansion or dispersion of soil particles. Application of wastewater with sodium (Na^+) content to soil increases sodicity, causes clay dispersion, changes pore geometry, and reduces K_s [7,8]. In contrast, [9] found no adverse impact on the hydraulic parameters while applying standard domestic effluents to soil in Israel. Soils in the arid region are generally calcareous with high pH in the upper soil horizons favoring the precipitation of most heavy metals and reduce the risk of groundwater pollution [10]. The primary goal of land application of wastewater is to maximize vegetative cover to increase the capacity of the soil to serve as a sink for wastewater contaminants, minimize salt accumulation in the root zone, and avoid NO_3^- leaching to the groundwater

[11,12]. In this context application of treated wastewater on common arid and semiarid shrubs could be more economical and environmentally beneficial. Soil chemical properties are one of the most researched aspects of wastewater irrigation. Changes due to irrigation vary greatly and are largely dependent on the quality of the irrigation water. However, little work has been conducted on the impact of wastewater irrigation in Chihuahuan desert ecosystem on the native vegetation. An earlier study conducted on part of the West Mesa irrigated site reported that the sprinkler distribution uniformity was low (53.7%) and could have caused the variability in soil chemical and physical properties between canopies and intercanopy areas [11]. In spite of the variability of application, the previous study did not report statistical differences in chemical and physical properties between vegetation canopies and intercanopy areas likely due to low sample size. Similarly, NO_3^- and OM content of wastewater listed by the Environmental Protection Agency (EPA) as a method of recycling nutrients and OM were not addressed in that study. The present study overcomes these limitations of the earlier study and provides a detailed account of the impact of wastewater on physical and chemical properties under different vegetation canopies and intercanopy areas within the entire irrigated site. The objectives of this study were to: (1) determine the influence of lagoon treated wastewater interception by shrub canopies on physical and chemical properties of canopy soil, and (2) compare physical and chemical properties among the canopy and intercanopy areas.

11.2 MATERIALS AND METHODS

11.2.1 EXPERIMENTAL SITE

The West Mesa industrial and municipal wastewater land application facility (West Mesa) is located near Las Cruces, NM(longitude W 106° 54.408' latitude N 32° 15.99', altitude 1298 m). This includes a wastewater treatment plant and a land application system. The untreated industrial and municipal wastewater generated from dairy processing and metal wire fabrication industries is treated in a 1,500 m^3d^{-1} capacity treatment plant, which can discharge 200 m^3d^{-1} of wastewater to the 36-ha study site. Additional details about the study locations could also found in [13,14]. Aerated lagoon effluent application on this site began on February 5, 2002 to the Chihuahuan Desert upland adjacent to the wastewater treatment plant by 1,243 fixed-head sprinklers operated by automated pumps [15]. The treated plots received variable amounts of effluent due to the temporal fluctuations in tenant-generated wastewater and the high evaporation losses from the wastewater lagoons through the peak summer months. During the late summer the application onto the treated site increased usually due to the decreased evaporation and increased tenant's wastewater discharge. From 2006 to 2008, the entire 36-ha received an average of 57.66 cm of water of which 34.68 cm came from the effluent application (Table 1). Total average non stressed ET for mesquite and creosote shrubs was 154.06 cm during 2006-08 and the ratio of total water applied to ET was about 0.37 ± 0.03. Overall, vegetation in the experimental site was water stressed because little or no wastewater was available for application during the summer months when ET demands were high. This area is dominated by woody perennials such as creosote (*Larreatridentata*, (DC) Cov.) and honey mesquite (*Prosopisglandulosa Torr. varglandulosa*) whose percent groundcover in 2002 were approximately 8.7 and 14.4%, respectively (Babcock et al. [11]). The visual observation during the spring and early summer months of 2008 revealed that approximately 80% of the irrigated area was covered with perennial vegetation including, desert daisy (*Bebbiajuncea Benth.*),snakeweed (*Gutierreiza Lag.*), pigweed (*Amaranthus L.*), spiderling (*Boerhavia L.*), sagebrush (*Artemisia L.*), and chinchweed (*Pectis L.*). Coppice dunes occur under mesquite canopies and occasionally under creosote canopies over most of the experimental site. Before the development of coppice dunes the area was level and surface horizons consisted of coarse textured materials [16] that provides better condition for infiltration and leaching of Na^+ and other soluble salts. Soil texture of the coppice dunes and the intercanopy areas varies from sand to light sandy loam with little or no gravel. Soil series identified in and around the West Mesastudy site are Onite (coarseloamy, mixed, superactive, thermic TypicCalciargids), Pintura (Mixed, thermic Typic Torripsamments), Bucklebar (TypicHaplargid), Pajarito (Coarse-loamy, mixed, superactive, thermic TypicHaplocambids), and Bluepoint (Mixed, thermic TypicTorripsamments) [16].

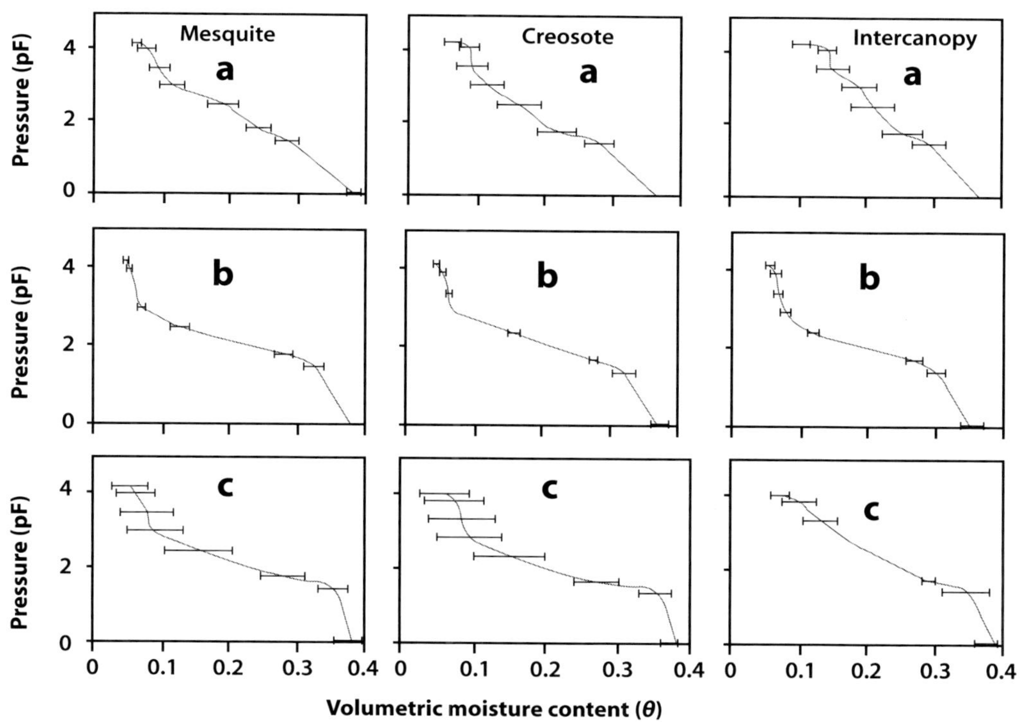

FIGURE 1: Soil moisture release curves of mesquite, creosote, and intercanopy areas at 0-20 cm depth by plot where pF is log of pressure in centimeters (a) irrigated plot-I (b) unirrigated (c) irrigated plot-II during 2007 [1-5].

TABLE 1: Amounts of wastewater, precipitation, and evapotranspiration (ET) during 2006-2008 at West Mesa land application site.

Year	Wastewater	Precipitation	Total water applied	Creosote ET	Mesquite ET	Average crop ET	Deficit
	cm						
2006	17.62	33.93	51.55	170.30	179.83	175.18	123.63
2007	36.79	20.45	57.24	177.66	143.63	158.53	101.29
2008	49.65	14.55	64.20	135.73	121.23	128.48	64.27
Ave.	34.68	22.97	57.66	161.23	148.23	154.06	96.39

11.2.2 SOIL SAMPLING AND ANALYSIS

Three plots were identified for soil sampling: an unirrigated plot, irrigated plot-I, and irrigated plot- II. The soils in unirrigated and irrigated plot-I were classified as Blue point loamy sand whereas in irrigated-II, it was Onite-Pajarito association. Amount of wastewater received was approximately 10 % higher in the Irrigated plot-I than the Irrigated plot-II due to the head differences from the wastewater holding point. Three mesquite and three creosote shrubs were selected randomly in each plot. Shrubs within the irrigated plot-I and II were located on the periphery of the sprinkler uniformity test site. Four sampling points were selected in the center of each canopy (four cardinal directions within the canopy) and three on each intercanopy area. Intact soil cores were taken by a core sampler (19 cm length and 5.5 cm diameter) from each sampling point at 0-20 cm and 20-40 cm depths. Similarly, bulk soil samples were taken by a metal auger (3 cm diameter) from each sampling point at 0-20, 20-40, 40-60, 60-80, 80- 100 and 100-150 cm depths. Thus, a total of 162 core and 486 bulk soil samples were collected from all three plots. Visual observations were made to detect the signs of stress and leaf burn caused by wastewater application. Particle size analysis (PSA) was performed by hydrometer method using air-dried sample < 2 mm [17]. Soil cores were trimmed and the BD was determined by soil core method [18]. Cores were saturated with tap water by slowly raising water level in the trough and K_s was determined by the constant head method [19]. Volumetric moisture content (θ) of each core was determined at 0, 0.003, 0.006 MPa suctions using tension table and 0.03, 0.1, 0.3, 1, 1.5 MPa using pressure plate apparatus [20].The difference in θ at 0 MPa and 0.006 MPa was calculated to estimate drainable porosity (θ_d) or soil macroporosity, the difference in θ at 0.03MPa (field capacity; FC) and 1.5 MPa (wilting point; WP) was used to estimate plant available water content (AWC). The van Genuchten (1980) model was fitted to the measured soil moisture retention [h(θ)] curves to obtain the air entry value ($1/\alpha$),the pore size distribution parameter(λ), and empirical parameters (n and m) using the retention curve (RETC) program of van Genuchten et al., (1991).

$$S_e = \frac{\theta - \theta_r}{\theta_s - \theta_r} = [1 + (\alpha h)^n]^{-m} \ldots . h < 0 \tag{1}$$

$$= \theta_s \ldots \ldots \ldots \ldots \ldots \ldots \ldots \ldots \ldots \ldots h \geq 0$$

Where S_e is the degree of saturation $0 \leq S_e \leq 1$, θs and θ_r are saturated and residual water contents. The RETC uses a non-linear least-square optimization approach to estimate the unknown model parameters and empirical constants affecting the shape of the retention curve. Chemical properties, like EC and pH were determined on 1:2 ratio of soil: water. NO_3^- concentration was measured using auto analyzer [21]. For NO_3^- concentration, 2.5 g of sieved soil sample was mixed with 25 ml of 2N sodium chloride (KCL) solution in 125 ml Erlenmeyer flask and shaken for one hour using mechanical shaker. The solution was filtered through Whatman no. 2 filter paper before analysis. The extract was used to analyze the amount of nitrate-nitrogen (NO_3^--N) through the Technicon auto analyzer [22]. The amount of NO_3^- was calculated from NO_3^--N. For Cl- analysis, about 5 g of soil and 25 ml of DI water was mixed in a centrifuge tube, shaken for an hour in a mechanical shaker, and centrifuged for15 minutes at 2000 rpm speed. A mixture consisting of 5-ml of final soil solution, 35 ml of DI water and 2 ml of nitric acid was titrated with the 0.1 N silver nitrate by 798 MPT Titrinotitrator. Only one sample was analyzed for OM, SAR, ESP and Na^+ from unirrigated plot because no wastewater was applied to this plot and an earlier study showed no significant differences in soil chemical properties between 2002 and 2006 for the unirrigated plot. In addition, 126 composite soil samples were analyzed for pH, EC, Cl^-, N_3^--, OM, ESP and SAR (Harris Lab, Columbus, Nebraska). Plant samples of mesquite, creosote and perennial weeds from intercanopy areas were collected from both irrigated and unirrigated plots. Each sample was washed, oven dried at 60°C, ground and analyzed for Na^+ and NO_3^- (Harris Lab, Columbus, Nebraska). Chemical properties including heavy metal concentrations of wastewater influent and effluent from 2006-2008 were provided by the City of Las Cruces, Water Quality Lab. All the wastewater analysis was conducted in the Continental Analytical Service Inc., Salina Kansas, following the United States Environmental Protection Agency (USEPA) guidelines. Sprinkler uniformity tests were conducted to determine the effectiveness of sprinklers to discharge the wastewater uniformly. The sprinklers in irrigated I were installed on a trapezoidal grid rather than on a square grid. The spacing of sprinklers was 11 m by 12.7 m and 11.5 m by 14.2 m in irrigated I and 11.9 m by 12.6 m and 12.0 m by 11.4 m in the irrigated II. Uniformity of wastewater application with sprinkler irrigation system was calculated by Christiansen's coefficient (Cu) (Christiansen, 1942) using the American Society of Agricultural Engineers standard #3301 (ASAE Standards, 1993).

$$Cu = 100\,(1.00 - \Sigma|dv|/nX) \tag{2}$$

Where Dv = deviations of volume of water collected in the catch funnel from the mean catch volume; n= number of catch funnels; X =mean volume collected in catch funnel.

11.2.3 STATISTICAL ANALYSIS

To assess differences in soil chemical and physical properties among plots, one-way analysis of variance (ANOVA) with contrasts was performed.

Similarly, ANOVA was also performed to assess differences in soil chemical and physical properties between the canopies within the plots. The SAS General Linear Model Procedure (Proc GLM) was used to assess plot, vegetation and plot x vegetation interaction due to the application of wastewater for soil physical and chemical properties at 0-20 and 20-40 cm depths. All statistical analyses were performed using SAS® software version 9.1.3 (SAS Institute Inc., 2002-2003). All statistical analyses were performed for a significance level of $P \leq 0.05$.

11.3 RESULTS AND DISCUSSION

11.3.1 WASTEWATER QUALITY AND APPLICATION

Evaporation losses at the experimental site ranged from 50 to 90% similar to the typical values reported for arid regions, which can result in 2 to 20 fold increases in soluble salt concentrations [23]. Water quality for the irrigation water was based on the SAR, total salinity, EC, and specific ion concentrations. Analysis of the wastewater showed higher amounts of TDS, Cl-, Na+, EC, and SAR in the effluent than influent primarily due to high rate of evaporation in the holding ponds (Table 2). Wastewater generated from meat and dairy processing industry is reported to contain elevated concentrations of Na^+, with SAR ranging between 4 and 50 [24]. The average SAR and Na^+ concentration of applied wastewater was 37.16 ± 2.48 and 1122.36 ± 87.39 mg L^{-1}, respectively. Irrigation with water having high Na^+ concentrations is reported to cause an accumulation of exchangeable Na^+ on soil colloids and affect the survival of vegetation in the long run [25]. Visual observations during field visits also indicated sign of stress e.g., leaf burn in creosote and wilting in the mesquite possibly due to the application of sodic wastewater. The EC tolerance limit for mesquite is 9.36 dS m^{-1} [26] and for creosote is 7.51 dS m^{-1} [27]. The highest measured wastewater EC from 2006 -2008 was 5.64 dS m^{-1}. Thus, with regard to EC of wastewater, there is no immediate danger for the sustainability of native shrubs in the area. However, shallow rooted annual and perennial weed mustard may be threatened due to higher SAR irrigation water (37.16 ± 2.48).

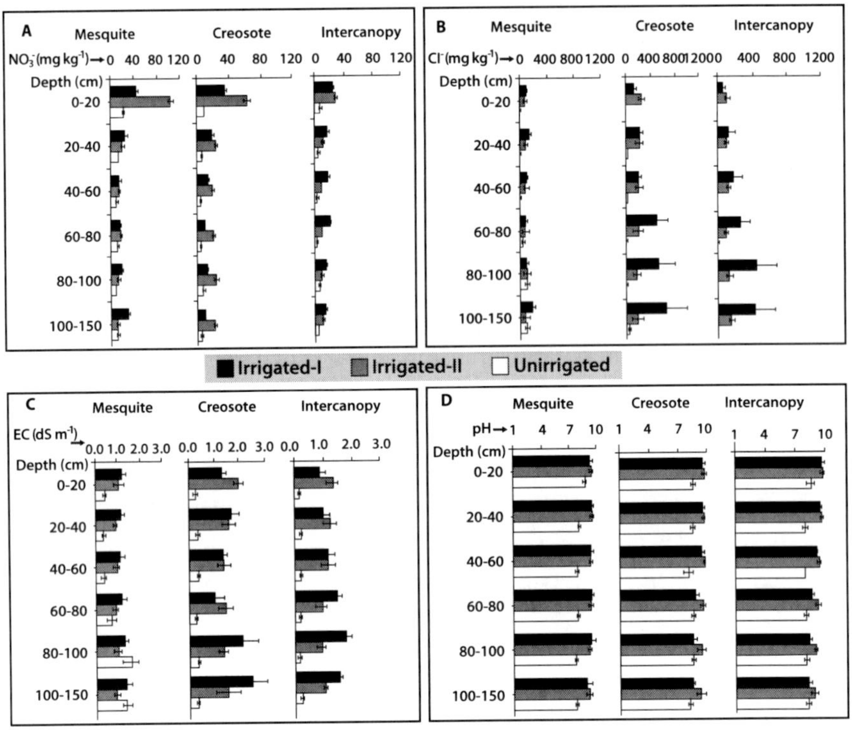

FIGURE 2: Concentration of (A) nitrate, NO3;(B) chloride, Cl-; (C) electrical conductivity, EC and (D) pH in three plots under the canopies of mesquite, creosote and intercanopy area during 2007 [1-5].

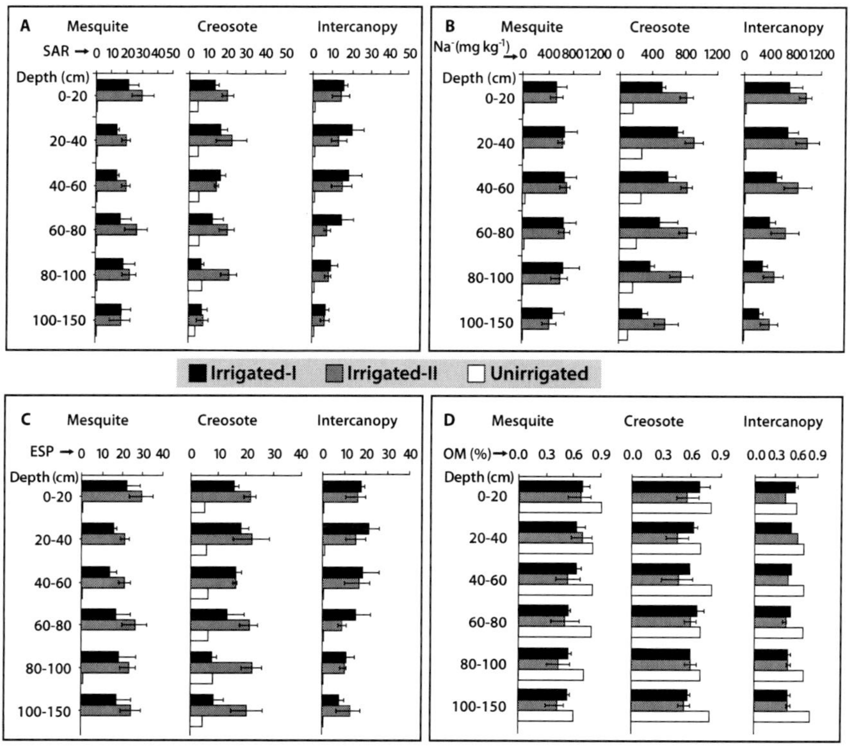

FIGURE 3: Concentration of (A) sodium adsorption ratio, SAR; (B) sodium, Na+; (C) exchangeable sodium percentage, ESP; and (D) organic matter, OM, in three plots under the canopies of mesquite, creosote and intercanopy area during 2007 [1-5].

TABLE 2: Influent and effluent chemical values means and standard errors from 2006-2008.

	TDS (mg L^{-1})		Chloride (mg L^{-1})	
Year	Influent	Effluent	Influent	Effluent
2006	1866.66 ± 450.41	3160.00 ± 900.68	320.66 ± 43.07	528.00 ± 169.00
2007	810.00 ± 28.86	3455.00 ± 293.72	247.00 ± 28.61	633.25 ± 67.71
2008	982.50 ± 221.74	3607.50 ± 455.60	252.50 ± 73.86	855.00 ± 127.19
Average	1219.72 ± 223.67	34075.50 ± 550.00	273.39 ± 145.54	672.08 ± 121.30
	Nitrate (mg L^{-1})		Sodium (mg L^{-1})	
2006	1.47 ± 0.47	13.49 ± 13.04	332.00 ± 83.57	1175.33 ± 149.69
2007	1.61 ± 0.97	1.19 ± 0.32	215.00 ± 47.61	1094.00 ± 17.87
2008	0.66 ±0.22	0.36 ± 0.10	184.75 ± 48.58	1097.75±94.63
Average	1.24 ± 0.55	5.01 ± 4.48	389.40 ± 64.76	1122.36 ± 87.39
	EC (dS m^{-1})		SAR	
2006	2.91 ± 1.21	4.93 ± 1.40	7.55 ± 1.91	41.47 ± 4.33
2007	1.26 ± 0.04	5.39 ± 0.45	5.46 ± 1.08	36.89 ± 1.60
2008	1.54 ± 0.34	5.64 ± 0.72	4.13 ± 0.89	33.14 ± 1.52
Average	1.90 ± 0.53	5.32 ± 2.57	5.71 ± 1.29	37.16 ± 2.48

Source-City of Las Cruces, Water Quality Lab

11.3.2 WATER TRANSPORT AND RETENTION PARAMETERS

There are several attributes of wastewater, such as SAR, EC and OM content that can affect the soil hydraulic properties. Soil porosity can change due to the blockage of the inter-soil spaces by suspended materials [6] and can also impact soil hydraulic conductivity [28,29]. A one -way ANOVA contrast detected significant difference for Ks and d between irrigated -I and unirrigated plots at 0-20 cm depth (Table 3).The plot and vegetation interactions were significant for K_s at both depths and plot x vegetation interaction at 0-20 cm depth (Table 4). The average K_s of canopies and intercanopy areas at 0-20 cm depth in the unirrgated plot was 15.18 ± 1.50 cm h^{-1}, irrigated plot -I was 11.16 ± 1.42cm h^{-1} and in irrigated plot-II was 12.33 ± 0.80cm h^{-1} (Table 4). The K_s was higher under mesquite canopies

(18.20 ± 1.29 cm h^{-1}) followed by creosote (14.20 ± 0.78 cm h^{-1}) and intercanopy areas (4.80 ± 0.34 cm h^{-1}) in all three plots (Table 3). Higher Ks under mesquite canopies than intercanopy areas and creosote canopies were likely due to higher sand content and higher amounts of macrospores associated with coppice dunes. In addition, differences in K_s between vegetation canopies might be due to the differences in morphological structure of the vegetation and differences in the interception of wastewater by vegetation canopies. A white coating on the soil surface was observed only in the intercanopy areas, which was due to the reprecipitation of salt due to evaporation and could have caused reductions in the K_s at the intercanopy areas. The water content at FC and AWC are reported to increase due to the application of wastewater [30]. In this study, significant differences for water content at FC were detected between irrigated and unirrigated plots at both depths; some differences were observed among vegetations but were not significant (Table 4). No significant plot, vegetation or plot x vegetation interactions were detected for AWC and θ_d at 0-20 cm depth (Table 5) but θ_d was significantly different among plots and vegetation at 20-40 cm depth (Table 5). The θ_d was higher under the mesquite than creosote canopy and was in accord with high macroporosity of coppice dunes.

Soil moisture content variations under vegetation and intercanopy areas in different plots expressed as standard errors were generally lower at most suctions for vegetation canopies as well as for intercanopy areas in unirrigated than in irrigated-I and irrigated-II plots at 0-20 cm depth (Figure 1). The coefficient of determination (R^2) between measured and [31] model fitted $h(\theta)$ ranged from 0.96 to 0.99 (Table 6). The bubbling pressure, which is the inverse of α, was higher under vegetation canopies and intercanopy areas of unirrigated plot than both the irrigated plots. The irrigated plots have higher SAR and EC and lower bubbling pressure, which could be due to the higher osmotic potential than the unirrigated plot.

11.3.3 SOIL NITRATE AND CHLORIDE CONCENTRATION

Significant plot, vegetation and plot x vegetation interaction effects were obtained for NO_3^- at 0-20 and 20-40 cm depths (Table 7). One-way ANOVA contrasts also detected differences for NO_3^- between creosote canopies and intercanopy areas at 0-20 cm depth, between mesquite and creosote canopie at 60-80 and 100-150 cm in the irrigated plot-I (Table 8). NO_3^- concentration was higher under mesquite canopies in both irrigated and unirrigated plots than under creosote canopies and intercanopy areas at 0-20 cm depth (Figure 2A). Mesquite is N fixing tree and that may be the reason for higher NO_3^- under mesquite canopies because nitrate concentration of effluent water was low. It is reported that mesquite can store soil nitrogen 3 to 7 times greater beneath its canopies than in the interspaces between species [32]. Higher NO_3^- at upper depths than deeper depths indicated no leaching of NO_3^-. Significant plot, vegetation and plot x vegetation interaction effects were observed for Cl^- at 0-20 cm and only plot interaction was significant at 20-40 cm depth (Tables 7). Chloride concentration was higher under creosote canopies than mesquite and intercanopy areas in irrigated plot-I at all depths (Figure 2B; Table 8). The Cl^- concentration almost linearly increased with depth under creosote and intercanopy areas. Higher Cl^- concentration under creosote canopies than intercanopy areas and mesquite canopies could be due to the higher wastewater interception by creosote canopies. Soil Cl^- accumulation was observed between 60 and 150 cm depth under creosote and intercanopy areas (Figure 3). However, a lower level of Cl^- under mesquite might be the effect of higher K_s that resulted in the deeper leaching of Cl^- below the sampling depths. This is also supported by larger errors in the Cl balance (total applied-available at 0-150 cm depth) under mesquite canopy than creosote or intercanopy area. An earlier study conducted on the same site reported high Cl^- concentration in the upper profile (0-15cm) of intercanopy areas due to wastewater ponding that could not be supported by this study and the white precipitate observed in the intercanopy areas were primarily due to Na^+. The NO_3^- and Cl^- are weakly held anions and can leach to greater depths with percolating water; however, most of the applied NO_3^- was accounted for within 0-150 cm depth. This study demonstrated that Cl^- but not the NO_3^- was leached below the sampling depths of 150 cm.

11.3.4 SOIL ELECTRICAL CONDUCTIVITY AND PH

Significant interactions in EC were obtained only among plots (Table 7). The EC was higher under creosote than mesquite canopies at 0-20 cm depth of the irrigated plot-I (Figure 2C). Higher EC under creosote canopies was also in accord with the higher wastewater interception by the canopies. The EC was similar under vegetation canopies at all sampled depths in the unirrigated plot. Similar to Cl^-, EC in irrigated-I increased by depth under both vegetation canopies and intercanopy areas.

TABLE 3: One-way ANOVA contrasts between vegetation canopies and intercanopy areas for particle size, bulk density (BD), hydraulic conductivity (Ks) available water content (AWC), field capacity (FC) and drainable porosity (θd) at 0-20 and 20-40 cm depth during 2007.

Properties	Unirrigated		Irrigated-I		Irrigated-II	
	0-20	20-40	0-20	20-40	0-20	20-40
				P-values		
Sand	(>.08)[1,2,3]	(>.08)[1,2,3]	(<.05)[1] *(>.08)[2,3]	(>.12)[1,2,3]	(>.07)[1,2,3]	(<.01)[1] *(>.06)[2,3]
Silt	(>.55)[1,2,3]	(>.18)[1,2,3]	(>.09)[1,2,3]	(>.07)[1,2,3]	(<.05)[1,3]*(>.48)[2]	(>.10)[1,2,3]
Clay	(>.17)[1,2,3]	(>.05)[1,2,3]	(>.69)[1,2,3]	(>.41)[1,2,3]	(>.18)[1,2,3]	(>.05)[1,2,3]
BD	(<.005)[2,3]* (>.08)[1]	(>.66)[1,2,3]	(>.05)[1,2,3]	(>.06)[1,2,3]	(>.14)[1,2,3]	(>.07)[1,2,3]
K_s	(<.001)[2,3]* (<.08)[1]	(0.09)[1,2,3]	(<.001)[2] *(>.09)[1,3]	(>.08)[1,2,3]	(<.005)[1,2,3]*	(>.48)[1,2,3]
AWC	(>.07)[1,2,3]	(<.005)[1]* (>.38)[2,3]	(>.61)[1,2,3]	(>.44)[1,2,3]	(>.52)[1,2,3]	(>.38)[1,2,3]
FC	(>.17)[1,2,3]	(<.05)[1] *(>.07)[2,3]	(>.21)[1,2,3]	(>.21)[1,2,3]	(>.05)[1,2,3]	(>.07)[1,2,3]
θ_d	(>.33)[1,2,3]	(>.27)[1,2,3]	(>.37)[1,2,3]	(>.39)[1,2,3]	(>.42)[1,2,3]	(>.25)[1,2,3]

*[1]= one-way ANOVA contrast between mesquite vs. creosote, [2]= mesquite vs. intercanopy, [3]= creosote vs. intercanopy. Numbers inside the parenthesis indicate the P-values * Indicates significant differences at $P < 0.05$*

Increased irrigation with salty water generally tended to increase soil EC with soil depth except at shallow (2.5-5 cm) depths because of the evaporation at the soil surface [33]. Similar patterns of increases in EC were observed except under mesquite canopies in irrigated plot-II. These values were lower in 2007 than those reported in 2005 [11]. This might be due to the time of the sampling, amount of wastewater application and precipitation. Samples were collected during July 2007 after some rainfall events and no application of wastewater was made during March 2007 to July 2007. Whereas in 2005 samples were collected during December and wastewater was continuously applied from September onwards with no precipitation recorded during the past three months.

TABLE 4: Values of F statistic and the probability (Pr) from analysis of variance (n=27) for sand, silt, clay, Ks, BD, available water content (AWC), field capacity (FC), and drainable porosity (θd) at 0-20 and 20-40 cm depth during 2007

Source	DF	F value	Pr>F	F Value	Pr>F
		0-20 cm		20-40 cm	
Sand					
Plot	2	1.07	0.365	2.20	0.227
Vegetation	2	2.45	0.114	2.14	0.233
Plot x vegetation	4	0.16	0.956	1.35	0.288
Silt					
Plot	2	1.49	0.328	2.06	0.156
Vegetation	2	0.56	0.611	4.62	0.024
Plot x vegetation	4	0.99	0.438	0.23	0.911
Clay					
Plot	2	3.73	0.121	2.53	0.194
Vegetation	2	4.62	0.091	2.40	0.206
Plot x vegetation	4	0.86	0.503	0.63	0.647
K_s					
Plot	2	5.28	<.05*	129.43	<.0005*
Vegetation	2	29.04	<0.0001*	22.83	<.005*
Plot x vegetation	4	2.64	<.05*	0.05	0.994
BD					
Plot	2	1.97	0.253	1.47	0.331
Vegetation	2	4.65	0.090	2.07	0.155
Plot x vegetation	4	1.89	0.156	1.00	0.434
AWC					
Plot	2	4.95	0.082	0.29	0.760
Vegetation	2	3.35	0.139	0.60	0.593

TABLE 4: *Cont.*

Source	DF	F value	Pr>F	F Value	Pr>F
		0-20 cm		20-40 cm	
Plot x vegetation	4	0.76	0.564	5.34	0.005*
FC					
Plot	2	20.19	<.005*	57.03	<.001*
Vegetation	2	6.66	0.053	2.66	0.069
Plot x vegetation	4	0.78	0.555	0.27	0.894
θ_d					
Plot	2	3.34	0.140	8.87	<.05*
Vegetation	2	1.21	0.065	7.28	0.05
Plot x vegetation	4	0.36	0.832	1.03	0.418

* *Indicates significant differences at $P < 0.05$*

Soil pH was similar (9.20 ± 0.01 to 9.80 ± 0.09) under vegetation canopies and intercanopy areas in both irrigated plots until 60 cm depth. Although plot interaction for pH was significant at 0-20 and 20-40 cm depths (Table 7), one-way ANOVA contrasts for pH did not detect differences between vegetation canopies and intercanopy areas in the irrigated plots (Table 8). Irrigation with wastewater with a pH of 9.70 ± 0.10 on soils in irrigated plots raised the soil pH to >9. Although mesquite and creosote are deep rooted bushes, it is difficult to assess the exact influence of high surface pH on their survival. However, high pH can certainly have an effect on survival and growth of native perennial and herbal vegetation by reducing the availability of certain micronutrients, particularly iron (Fe) and manganese (Mn).

11.3.5 SODIUM ADSORPTION RATIO AND EXCHANGEABLE SODIUM PERCENTAGE

Application of high SAR wastewater raised soil SAR in both irrigated plots and the SAR was higher in irrigated than unirrigated plots at most depths (Figure 4A). Significant plot interactions for SAR were observed at 0-20 cm depth alone (Table 7). One way ANOVA contrasts for SAR did not detect differences between vegetation canopies and intercanopy areas among the plots (Table 8). The SAR under vegetation canopies and intercanopy areas was >15 and pH> 8.5 within 0-100 cm depth which is characterized by reduced nutrient and micronutrient availability (Brady and Weil, 2000). Mesquites are deep-rooted plants which can survive with less moisture [32]. The rooting depth is about 12 m for mesquite and 3 m for creosote. However, majority of mesquite roots are distributed within 0-100 cm depth [34] and creosote within 0-25 cm depth [35]. Therefore high SAR and Na^+ concentration would affect the survival of mesquite and creosote bushes along with other perennial vegetation. Significant plot interactions were observed for Na^+ concentration at 0-20 and 20-40 cm depths (Table 7). The Na^+ concentration was higher in the intercanopy areas at 0-20 cm depth than under vegetation canopies in both irrigated plots (Figure 3B). Higher Na^+ in upper depths in the intercanopy areas were likely due to lower K_s and θ_d at the intercanopy areas than under the vegetation canopies which accumulated Na^+ in the upper depths. The ESP showed a similar trend as SAR and only plot interaction was significant (Figure 3C, Table 7). Differences in ESP were also detected between creosote canopies and the intercanopy areas at 0-20 and 80-100 cm depth in the irrigated plot-II (Table 8). However, no significant plots, vegetations and plot x vegetation interactions were observed for ESP at 0-20 and 20-40 cm depth (Table 7).

11.3.6 SOIL ORGANIC MATTER

Few differences were detected for OM between mesquite canopies and intercanopy areas, between creosote canopies and intercanopy areas at 20-40, 40-60, 80-100 and100-150 cm depth of the irrigated plots (Table 7). The EPA has recommended wastewater application as a method of recycling nutrients and organic matter. However, organic matter content was lower in both irrigated plots than in the unirrigated plot. Soil microorganisms and plants prefer a near neutral pH range of 6 to 7 for better performance [36]. Since the pH of irrigated plot is >9 at upper depths, it may have decreased the performance of microorganisms and the decomposition of OM in the irrigated plots. This study did not support that land application of solid organic residuals increases the OM content and soil moisture retention [3].

TABLE 5: Mean, standard errors and one-way ANOVA contrasts between plots for particle size, bulk density (BD) and hydraulic conductivity (K_s) available water content (AWC), field capacity (FC), and drainable porosity (θ_d) at 0-20 cm depth during 2007.

Vegetation	Sand (%)	Silt (%)	Clay (%)	BD (Mg m^{-3})	K_s (cm h^{-1})	AWC (cm^3 cm^{-3})	FC (cm^3 cm^{-3})	θ_d (cm^3 cm^{-3})
Unirrigated								
Mesquite	89.77 ± 0.31	3.61 ± 0.24	6.62 ± 0.37	1.52 ± 0.00	22.20 ± 2.82	1.85 ± 0.13	0.11 ± 0.00	0.14 ± 0.01
Creosote	89.69 ± 0.41	3.83 ± 0.41	6.48 ± 0.72	1.57 ± 0.01	12.35 ± 0.30	2.02 ± 0.15	0.13 ± 0.00	0.11 ± 0.14
Intercanopy	88.64 ± 1.15	4 .00 ± 0.57	7.36 ± 0.57	1.59 ± 0.03	11.00 ± 1.40	1.27 ± 0.19	0.11 ± 0.00	0.12 ± 0.00
Average	89.37 ± 0.62	3.81 ± 0.40	6.82 ± 0.55	1.56 ± 0.01	15.18 ± 1.50	1.71 ± 0.15	0.12 ± 0.00	0.12 ± 0.05
Irrigated-I								
Mesquite	89.19 ± 0.06	4.67 ± 0.05	7.14 ± 0.13	1.54 ± 0.01	13.54 ± 1.58	2.06 ± 0.25	0.16 ± 0.01	0.12 ± 0.03
Creosote	88.94 ± 0.16	3.41± 0.22	7.62 ± 0.08	1.49± 0.00	11.65± 1.97	2.90 ± 0.19	0.21 ± 0.01	0.11 ± 0.01
Intercanopy	87.98 ± 0.57	4.2 ± 0.33	7.84 ± 0.09	1.57± 0.01	8.20 ± 0.72	2.21 ± 0.29	0.17 ± 0.01	0.10 ± 0.01
Average	88.70 ± 0.26	3.76 ± 0.20	7.53 ± 0.10	1.53 ± 0.01	11.16 ± 1.42	2.39 ± 0.24	0.18 ± 0.01	0.11 ± 0.01
Irrigated-II								
Mesquite	89.35 ±0.66	3.67 ± 0.72	6.98 ± 0.21	1.51 ± 0.01	18.20 ± 1.29	2.08 ± 0.21	0.17 ± 0.01	0.13 ± 0.00
Creosote	88.98 ± 0.43	3.90 ± 0.36	7.12 ± 0.16	1.50 ± 0.03	14.00 ± 0.78	2.37 ± 0.14	0.20 ± 0.01	0.10 ± 0.00
Intercanopy	89.12 ± 1.33	2.83 ± 1.20	8.05 ± 0.33	1.55 ± 0.01	4.80 ± 0.34	2.07 ± 0.53	0.16 ± 0.00	0.10 ± 0.02
Average	89.15 ± 0.80	3.47 ± 0.76	7.38 ± 0.23	1.52 ± 0.01	12.33 ± 0.80	2.17 ± 0.29	0.17 ± 0.00	0.11 ± 0.02

TABLE 5: ***Cont.***

Vegetation	Sand (%)	Silt (%)	Clay (%)	BD (Mg m^{-3})	K_s (cm h^{-1})	AWC (cm^3 cm^{-3})	FC (cm^3 cm^{-3})	θ_d (cm^3 cm^{-3})
One way ANOVA Contrast								
Irri-I vs. Uni	0.055	0.315	0.201	0.074	<0.001*	0.823	0.047*	0.029*
Irri-II vs. Uni	0.093	0.106	0.319	0.285	0.496	0.446	0.005*	0.094
Irri-I vs. Irri-II	0.085	0.523	0.057	0.603	0.459	0.62	0.39	0.29

**Indicates significant differences at P< 0.05*

TABLE 6: Mean and standard errors for the van Genuchten (1980) parameters at 0-20 cm depth in both irrigated and unirrigated plots during 2007.

Plots	Vegetation	θ_r	θ_s	α	η	R^2	α-1cm
Irrigated-I	Mesquite	0.03 ± 0.02	0.38 ± 0.00	0.65 ± 0.15	1.35 ± 0.03	0.98	1.54
	Creosote	<0.001	0.36 ± 0.01	0.94 ± 0.47	2.10 ± 0.89	0.98	1.06
	Intercanopy	<0.001	0.35 ± 0.05	0.83 ± 0.47	1.13 ± 0.00	0.99	1.22
Unirrigated	Mesquite	0.04 ± 0.00	0.37 ± 0.00	0.17± 0.05	1.93 ± 0.13	0.99	5.88
	Creosote	0.03 ± 0.01	0.36 ± 0.00	0.17 ± 0.05	1.77 ± 0.19	0.98	5.56
	Intercanopy	0.04 ± 0.00	0.36 ± 0.00	0.18 ± 0.00	1.79 ± 0.05	0.99	5.56
Irrigated-II	Mesquite	0.05 ± 0.01	0.37 ± 0.00	0.38 ± 0.04	1.35 ± 0.04	0.98	2.63
	Creosote	0.09 ± 0.02	0.39 ± 0.00	0.44 ± 0.04	1.39 ± 0.08	0.96	2.27
	Intercanopy	0.01± 0.01	0.37 ± 0.02	0.50 ± 0.04	1.21 + 0.02	0.99	2.00

Where θr is residual soil moisture, θs is saturation soil moisture, α and η are equation parameters, R2 is coefficient of determination

11.3.8 VEGETATION ANALYSIS

The analysis of plant samples showed higher amount of Na^+ in the vegetation of irrigated than the unirrigated plots. The Na^+ content of creosote was eleven times higher in irrigated plots (880 mg.kg^{-1}) than the unirrigated plot (80 mg kg^{-1}), mesquite Na^+ content was two times higher in irrigated

(1600 mg kg^{-1}) than unirrigated (800 $mg.kg^{-1}$) and perennial vegetation Na^+ content was three times higher in irrigated (240 $mg.kg^{-1}$) than unirrigated (80 mg kg^{-1}). Total percentage N in irrigated mesquite was 3.5 %, unirrigated mesquite was 2.9%, irrigated creosote 2.5% and unirrigated creosote 1.9%. The N percentage in the irrigated perennial vegetation was three times higher (4.952) then in the unirrigated weeds. Thus native vegetations were taking up chemical constituents from the soil added through wastewater. The SAR under vegetation canopies and intercanopy areas was >15 and pH > 8.5 within 0-100 cm depth which is characterized by reduced nutrient and micronutrient availability (Brady and Weil, 2000). As the primary vegetation in the study areas are mesquite and creosote with rooting depths of about 12m and 3m, respectively. Majority of mesquite roots are distributed within 0-100 cm depth [34] and creosote within 0-25 cm depth [35]. Therefore high SAR and Na+ concentration would affect the survival of mesquite and creosote bushes along with other perennial vegetation.

TABLE 7: Values of F statistic and the probability (Pr) from analysis of variance (n=27) for nitrate (NO_3^-), chloride (Cl^-), electrical conductivity (EC), pH, sodium adsorption ratio (SAR), sodium (Na+), exchangeable sodium percentage (ESP), and organic matter (OM) at 0-20 and 20-40 cm depth during 2007

Source	DF	F value	Pr>F	F Value	Pr>F
		0-20 cm		20-40 cm	
NO_3^-					
Plot	2	16.33	<.0001*	16.24	<0.0001*
Vegetation	2	8.12	<.005*	8.5	<.005*
Plot x vegetation	4	4.7	<.005*	4.3	<.05*
Cl					
Plot	2	24.45	<0.0001*	9.3	<.05*
Vegetation	2	10.67	<.0005*	2.84	0.177
Plot x vegetation	4	4.84	<.005*	2.09	0.124
EC					
Plot	2	11.92	<.05*	13.96	<.05*
Vegetation	2	2.08	0.240	3.07	0.155
Plot x vegetation	4	2.14	0.117	1.62	0.213
pH					

TABLE 7: *Cont.*

Source	DF	F value	Pr>F	F Value	Pr>F
		0-20 cm		20-40 cm	
Plot	2	45.69	<0.0001*	66.82	<.005*
Vegetation	2	9.57	<.05*	1.87	0.267
Plot x vegetation	4	0.25	0.908	1.31	0.303
SAR					
Plot	2	7.14	<.001*	3.47	0.133
Vegetation	2	1.66	0.298	0.06	0.946
Plot x vegetation	4	0.68	0.61	2.11	0.141
Na^+					
Plot	2	19.53	<.005*	18.52	<.005
Vegetation	2	1.5	0.327	1.08	0.421
Plot x vegetation	4	0.51	0.731	0.58	0.684
ESP					
Plot	2	9.48	<.005*	5.21	0.076
Vegetation	2	0.93	0.420	0.06	0.946
Plot x vegetation	4	0.64	0.645	2.01	0.157
OM					
Plot	2	0.1	0.905	0.31	0.738
Vegetation	2	0.96	0.456	2.91	0.083
Plot x vegetation	4	3.04	0.05*	3.04	0.05

* *Indicates significant differences at* $P < 0.05$

11.4 CONCLUSIONS

Chemical parameters were higher in the effluent than in the influent primarily due to evaporation in the holding pond. Low sprinkler uniformity in both irrigated plots was observed primarily due to the non uniform sprinkler distances, wind velocities and wastewater interception by vegetation canopies. Application of wastewater containing high EC, SAR, and Na^+ concentration decreased the K_s of the irrigated west mesa soil. NO_3^- did not leach to the deeper depths but Cl_- did leach below the sampling depths. High Na^+ concentration (>693 mg kg-1), SAR (>15) and pH (>8.5) at 0-100 cm depth of the irrigated plots threaten the survival of woody as well as annual and perennial forbs and grass in the study areas as can be seen from high Na^+ content of vegetation of the irrigated area. Necessary steps should be taken to schedule uniform application of wastewater all around the year and measures should be taken to reduce the evaporation in the holding pond. Wastewater application in the site should also take into account the relative differences and importance of intercanopy and under the canopy soils.

REFERENCES

1. Hamilton AJ, Stagnitti F, Xiong X, Kredil SL, Benke KK, Maher P (2007) Wastewater irrigation: The State of Play. Vadose Zone J 6: 823-840.
2. Magesan GN (2001) Changes in soil physical properties after irrigation of two forested soils with municipal wastewater. N Z J For Sci 31: 188-195.
3. Magesan GN, Wang H (2003) Application of municipal and industrial residuals in New Zealand forests. Aust J Soil Res 41: 557-569.
4. Adhikari P, Shukla MK, Mexal JG (2012-a) Spatial variability of infiltration rate and soil chemical properties of desert soils: implications for management of irrigation using treated wastewater. Transaction of the ASABE 55: 1711-1721.
5. Mamedov AI, Shainberg I, Levy GJ (2000) Irrigation with effluent water: effects of rainfall energy on soil infiltration. Soil Sci Soc Am J 64: 732-737.
6. Abedi-Koupai, Mostafazadeh JB, Bagheri MR (2006) Effect of treated wastewater on soil chemical and physical properties in an arid regions. Plant Soil Environ 52: 335-344.
7. Halliwell DJ, Barlow MK, Nash MD (2001) A review of the effect of wastewater sodium on soil physical properties and their implications for irrigation systems. Aust J Soil Res 39: 1259-1267.
8. Sparks DL (2003) Environmental soil chemistry.2nd edition, Academic Press, San Diego, California, USA.
9. Agassi M, Tarchitzky J, Keren R, Chen Y, Goldstein D, et al. (2003) Effects of prolonged irrigation with treated municipal effluent on runoff rate. J Environ Qual 32: 1053-1057.
10. Rostango CM, Sosebee RE (2001) Biosolids application in the Chihuahuan desert: Effects on runoff water quality. J Environ Qual 30: 160-170.
11. Babcook M, Shukla MK, Picchioni GA, Mexal JG, Daniel D (2009) Chemical and physical properties of Chihuahuan desert soils irrigated with industrial effluent. Arid Land Research and Management 23: 47-66.
12. Adhikari P, Shukla MK, Mexal JG (2012-b) Treated wastewater application in Southern New Mexico: Effect on soil chemical properties and surface vegetation. New Mexico Journal of Science 46: 105-120.

13. Adhikari P, Shukla MK, Mexal JG (2011-a) Spatial variability of electrical conductivity of desert soil irrigated with treated wastewater: implications for irrigation management. Applied and Environmental Soil Science 2011: 1-11.
14. Adhikari P, Shukla MK, Mexal JG, Sharma P (2011-b) Assessment of the soil physical and chemical properties of desert soils irrigated with treated wastewater using principal component analysis. Soil Sci 176 : 356-366.
15. Adhikari P, Shukla MK, Mexal JG (2012-c) Spatial variability of soil properties in an arid ecosystem irrigated with treated municipal and industrial wastewater. Soil Sci 177: 458-469.
16. Gile LH, Hawley JW, Grossman RB (1981) Soils and geomorphology in the basin and range area of southern New Mexico-Guide book to the Desert Project. New Mexico Bureau of Mines and Mineral Resources, Memoir 39: 1-222.
17. Gee GW, Bauder JW (1986) Particle size analysis. Methods of soilanalysis Part 1 (2nd edition), ASA and SSSA, Madison, WI.
18. Blake GR, Hartge KH (1986) Bulk density. Methods of soil analysis. Part 1, 2nd edition. ASA and SSSA, Madison, WI.
19. Klute A, Dirksen C (1986) Hydraulic conductivity and diffusivity. Methods of soil analysis Part I (2nd edition): WI, USA.
20. Klute A (1986) Water retention. Methods of soil analysis Part I (2nd edition): WI, USA.
21. Black CA (1965) Methods of soil analysis. Chemical and microbiological properties, ASA Inc, Madison, WI.
22. Maynard DG, Kalra YP (1993) Nitrate and exchangeable ammonium nitrogen. Soil sampling and methods of analysis. CSSS, Lewis Publishers, USA.
23. Sparks DL (2003) Environmental soil chemistry.2nd edition, Academic Press, San Diego, California, USA.
24. Menner JC, McLay CDA, Lee R (2001) Effects of sodium contaminated wastewater on soil permeability of two New Zealand soils. Aust J Soil Res 39: 877-891.
25. Jalali M, Merrikhpour H (2008) Effects of poor quality irrigation waters on the nutrient leaching and groundwater quality from sandy soil. Environ Geology 53: 1289-1298.
26. Felker P, Clark PR, Laag AE, Pratt PF (1981) Salinity tolerance of the tree legumes: mesquite (Prosopisglandulosa var. torreyana, P. velutina and P. articulata), algarrobo (P. chilensis), Kiawe (P. pallida), and Tamarugo (P. tamarugo) grown in sand culture on nitrogen free media. Plant Soil 61: 311-317.
27. Al-Jibury LK (1972) Salt tolerance of some desert shrubs in relation to their distribution in the southern desert of North America. Arizona State University, Tempe AZ, USA.
28. Coppola A, Santini A, Boti P, Vacca S (2003) Urban wastewater effects on water flow and solute transport in soils. J Environ Sci Health 38: 1469-1478.
29. Al-Haddabi M, Ahmed M, Kacimov A, Rahman S, Al-Rawahy S (2004) Impact of treated wastewater from oil extraction process on soil physical properties. Commun Soil Sci Plan 35: 751-758.
30. Ibrahim A, Agunwamba JC, Idike FI (2005) Effects of wastewater effluent re-use on irrigated agricultural soils. J Sust Agri and the Environ 7: 60-80.

31. VanGenuchten MT (1980) A closed-form equation for predicting the hydraulic conductivity of unsaturated soils. Soil Sci Soc Am J 44: 892-898.
32. Ansley RJ, Huddle JA, Kramp BA (1997) Mesquite ecology. Texas Agricultural Experiment Station, Vernon, California, USA.
33. Costa JL, Prunty L, Montgomery BR, Richardson JL, Alessi RS (1991) Water quality effects on soils and alfalfa: II. Soil Physical and Chemical Properties. Soil Sci Soc Am J 55: 203-209.
34. Heitschmidt RK, Ansley RJ, Dowhower SL, Jacoby PW, Price DL (1988) Some observations from the excavation of honey mesquite root systems. Journal of Range Management Archives 41: 227-231.
35. Baynham P (2004) Sonoran originals: The unappreciated smell of rain. Master Garden Journal: 22-24.
36. Sylvia MD, Fuhrmann JJ, Hartel PG, Zuberer DA (2005) Principles and applications of soil microbiology. (2nd edition), Prentice Hall, Upper Saddle River, New Jersey, USA.

Table 8 is not available in this version of the article. To view this additional information, please use the citation on the first page of this chapter.

CHAPTER 12

Spatial Distribution of Fecal Indicator Bacteria in Groundwater beneath Two Large On-Site Wastewater Treatment Systems

CHARLES HUMPHREY, MICHAEL O'DRISCOLL, AND JONATHAN HARRIS

12.1 INTRODUCTION

On-site wastewater treatment systems (OWS) are a common method of wastewater treatment in many countries including the United States, Australia, Canada, New Zealand, and Ireland [1,2,3,4,5]. OWS treat wastewater that contains high concentrations of pathogenic microorganisms such as bacteria, viruses, and protozoa [1,6]. Most OWS include a septic tank, effluent distribution device, drainfield trenches, and soil beneath the trenches. The septic tank provides primary treatment via sedimentation and anaerobic digestion of organic matter. The effluent distribution device distributes wastewater to the drainfield trenches, where effluent is stored until infiltrating the soil. The soil beneath the drainfield trenches provides most of the physical, chemical and biological treatment of wastewater. If

OWS do not reduce the concentration of pathogens in wastewater effluent, then groundwater and surface water quality may be degraded and public health may be compromised. For example, studies have shown that concentrations of fecal indicator bacteria (FIB) including *E. coli*, enterococci, total coliform, and/or fecal coliform can be elevated in groundwater adjacent to OWS if there is not sufficient vertical separation between the OWS trenches and water table [7,8,9,10]. Other research has shown that surface water FIB concentrations were elevated in areas adjacent to high densities of OWS [11] and when groundwater levels beneath OWS were elevated surface water FIB concentrations were also elevated [12]. High densities of OWS have been associated with endemic diarrheal illness, with higher OWS densities correlating to higher incidences of bacterial and viral diarrhea in a study in central Wisconsin [13].

Other factors associated with OWS wastewater treatment performance include soil type, loading rates, and system type. OWS installed in soils with higher percentages of fine-textured materials such as silt and clay, typically are more efficient at reducing microbial concentrations than OWS installed in sandy soils, due to the increased reactive surface area and residence time [7,14,15]. Research has also suggested that higher wastewater loading rates may lead to decreased treatment [8,16,17,18] and that OWS that utilize low pressure pipe (LPP) distribution can more efficiently treat wastewater constituents such as bacteria relative to gravity flow effluent distribution systems [16,19,20]. LPP systems include a septic tank and pump tank, a pump that delivers effluent to the drainfield trenches via a pressure manifold connected to small diameter (2.54 to 5.08 cm) PVC conveyance pipes (laterals) that extend the length of each trench (Figure 1). The laterals for each trench are typically the same diameter, and have small holes (0.31 to 0.39 cm) drilled along the lateral about every 1.5 m (in sandy soils) to deliver wastewater to the trench. The ends of the laterals include a "turn-up" that is capped, so that when the pump is activated, the manifold and then all laterals should pressurize and thus deliver equal distribution of effluent across the drainfield. This is in contrast to OWS that use distribution boxes. Effluent distribution with distribution boxes is facilitated by gravitational forces, not pressure, even if the OWS uses a pump. The distribution box is installed at a higher elevation relative to the trenches. The distribution boxes are connected to the drainfield trenches with conveyance pipes that have larger diameter weep holes (1.3 to 1.9 cm) that are more closely spaced (10 cm apart) together than LPP systems [21] (Figure 1). The conveyance pipes are not capped at the end, and thus often deliver more wastewater to the front of the trenches than an LPP system. Because some of the earlier studies suggested that LPP systems could perform better than conventional OWS [16,19,20], there are less stringent siting requirements for LPP systems relative to conventional systems with distribution boxes in North Carolina. For example, in sandy areas, LPP systems can be installed in soils with 15 cm less soil depth and separation to groundwater than conventional OWS [21], and LPP systems typically require about 25%–30% less overall drainfield area than conventional trench systems with distribution boxes [21]. If LPP systems are more effective at distributing wastewater across the entire drainfield area than gravity flow distribution systems, then groundwater characteristics (as influenced by infiltrating wastewater) beneath the LPP systems may also be more homogeneous. However, there is a lack of field research that compares FIB groundwater quality uniformity beneath OWS that use LPP and gravity flow distribution. If the LPP systems are not more effective at distributing effluent across the entire drainfield area than OWS that use distribution boxes, then the overall treatment of wastewater constituents such as pathogenic microorganisms may not be as effective as prior work has suggested.

The objectives of this research were to: (1) compare the spatial variability of groundwater FIB concentrations beneath OWS that use LPP and distribution boxes for effluent distribution; and (2) evaluate the efficiency of the LPP and pump to distribution box OWS in reducing FIB concentrations.

12.2 MATERIALS AND METHODS

12.2.1 SITE SELECTION

Sites were selected at two schools in Craven County North Carolina, USA including James W. Smith Elementary (JWS) and West Craven High School (WCH). The sites were chosen because they were located in coastal North Carolina where issues regarding FIB contamination of surface waters have been documented [8,11]. Both schools have OWS that include large capacity septic tanks, pump tanks with effluent pumps and dual alternatingly dosed disposal fields (Table 1). The OWS at JWS and WCH were routinely maintained by a certified operator with oversight from the Craven County Environmental Health Department. Both systems were in operation during the study and each had 32 drainfield trenches. The OWS at WCH was a replacement system that was installed in 1997. The OWS at JWS was the original system that has been in use since 1987. The OWS at JWS uses a pump to distribution box system and WCH uses a low pressure pipe (LPP) system. Other system characteristics are listed in Table 1.

The soil survey for Craven County listed the soil series as Autryville loamy sand (Loamy siliceous, thermic, Arenic Paleudult) at JWS, and Tarboro sand (Mixed, thermic, typic Udipsamments) at WCH [22]. The soil series at the sites were confirmed by a NC Licensed Soil Scientist working on the project. Soil cores were also collected to depths of 5 m at each site within the drainfield area for descriptive analysis. Grain-size analysis was completed on the soil samples using the GRADISTAT 4.0 method [23]. The mean % sand from soil samples collected from the cores was >90% for both sites.

TABLE 1: Site and wastewater system characteristics.

Site	Install Date	Septic Tank Capacity (L)	Max Design Flow (L/d)	Grease Trap (L)	Pump Tank (L)	Distribution Device	Dispersal Area (m^2)	Vertical Separation (m)	USDA Soil Series
JWS	1987	37,800	37,800	3780	18,144	D-box (2)	892	>4 m	Autryville
WCH	1997	73,827	73,827	11,340	11,340	LPP (2)	1115	>1 m	Tarboro

12.2.2 GROUNDWATER MONITORING NETWORK

Monitoring wells constructed with 3.2 or 5 cm diameter, solid PVC pipe coupled to 0.90 m of well screen were installed near (within 3 m) the corners of both drainfields at each site (~25 m apart) (Figure 2 and Figure

3). Therefore, there were a total of four "front" wells (closest to distribution box/mainfold) and four "end" wells (farthest from the distribution box/manifold). Soil augers and a geoprobe were used to create bore holes to depths below the water table (range: 1 m to 10.5 m) for well placement. Once the wells were placed and driven in the boreholes, the annular space adjacent to the screen was filled with sand, and the annular space above the screen was filled with a mixture of native sand and bentonite. The wells were capped, labeled, and enclosed in valve boxes. The relative elevation of the monitoring wells was determined using a laser level, and the coordinates of the wells were determined using a Trimble 6000 series GPS (Trimble Navigation Limited, Westminster, CO, USA). Site maps with the monitoring well locations overlain on aerial photographs were created for both sites. Depth to groundwater was determined using a Solinst TLC meter (Solinst, Ontario, Canada), and the relative elevation of the groundwater was labeled on the maps. Groundwater flow direction was determined using the groundwater data, and the 3-pt contouring method [24]. Monitoring wells were installed up-gradient and down-gradient from the systems at both sites, after the groundwater flow direction was determined. A spring and stream and were also sampled at JWS down-gradient from the OWS.

12.2.3 GROUNDWATER CHARACTERIZATION AND ANALYSIS

Environmental readings and water samples were collected bi-monthly (4 times) during the study period (November 2012 to May 2013) from each well, septic tank, and adjacent surface water. Samples were collected during periods without rainfall to help reduce the potential for sample contamination. The depth to groundwater at each well was determined using a Solinst TLC meter (Solinst, Ontario, Canada). The well was then purged using a sterile (unwrapped immediately prior to use) bailer, allowed to recharge, and a sample was transferred from the bailer to sterile sample bottle. New, neoprene gloves were used for each sample location. Sample bottles were labeled and placed in ice filled coolers for later transport to the East Carolina University Environmental Health Sciences Water Laboratory for analysis. Samples were prepared and incubated within 6 h of collection. Samples from the bailers were also transferred to the multi-meter sensor cup. An YSI 556 multi-meter (YSI Inc., Yellow Springs, OH, USA) was used to determine sample pH, specific conductivity (SC), and temperature in the field. The meter was calibrated prior to each field day. Septic effluent samples were obtained from the distribution box at JWS and from the septic tank at WCH by removing the manhole covers and lowering bailers into the box and tank for sample collection. Effluent samples were also transferred to the sensor cups for environmental readings. Water samples were analyzed for total coliform and *E. coli* using the IDEXX Colilert method, and for Enterococci using IDEXX Enterolert method (IDEXX Laboratories, Westbrook, Maine). Total coliform and *E. coli* samples were incubated at 35 °C for 24 h and enterococcus samples were incubated at 41 °C for 24 h and then read. Septic effluent samples were diluted (dilution factors of 10–2000×) because of elevated FIB concentrations. For quality control and assurance, field blanks were inserted for approximately 10% of samples.

12.2.4 SPATIAL DISTRIBUTION OF FECAL INDICATORS

Groundwater FIB concentrations near the front of each drainfield were compared to groundwater FIB concentrations near the end of each drainfield to determine if there were significant differences in distribution of wastewater and FIB along the length of the drainfield trenches. The data were pooled for each field (Field 1 fronts in comparison to Field 1 ends) and between fields (Field 1 and Field 2 fronts in comparison to Field 1 and Field 2 ends). Groundwater FIB concentrations beneath Field 1 and Field 2 were compared at each site to determine if significant differences were observed between the fields. Mann Whitney tests were performed with Minitab16 statistical software to determine if differences in concentrations were statistically significant. Groundwater physical and chemical properties including pH, specific conductivity (SC), depth from surface, and temperature were also summarized for Fields 1 and 2, and for proximal and distal ends of the drainfields. FIB data were log transformed and a Pearson Correlation test was used to determine if there were significant correlations between the different FIB concentrations, and between FIB concentrations and SC. Correlations between FIB and SC were performed because SC is often elevated in fresh groundwater influenced by effluent [7,25,26].

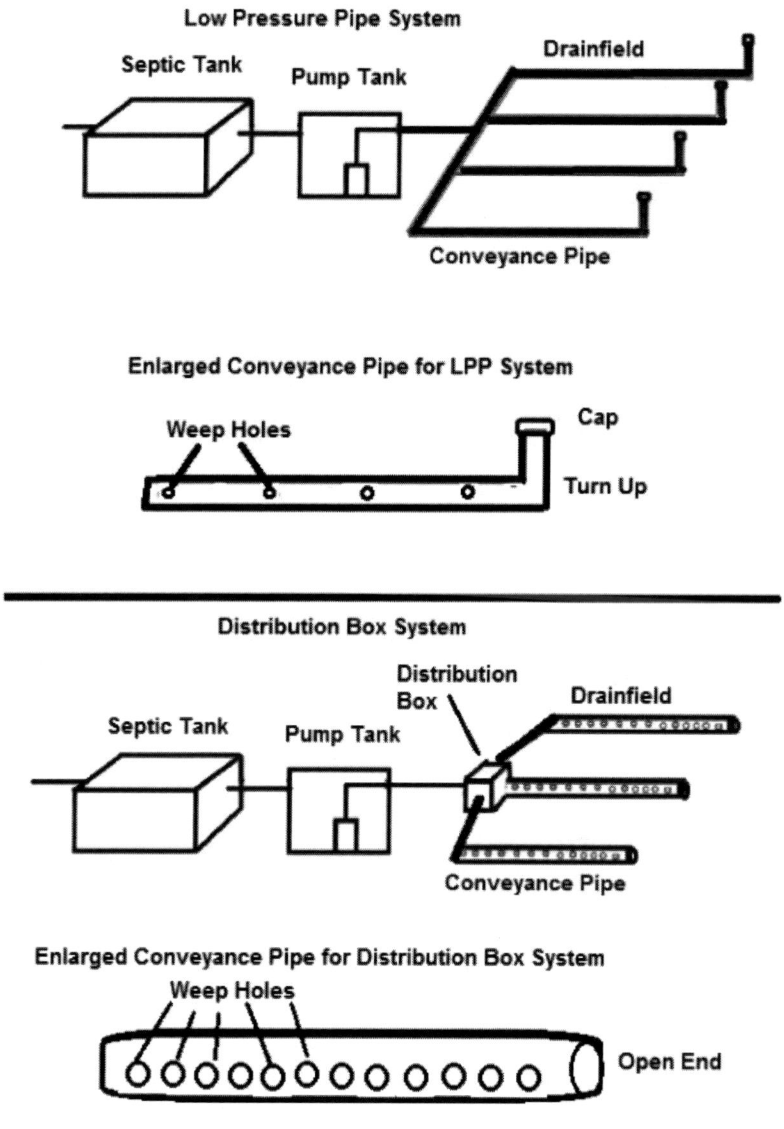

FIGURE 1: Low pressure pipe (LPP) system (top) and pump to distribution box system (bottom) showing the differences in pipe diameter, weep hole size and spacing, and end of the pipes (capped for LPP, open for distribution box).

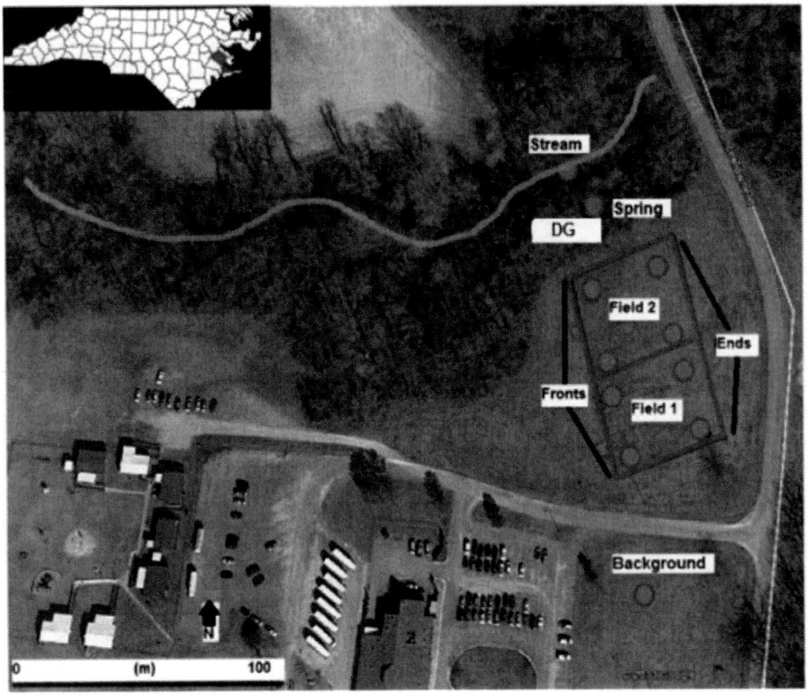

FIGURE 2: James Smith Elementary School (JWS) in Craven County, North Carolina. The blue circles show monitoring wells, (DG) is down-gradient wells, the red boxes show the boundaries of drainfields 1 and 2. The fronts of the drainfields are located on the west side of the boxes, while the ends of the drainfields are located on the east side of the boxes. Groundwater flow direction is to the north.

12.2.5 TREATMENT EFFICIENCY

FIB concentrations in septic effluent and groundwater beneath the drainfield with the highest FIB concentrations (Field 1 or Field 2 for each site) were compared to determine the effectiveness of the OWS in reducing FIB concentrations in the vadose zone. Septic effluent concentrations of FIB were also compared to groundwater down-gradient and within the flow path of the OWS to determine the effects on groundwater quality down-gradient from the OWS. Groundwater FIB concentrations beneath the drainfield trenches and within the flow paths of the OWS were compared to up-gradient groundwater FIB concentrations to determine if the OWS were influencing groundwater quality. Surface water was also sampled at JWS from a spring located down-gradient and within the groundwater flow path of the OWS, and from an adjacent stream that received the spring discharge. A piezometer adjacent to the spring at JWS was also sampled. Groundwater and surface water FIB concentrations were compared to surface waters standards for the specific indicators. The OWS treatment efficiencies of JWS and WCH were compared. The geometric mean FIB concentrations we observed were not compared to the geometric mean FIB water quality standards for FIB because the sampling frequency used in this study did not conform to the frequency recommended by the US United States Environmental Protection Agency (EPA). The geometric mean FIB concentrations were instead compared to the single sample FIB standard for infrequently used surface waters suggested by the US EPA.

12.3 RESULTS

12.3.1 GROUNDWATER ENTEROCOCCI DISTRIBUTION

The geometric mean groundwater enterococci concentrations at JWS (MPN/100 mL) were higher towards the end of the drainfield trenches relative to the front of the trenches for Field 1 (Front: 336, End: 1329) and Field 2 (Front: 36; End:149), and for the pooled data (All Fronts: 110, All Ends: 445) (Figure 4). Statistically significant differences in groundwater enterococci concentrations were observed between the end and front of the trenches for Field 1 ($p = 0.05$) and for the pooled (end vs. front) data ($p = 0.004$). Groundwater enterococci concentrations were significantly ($p = 0.0035$) higher beneath Field 1 (667 MPN/100 mL) relative to Field 2 (73 MPN/100 mL). The groundwater geometric mean enterococci concentrations at WCH were elevated toward the end of the trenches beneath Field 1 (End: 1196 MPN/100 mL; Front: 562 MPN/100 mL) and lower toward the end of the trenches beneath Field 2 (Front: 123 MPN/100 mL; End:

92 MPN/100 mL), but the differences were not statistically significant ($p > 0.10$) (Figure 4). When pooling the data, groundwater towards the end of drainfields (geometric mean enterococci concentrations: 331 MPN/100 mL) was not significantly different than groundwater towards the front of the trenches (geometric mean: 262 MPN/100 mL). Groundwater enterococci concentrations were significantly ($p = 0.001$) higher beneath Field 1 (820 MPN/100 mL) relative to Field 2 (106 MPN/100 mL). All field blanks ($n = 12$) were negative for FIB.

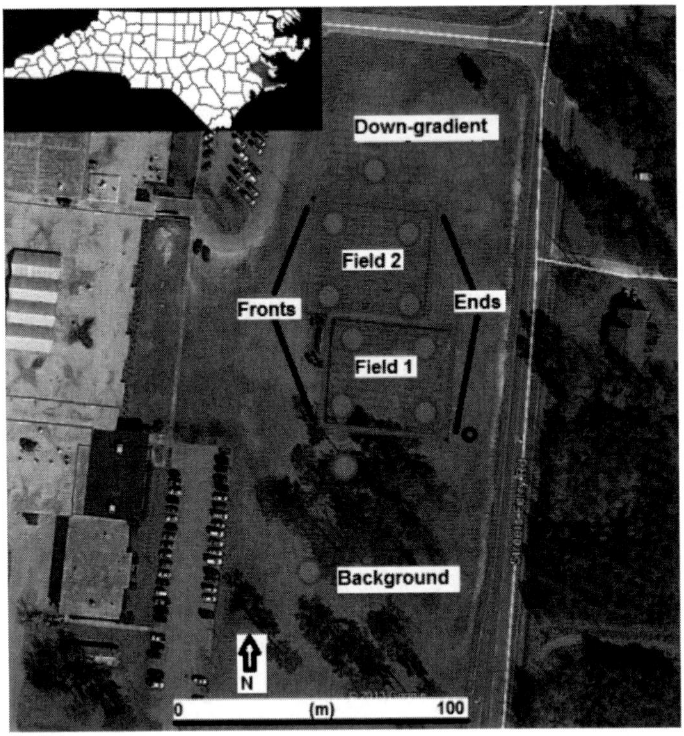

FIGURE 3: West Craven High School (WCH) in Craven County, North Carolina. The blue circles show the location of monitoring wells, the red boxes show drainfields 1 and 2. The fronts of the drainfields are located on west side of the boxes, while the ends of the drainfields are located on the east side of the boxes. Groundwater flow direction is to the north.

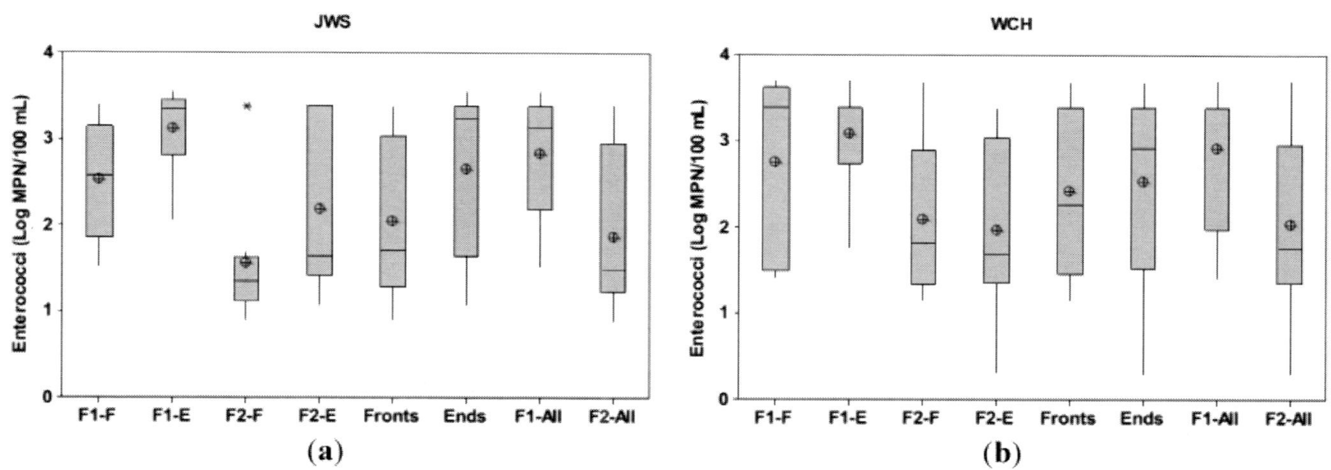

FIGURE 4: Enterococci concentrations (Log MPN/100 mL) in groundwater beneath the drainfields at James W. Smith (a) and West Craven High (b). Field 1 is F1, Field 2 is F2, front of the field is F, end of the field is E. Data indicate that groundwater enterococci concentrations were significantly higher near the ends of the trenches relative to fronts of the trenches at James Smith, and beneath Field 1 relative to Field 2 at James Smith and West Craven.

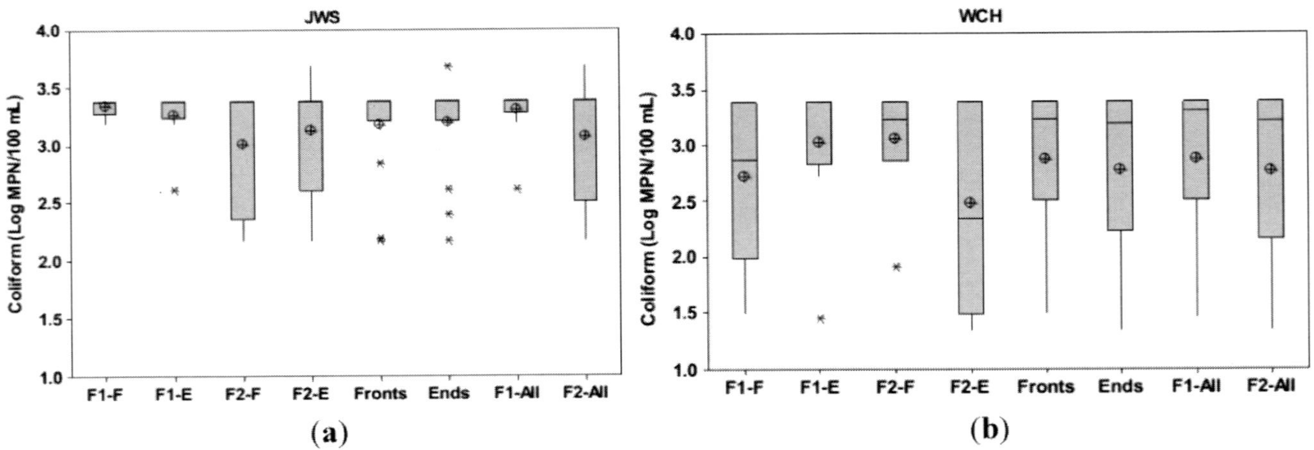

FIGURE 5: Coliform concentrations (Log MPN/100 mL) in groundwater beneath the drainfields at James W. Smith (a) and West Craven High (b). Field 1 is F1, Field 2 is F2, front of the field is F, end of the field is E. Data indicate that groundwater coliform concentrations were evenly distributed beneath the systems at both sites.

12.3.2 GROUNDWATER TOTAL COLIFORM DISTRIBUTION

The geometric mean total coliform concentrations in groundwater towards the end of Field 2 (1341 MPN/100 ML) was similar to groundwater coliform concentrations near the front (1034 MPN/100 mL), and for the pooled data (Fronts: 1507, Ends: 1568) (Figure 5). The groundwater total coliform concentrations were elevated beneath Field 1 (2007 MPN/100 mL) in comparison to Field 2 (1178 MPN/100 mL) and the differences were statistically significant (at $p < 0.10$). There was a weak, but significant correlation between enterococci and total coliform concentrations at JWS ($r = 0.446$, $p = 0.011$). Groundwater total coliform concentrations near the end and front of trenches beneath Field 1 (End: 1071 MPN/100 mL; Front: 518 MPN/100 mL) and Field 2 at WCH (Front: 1140 MPN/100 mL; End: 300 MPN/100 mL) were not significantly different (Figure 5). Overall, pooled data revealed that groundwater near the front of the drainfield trenches had similar geometric mean total coliform concentrations (748 MPN/100 mL) in comparison to groundwater near the end of the trenches (592 MPN/100 mL). Groundwater beneath Field 1 had similar geometric mean coliform concentrations (744 MPN/100 mL) in relation to groundwater beneath Field 2 (585 MPN/100 mL). There was a weak correlation between enterococci and total coliform concentrations at WC ($r = 0.328$, and $p = 0.07$).

12.3.3 GROUNDWATER E. COLI DISTRIBUTION

The mean groundwater *E. coli* concentrations were higher at JWS near the end of the trenches beneath Field 1 (7 MPN/100 mL) and Field 2 (2 MPN/100 mL) relative to the end of trenches (Field 1 Front: 1 MPN/100 mL, Field 2 Front: 1 MPN/100 mL) (Figure 6). Differences were significant at $p = 0.097$. Groundwater beneath Field 1 had similar average *E. coli* concentrations (4 MPN/100 mL) to groundwater beneath Field 2 (2 MPN/100 mL). The mean groundwater *E. coli* concentrations at WCH towards the front of the trenches for Field 1 (8 MPN/100 mL) and Field 2 (4 MPN/100 mL) were similar to concentrations near the end of the trenches (Field 1 End: 4 MPN/100 mL; Field 2 End: 1 MPN/100 mL) (Figure 6).

The pooled data also showed similar mean concentrations of *E. coli* in groundwater towards to the front of the trenches (6 MPN/100 mL) in comparison to the end (3 MPN/100 mL), and beneath Field 1 (6 MPN/100 mL) than Field 2 (3 MPN/100 mL).

12.3.4 MICROBIAL TREATMENT

The OWS at JWS was efficient (>99%) at reducing all FIB concentrations before discharge to groundwater and adjacent surface water. Wastewater geometric mean FIB concentrations from the septic tank (enterococci: 296,669 MPN/100 mL; total coliform: 1,104,391 MPN/100 mL; and *E. coli*: 457,809 MPN/100 mL) were significantly higher (all $p < 0.05$) than all other sampling locations. Between the JWS septic tank and groundwater beneath the field with the highest FIB concentrations (Field 1), there were significant reductions in the concentrations of enterococci (log 2.65 reduction or 99.78%), total coliform (log 2.74 reduction or 99.82%) and *E. coli* (log 5.06 reduction or >99.99%) (Figure 7). Groundwater enterococci and total coliform (but not *E. coli*) concentrations beneath portions of Field 1 were significantly elevated relative to background groundwater (enterococci: 105 MPN/100 mL; total coliform: 267 MPN/100 mL; *E. coli*: 0 MPN/100 mL). Microbe concentrations at JWS were further reduced as groundwater moved 24 m down-gradient from the OWS (enterococci: 57 MPN/100 mL; total coliform: 1418 MPN/100 mL; *E. coli*: < 1 MPN/100 mL). Surface water microbe concentrations at the spring down-gradient from the OWS (enterococci: 4 MPN/100 mL; total coliform: 149 MPN/100 mL; *E. coli*: 3 MPN/100 mL) were less than or similar to background groundwater concentrations, and surface water standards. The overall FIB concentration reductions at JWS from septic tank to the spring were significant (enterococci: log 4.86 or 99.98% reduction; total coliform: log 3.87 or 99.87% reduction; and *E. coli*: log 5.18 or >99.99% reduction). Creek water upstream from where the spring discharges into the stream contained elevated concentrations of all microbial indicators (enterococci: 71 MPN/100 mL; total coliform: 1221 MPN/100 mL; and *E. coli*: 291 MPN/100 mL).

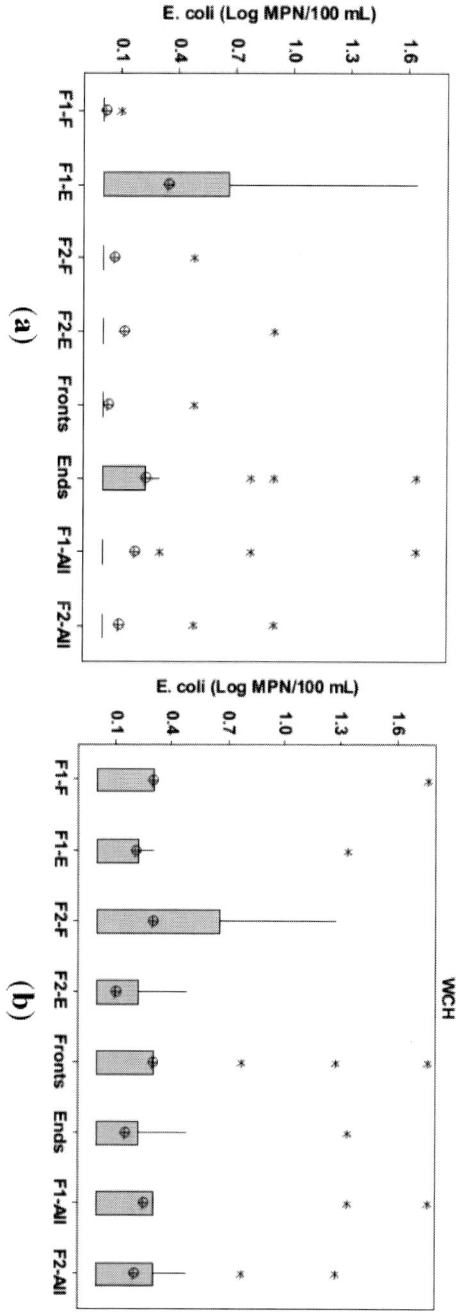

FIGURE 6: *E. coli* concentrations (Log MPN/100 mL) in groundwater beneath the drainfield trenches at James W. Smith (a) and West Craven High (b). Field 1 is F1, Field 2 is F2, front of the field is F, end of the field is E. Data indicate that the *E. coli* concentrations were more variable at James W. Smith relative to West Craven.

The OWS at WCH was also efficient at reducing all measured wastewater FIB concentrations before discharge to groundwater beneath the drainfields (Figure 7). There was a log 2.16 reduction (99.30%) in enterococci concentrations from the septic tank (117,532 MPN/100 mL) to groundwater beneath Field 1 (820 MPN/100 mL). Groundwater enterococci concentrations beneath Field 1 were still significantly ($p = 0.03$) higher than background groundwater (52 MPN/100 mL). Similar to JWS, the OWS at WCH also had elevated geometric mean enterococci (820 MPN/100 mL) relative to the single sample standard for infrequently used recreational waters (151 MPN/100 mL) [27]. However, there was a log 3.75 reduction (99.98%) between the septic tank and groundwater 10 m down-gradient (21 MPN/100 mL) from the drainfield, and at that distance, enterococci concentrations were not significantly different than background groundwater.

There was a log 3.87 reduction (99.98%) in total coliform at WCH between the septic tank (4,859,918 MPN/100 mL) and groundwater beneath Field 1 (744 MPN/100 mL). Total coliform concentrations did not change much 10 m down-gradient from the OWS (949 MPN/100 mL). Background groundwater total coliform concentrations (65 MPN/100 mL) were lower than all other sampling locations, but statistically significant differences were only observed when comparing septic wastewater to background groundwater. Septic wastewater coliform concentrations were significantly ($p < 0.05$) higher than all other sampling locations. The geometric mean total coliform concentrations beneath Field 1 and down-gradient from the OWS were lower than total coliform surface water standards (1000 MPN/100 mL) used by some countries [28].

There was a log 5.63 reduction (>99.99%) in *E. coli* concentrations between the septic tank (2,573,800 MPN/100 mL) and groundwater beneath Field 1 (6 MPN/100 mL), and a log 6.11 reduction (>99.99%) between the tank and groundwater 10 m down-gradient (2 MPN/100 mL) from the OWS at WCH. Septic tank *E. coli* concentrations were significantly higher than all other sampling locations. Groundwater *E. coli* concentrations beneath the OWS were lower than surface water standards for *E. coli* (geometric mean: 126 MPN/100 mL; single sample: 576 MPN/100 mL) [27].

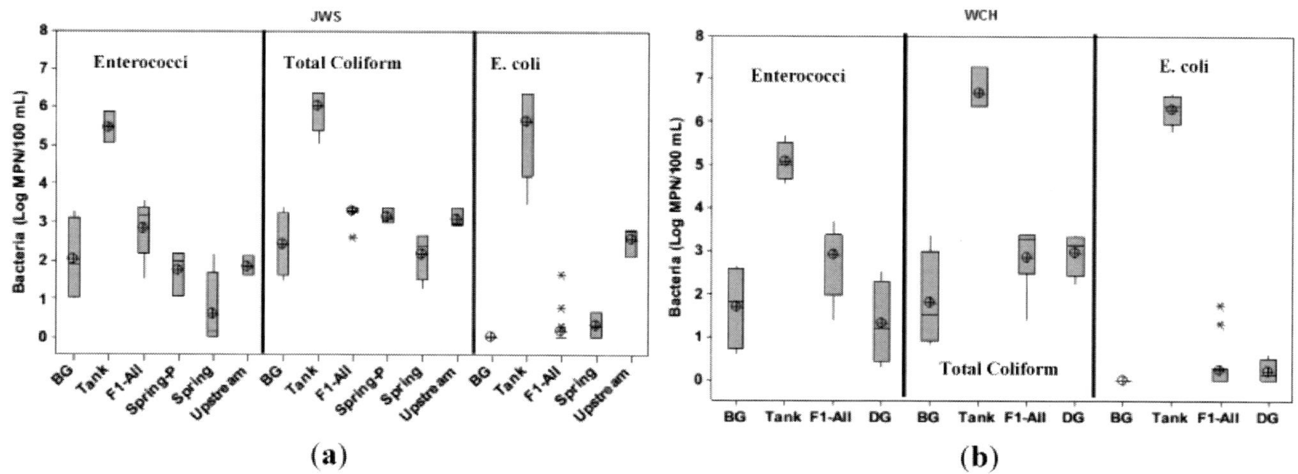

FIGURE 7: Fecal indicator bacteria concentrations (Log MPN/100 mL) data at James W. Smith (a) and West Craven High (b) for background groundwater (BG), septic tank effluent (Tank), groundwater beneath the drainfield with the highest microbial concentrations (F1-All), down-gradient groundwater (DG), spring water (Spring) and stream (Up-stream). Data indicate excellent treatment (>99% reduction) for all indicator bacteria from the septic tanks to groundwater, and further treatment as indicator bacteria concentrations decrease in groundwater down-gradient from the systems. Enterococci reductions are the lowest.

12.3.5 PHYSICAL AND CHEMICAL PARAMETERS

The mean SC at each site was highest for wastewater samples from the septic tanks (JWS: 1057 ± 387 μS/cm; WCH: 1196 ± 432 μS/cm) (Table 2). Groundwater beneath the drainfields and down-gradient from the OWS had SC significantly ($p < 0.05$) higher than background groundwater for both sites (Table 2). The spring at JWS also had elevated SC (620 ± 69 μS/cm) relative to background groundwater (98 ± 51 μS/cm) and stream water (144 ± 12 μS/cm). Groundwater SC trends were similar to microbial concentration trends at JWS. More specifically, the average groundwater SC was higher beneath Field 1 (755 μS/cm) relative to Field 2 (547 μS/cm), and higher towards the end of the trenches (754 μS/cm) relative to the front (547 μS/cm). At WCH, groundwater SC trends were similar to microbe trends because the average SC was elevated near the front of the trenches (521 μS/cm) relative to the ends of the trenches (340 μS/cm). However, there were some dissimilarities as mean groundwater SC was elevated beneath Field 2 (550 μS/cm) relative to Field 1 (312 μS/cm), while microbe concentrations were typically higher beneath Field 1 at WCH. There were weak but significant correlations between SC and enterococci concentrations at JWS ($r = 0.428$, $p = 0.007$) and WCH ($r = 0.387$, $p = 0.014$) and between SC and total coliform concentrations at WC ($r = 0.534$, $p = 0.001$).

Sample pH values were highest for wastewater at each site (JWS: 7.31 ± 0.26; WCH: 6.91 ± 0.28), and were generally similar to or elevated in groundwater beneath the drainfields (JWS: 6.37 to 6.89; WCH: 5.97 to 6.53) relative to background groundwater (JWS: 5.12 ± 0.65; WCH: 6.56 ± 0.99).

Water temperature at JWS was similar for the septic tank wastewater (17.8 ± 2.9 °C), groundwater beneath the drainfields (18.3 to 18.9 °C), and spring (18.2 ± 0.6 °C). The average stream water temperature at JWS was lower (13.7 ± 1.1 °C) than the groundwater and wastewater temperatures.

The average depth to groundwater near the drainfield area at JWS (4.56 to 6.21 m) was greater than at WCH (1.62 to 1.76 m). There was also more variability in groundwater depths between Fields 1 and 2 at JWS (mean difference of 1.55 m) in comparison to WCH (mean difference of 0.14 m).

TABLE 2: Mean and standard deviation (in brackets) physical and chemical properties of water at various sampling locations at James W. Smith and West Craven High. Data indicate higher mean specific conductivity values for groundwater influenced by wastewater.

Site	Sampling Location	Specific Conductivity (μS/cm)	pH	Temp (°C)	Depth to Water (m)
JWS	Field 1	753 (246)	6.37 (0.57)	18.3 (1.1)	4.56 (0.91)
	Field 2	547 (237)	6.89 (0.75)	18.9 (1.1)	6.21 (1.02)
	Front	547 (204)	6.45 (0.80)	18.4 (0.9)	4.92 (1.25)
	Ends	754 (268)	6.78 (0.58)	18.8 (1.2)	5.85 (1.15)
	Background	98 (51)	5.12 (0.65)	18.1 (0.5)	4.32 (0.31)
	Tank	1057 (387)	7.31 (0.26)	17.8 (2.9)	-
	Down-gradient	620 (69)	7.23 (0.27)	17.5 (0.8)	0.34 (0.02)
	Spring	445 (29)	6.91 (0.37)	18.2 (0.6)	-
	Upstream	144 (12)	7.22 (0.25)	13.7 (1.1)	-
WCH	Field 1	312 (401)	5.97 (0.65)	16.4 (2.1)	1.76 (0.16)
	Field 2	550 (302)	6.53 (0.34)	16.7 (1.8)	1.62 (0.11)
	Front	521 (327)	6.32 (0.55)	16.6 (2.0)	1.69 (0.16)
	Ends	340 (397)	6.17 (0.63)	16.4 (1.9)	1.69 (0.16)
	Background	49 (12)	6.56 (0.99)	15.5 (2.2)	1.53 (0.31)
	Tank	1196 (432)	6.91 (0.28)	17.9 (3.5)	-
	Down-gradient	710 (212)	6.53 (0.22)	17.2 (1.5)	1.44 (0.13)

12.4 DISCUSSION

12.4.1 MICROBIAL DISTRIBUTION

Overall, the OWS at WCH had a more uniform distribution of FIB in groundwater beneath the system than the OWS at JWS. More specifically, statistically significant differences ($p < 0.05$) were observed when comparing the groundwater enterococci concentrations near the end of the trenches relative to the front of the trenches at JWS, but not at WCH. Higher concentrations of enterococci concentrations in groundwater towards the ends of the trenches at JWS may have been because of the formation of a biomat near the front of the trenches that reduced the infiltration rate of effluent and bacteria at JWS near the front [17,29,30,31]. Biomats form along the trench bottom and trench sidewalls as solids, microorganisms, and secretions from microorganisms accumulate and begin to clog soil pores and reduce the infiltration rate of wastewater into the soil. Biomat formation is influenced by, dosing characteristics, hydraulic and organic loading rate, age of the system, and aeration status of the infiltrative surface [30,32,33]. While both OWS used pump stations, and thus had effluent dosing and resting times, the OWS at WCH used low-pressure pipe (LPP) distribution. The LPP system may have provided more even distribution of wastewater across the drainfield area, and limited the development of a biomat relative to JWS, or allowed a more even development of a biomat along the trenches. Prior studies have shown that OWS that use LPP can be effective at reducing microbial concentrations, in theory, because of the uniform distribution of effluent and the resting and dosing cycles which promote unsaturated flow [19,20]. Research has also shown that effluent distribution via distribution box, like at JWS, can often be unequal [5,9]. When effluent leaves the distribution boxes and enters the drainfield trenches at JWS, effluent initially leaks out of the closely spaced, relatively large holes in the pipes at the front of the trenches, thus dosing the front the of the trenches first, and resulting in a relatively high hydraulic and organic loading rate near the front of the trenches [30]. Overtime, a biomat forms at the front of the trenches and gradually extends towards the end, as effluent infiltration is reduced via the biomat and effluent is "pushed" along the length of the trench [30]. The OWS at JWS had been in use for 25 years prior to the start of this study. This may explain why groundwater enterococci, total coliform, *E. coli* concentrations and SC were higher towards the end of the trenches at JWS, because of the biomat formation at the front of the trenches that limited infiltration of wastewater near the front. Biomat formation can enhance the treatment of effluent by reducing the infiltration rate of wastewater into soil, creating unsaturated flow conditions beneath the OWS trenches, and increasing the residence time of wastewater in the vadose zone [30,31,32]. However, if the biomat does not extended to the distal end of the trenches, then uneven distribution and treatment of effluent along the trench may occur, potentially influencing groundwater quality dynamics beneath OWS. Biomat formation takes time. The OWS at WCH and JWS had been is use for 15 years and 25 years respectively, prior to the start of this project, so there was more time for biomat formation at JWS. However, the maximum design wastewater loading rate to drainfield trenches at WCH (66.2 L/m^2/d) was greater than at JWS (42.4 L/m^2/d), thus increasing the potential for biomat development at WCH. The higher loading rates at WCH were a reflection of the NC regulations for the design and installation of OWS [21].

There were consistent patterns of elevated enterococci, total coliform and *E. coli* concentrations beneath Field 1 relative to Field 2 at both sites. For the JWS site, this is somewhat expected because there was a significant difference in the depth to groundwater beneath Field 1 (4.56 ± 0.91 m) relative to Field 2 (6.21 ± 1.02 m). Therefore, the vadose zone was on average over 1.6 m thicker beneath Field 2 relative to Field 1 at JWS. Prior studies have shown that bacteria treatment typically improves as the unsaturated zone thickness increases [7,8,12,14,15,26]. Therefore, there was likely more bacterial filtration beneath Field 2, relative to Field 1. At WCH, the depth to groundwater was very similar beneath Field 1 (1.76 ± 0.16 m) and Field 2 (1.62 ± 0.11 m). The groundwater samples were collected during a period (November–May) when the water table is typically highest in eastern NC [22]. Therefore, it is possible the treatment efficiencies of the OWS may improve during the summer, when the water table is lower, and vadose zone is thicker.

There were significant differences in groundwater SC near the end of the drainfield trenches at JWS (754 ± 268 μS/cm) relative to the front of the trenches (547 ± 204 μS/cm), and beneath Field 1 (753 ± 246 μS/cm) in comparison to Field 2 (547 ± 237 μS/cm). Because wastewater contains elevated concentration of dissolved salts and solids, the SC of wastewater is elevated relative to most fresh groundwater, and SC has been used as a tracer for wastewater impacted groundwater [34,35]. The SC patterns were very similar to the microbial indicator patterns at JWS (both higher for Field 1 relative to Field 2, and higher towards the end of the trenches relative to front). At WCH, there were also statistically significant differences ($p < 0.05$) in groundwater SC beneath the front (521 ± 327 μS/cm) and ends of the trenches (340 ± 397 μS/cm), and beneath Field 1 (312 ± 401 μS/cm), relative to Field 2 (550 ± 302 μS/cm). However the differences at WCH (Fronts > Ends: $p = 0.050$; and Field 2 > Field 1: $p = 0.011$) were not as significant as at JWS (Ends > Fronts: $p = 0.0167$; Field 1 >

Field 2: p = 0.0070). Therefore, the groundwater SC also showed a more even distribution at WCH with the LPP system in comparison to JWS with the pump to distribution box system.

While some of the differences in microbial concentrations between Fields 1 and 2 at JWS may have been influenced by the differences in vadose zone thickness, there was still more variation in groundwater microbe concentrations between the front and end of the trenches for each individual field (less variability in vadose zone thickness) at JWS in relation to WCH, indicating that effluent distribution was a key factor in groundwater microbial concentrations.

12.4.2 TREATMENT

There was significant attenuation (>99%) of all microbial indicators between the septic tank and groundwater for both systems. Total coliform were most abundant in all sampling locations at both sites. The OWS at WCH was on average, more efficient (Log 3.87 or 99.98%) at reducing total coliform concentrations prior to discharge to groundwater than the OWS at JWS (Log 2.74 or 99.82%). However, the OWS at JWS was more efficient at reducing enterococci (Log 2.65 or 99.78%) in comparison to WCH (Log 2.16 or 99.30%). The average *E. coli* treatment at the two sites were similar (JWS: 5.06 or >99.99%; WCH: 5.63 or >99.99%). Overall, the OWS at both sites were less efficient at reducing enterococci, relative to the other bacteria, and thus enterococci was the most resilient of the microbial indicators.

While FIB concentration reduction was efficient, groundwater with elevated concentrations of enterococci and total coliform were still observed beneath the OWS trenches at JWS. For example the geometric mean enterococci concentration beneath F1 (667 MPN/100 mL) was elevated relative to the single sample recreational surface water standard for infrequently used waters (151 MPN/100 mL) [27]. The United States Environmental Protection Agency has not set a surface water standard for total coliform, because coliform bacteria grow in soil and are not specific to fecal material, however, other countries such as Columbia, Cuba, and Japan have total coliform standards for recreational waters of 1000 MPN/100 mL [28]. The geometric mean total coliform concentration beneath JWS (Field 1) was 2007 MPN/100 mL, and thus would have exceeded standards established by other countries. However, FIB concentrations at JWS were further reduced as groundwater moved 24 m down-gradient from the OWS (enterococci: 57 MPN/100 mL; total coliform: 1418 MPN/100 mL; *E. coli*: < 1 MPN/100 mL). Surface water microbe concentrations at the spring down-gradient from the OWS (enterococci: 4 MPN/100 mL; total coliform: 149 MPN/100 mL; *E. coli*: 3 MPN/100 mL) were less than or similar to background groundwater concentrations, and surface water standards. Therefore, FIB transport to surface water was limited at JWS because of the combined effects of the vertical separation from OWS trenches to groundwater and the horizontal setback distances from surface water, resulting in efficient overall treatment (enterococci: Log 4.86 or 99.98% reduction; total coliform: Log 3.87 or 99.87% reduction; and *E. coli*: log 5.18 or >99.99% reduction). Creek water upstream from where the spring discharges into the stream contained elevated concentrations of all FIB (enterococci: 71 MPN/100 mL; total coliform: 1221 MPN/100 mL; and *E. coli*: 291 MPN/100 mL) indicating another source of surface water FIB contamination in the watershed.

Groundwater typically had higher enterococci concentrations than *E. coli*, but surface waters typically had higher *E. coli* relative to enterococci. Prior studies have also reported that OWS typically are not as effective at reducing enterococci relative to *E. coli* [8,9,10]. Transport of FIB beyond 10 m from the OWS was very limited, and groundwater FIB concentrations were typically at or below background concentrations 10 to 24 m down-gradient from the OWS. While the spring at JWS did contain detectable levels of all FIB (Enterococci: 4 MPN/100 mL; Total Coliform: 222 MPN/100 mL; *E. coli*: 2 MPN/100 mL), the mean spring FIB concentrations were lower than surface water sampled 5 m upstream from where the spring discharged into the creek (Enterococci: 71 MPN/100 mL; Total Coliform: 1221 MPN/100 mL; *E. coli*: 291 MPN/100 mL). Therefore, other sources of FIB pollution may have been influencing stream water quality. Research has shown that pets [36], livestock [37], wildlife [38], and various other non-point sources of pollution [39] can contribute significant concentrations of FIB to surface waters. There are some livestock farms upstream from JWS, sparse residential development with OWS, and wildlife such as raccoons, birds, and deer have been seen in riparian areas adjacent to the stream. Studies have also shown that surface waters near OWS may have FIB concentrations that are elevated relative to groundwater near properly functioning and malfunctioning systems [8,9,10].

12.5 CONCLUSIONS

Overall, there was a more even distribution of FIB in groundwater beneath the LPP system at WCH in comparison to the pump to distribution box system at JWS. The groundwater SC was also more similar beneath the drainfield at WCH in relation to JWS. Both systems were very effective (>99%) at reducing FIB concentrations before discharge to groundwater, and FIB concentrations were typically at background groundwater concentrations 10 to 24 m down-gradient from the OWS. The OWS at both sites were less efficient at reducing enterococci concentrations in relation to *E. coli* and total coliform treatment.

REFERENCES

1. United States Environmental Protection Agency (USEPA). On-site Wastewater Treatment Systems Manual; EPA/625/R-00/008. USEPA: Washington, DC, USA, 2002; pp. 1–11.
2. Carroll, S.; Hargreaves, M.; Goonetilleke, A. Sourcing faecal pollution from onsite wastewater treatment systems in surface waters using antibiotic resistance analysis. J. Appl. Microbiol. 2005, 99, 471–482.
3. Harman, J.; Robertson, W.D.; Cherry, J.A.; Zanini, L. Impacts on a sand aquifer from an old septic system: Nitrate and phosphate. Groundwater 1996, 34, 1105–1114.
4. Pang, L.; Close, M.; Goltz, M.; Sinton, L.; Davies, H.; Hall, C.; Stanton, G. Estimation of septic tank setback distances based on transport of *E. coli* and F-RNA phages. Environ. Int. 2003, 29, 907–921.
5. Patel, T.; O'Luanaigh, N.O.; Gill, L.W. A Comparison of gravity distribution devices used in On-site domestic wastewater treatment systems. Water Air Soil Pollut. 2008, 191, 55–69.
6. Lowe, K.S.; Rothe, N.K.; Tomaras, J.M.B.; DeJong, K.; Tucholke, M.B.; Drewes, J.; McCray, J.E.; Munakata-Marr, J. Influent constituent characteristics of the modern waste stream rom single sources. In Literature Review; Water Environment Research Foundation: Alexandria, VA, USA, 2007; pp. 3–19.

7. Humphrey, C.P.; O'Driscoll, M.A.; Zarate, M.A. Evaluation of on-site wastewater system Escherichia coli contributions to shallow groundwater in coastal North Carolina. Water Sci. Technol. 2011, 63, 789–795.
8. Conn, K.E.; Habteselassie, M.Y.; Blackwood, A.D.; Noble, R.T. Microbial water quality before and after the repair of a failing onsite wastewater treatment system adjacent to coastal waters. J. Appl. Microbiol. 2011, 112, 214–224.
9. Habteselassie, M.Y.; Kirs, M.; Conn, K.E.; Blackwood, A.D.; Kelly, G.; Noble, R.T. Tracking microbial transport through four onsite wastewater treatment systems to receiving waters in eastern North Carolina. J. Appl. Microbiol. 2011, 111, 835–847.
10. Harris, J.; Humphrey, C.; O'Driscoll, M. Transport of indicator microorganisms from an onsite wastewater system to adjacent stream. Univers. J. Environ. Res. Technol. 2013, 3, 423–426.
11. Cahoon, L.B.; Hales, J.C.; Carey, E.S.; Loucaides, S.; Rowland, K.R.; Nearhoof, J.E. Shellfishing closures in southwest Brunswick County, North Carolina: Septic tanks vs. stormwater runoff as fecal coliform sources. J. Coast. Res. 2006, 22, 319–327.
12. Meeroff, D.; Bloetscher, F.; Bocca, T.; Morin, F. Evaluation of water quality impacts of on-site wastewater treatment and disposal systems on urban coastal waters. Water Air Soil Pollut. 2008, 192, 11–24.
13. Borchardt, M.A.; Po-Huang, C.; DeVries, E.O.; Belongia, E.A. Septic system density and infectious diarrhea in a defined population of children. Environ. Health Perspect. 2003, 111, 742–748.
14. Scandura, J.E.; Sobsey, M.D. Viral and bacterial contamination of groundwater from on-site sewage treatment systems. Water Sci. Technol. 1997, 35, 141–146.
15. Karathanasis, A.D.; Mueller, T.G.; Boone, B.; Thompson, Y.L. Effect of soil depth and texture on fecal bacteria removal from septic effluents. J. Water Health 2006, 4, 395–404.
16. Ijzerman, M.M.; Hagedorn, C.; Reneau, R.B. Microbial tracers to evaluate an on-site shallow-placed low pressure distribution system. Water Res. 1993, 27, 343–347.
17. O'Luanaigh, N.D.; Gill, L.W.; Misstear, B.D.R.; Johnston, P.M. The attenuation of microorganisms in on-site wastewater effluent discharged into highly permeable subsoils. J. Contam. Hydrol. 2012, 142–143, 126–139.
18. Motz, E.C.; Cey, E.; Ryan, M.C.; Chu, A. Vadose zone microbial transport below at-grade distribution of wastewater effluent. Water Air Soil Pollut. 2012, 223, 771–785.
19. Carlile, B.L.; Cogger, C.G.; Sobsey, M.D.; Scandura, J.; Steinbeck, S.J. Movement and Fate of Septic Tank Effluent in Soils of the North Carolina Coastal Plain; Report to the Coastal Plains Regional Commission: Raleigh, NC, USA, 1981; p. 37.
20. Ijzerman, M.M.; Hagedorn, C.; Reneau, R.B. Fecal indicator organisms below an on-site wastewater system with low pressure pipe distribution. Water Air Soil Pollut. 1992, 63, 201–210.
21. North Carolina Division of Environmental Health: On-site Wastewater Section. Laws and Rules for Sewage Treatment and Disposals Systems; Raleigh, NC, USA, 1999. Available online: http://ehs.ncpublichealth.com/oswp/resources.htm (accessed on 11 March 2014).
22. U.S. Department of Agriculture, Natural Resources Conservation Service. Soil Survey of Craven County, North Carolina, USA. 1989. Available online: http://soils.usda.gov/survey/printed_surveys/ (accessed on 11 March 2014).

23. Blott, S.; Pye, K. Gradistat: A grain size distribution and statistics package for the analysis of uncosolidated sediments. Earth Surf. Process. Landf. 2001, 26, 1237–1248.
24. Domenico, P.A.; Schwartz, W. Physical and Chemical Hydrogeology, 2nd ed.; John Wiley & Sons, Inc.: New York, NY, USA, 1998; p. 36.
25. Del Rosario, K.L.; Humphrey, C.P.; Mitra, S.; O'Driscoll, M.A. Nitrogen and carbon dynamics beneath on-site wastewater treatment systems in pitt county, North Carolina. Water Sci. Technol. 2014, 69, 663–671.
26. Humphrey, C.P.; O'Driscoll, M.A. Biogeochemistry of groundwater beneath on-site wastewater systems in a coastal watershed. Univers. J. Environ. Res. Technol. 2011, 1, 320–328.
27. United States Environmental Protection Agency (USEPA). Managing Urban Watershed Pathogen Contamination; EPA/600/R-03/111. USEPA: Cincinnati, OH, USA, 2003.
28. United States Environmental Protection Agency. Ambient Water Quality Criteria for Bacteria; EPA440/5–84–002. USEPA: Washington, DC, USA, 1986.
29. Stevik, T.K.; Aa, K.; Ausland, G.; Hanssen, J.F. Retention and removal of pathogenic bacteria in wastewater percolating through porous media: A review. Water Res. 2004, 38, 1355–1367.
30. Hoover, M.T.; Disy, T.A.; Pfieffer, M.A.; Dudley, N.; Mayer, R.B.; Buffington, B. North Carolina Subsurface Wastewater Operators Training School Manual; 1996.
31. Beal, C.D.; Gardner, E.A.; Kirchhof, G.; Menzies, N.W. Long-term flow rates and biomat zone hydrology in soil columns receiving septic tank effluent. Water Res. 2006, 40, 2327–2338.
32. Siegrist, R.; Boyle, W.C. Wastewater-induced soil clogging development. J. Environ. Eng. 1987, 113, 550–566.
33. Stevik, T.K.; Ausland, G.; Jenssen, P.D.; Siegrist, R. Removal of *E. coli* during intermittent filtration of wastewater effluent as affected by dosing rate and media type. Water Res. 1999, 33, 2088–2098.
34. Alhajjar, B.J.; Chesters, G.; Harkin, J.M. Indicators of Chemical Pollution from Septic Systems. Groundwater 1990, 28, 559–568.
35. Humphrey, C.P.; Deal, N.E.; O'Driscoll, M.A.; Lindbo, D.L. Characterization of on-site wastewater nitrogen plumes in shallow coastal aquifers, North Carolina. In Proceedings of the 2010 World Environmental & Water Resources Congress, Providence, RI, USA, 16–20 May 2010; pp. 949–958.
36. Wright, M.E.; Solo-Gabriele, H.M.; Elmir, S.; Fleming, L.E. Microbial load from animal feces at a recreational beach. Mar. Pollut. Bull. 2009, 58, 1649–1656.
37. Liwimbi, L.; Graves, A.K.; Israel, D.W.; van Heugten, E.; Robinson, B.; Cahoon, C.W.; Lubbers, J.F. Microbial source tracking in a watershed dominated by swine. Water 2010, 2, 587–604.
38. Whitlock, J.E.; Jones, D.T.; Harwood, V.J. Identification of the sources of fecal coliforms in an urban watershed using antibiotic resistance analysis. Water Res. 2002, 36, 4237–4282.
39. Sanders, E.C.; Yuan, Y.; Pitchford, A. Fecal coliform and *E. coli* concentrations in effluent-dominated streams of the Upper Santa Cruz Watershed. Water 2013, 5, 243–261.

CHAPTER 13

Detection of Retinoic Acid Receptor Agonistic Activity and Identification of Causative Compounds in Municipal Wastewater Treatment Plants in Japan

KAZUKO SAWADA, DAISUKE INOUE, YUICHIRO WADA, KAZUNARI SEI, TSUYOSHI NAKANISHI, AND MICHIHIKO IKE

13.1 INTRODUCTION

Retinoic acid (RA) receptors (RARs) are nuclear receptors whose specific natural ligands are all-trans RA (atRA) and 9-cis RA (9cRA) derived from retinoid (vitamin A) precursors [1]. Retinoic acid receptors control aspects of vision, cell differentiation, immune response, and embryonic development in vertebrates [1, 2]. By contrast, ligands of RARs are known as potential teratogens in developing vertebrate embryos. Previous studies have demonstrated that excess RAs cause a spectrum of teratogenesis in many animals, including fish [3, 4], amphibians [5–7], and mammals [8, 9].

Detection of Retinoic Acid Receptor Agonistic Activity and Identification of Causative Compounds in Municipal Wastewater Treatment Plants in Japan. © *Sawada K, Inoue D, Wada Y, Sei K, Nakanishi T, and Ike M.* Environmental Toxicology and Chemistry ***31****,2 (2012), DOI: 10.1002/etc.741. Reprinted with permission from the authors.*

Accordingly, the environmental occurrence of RAR agonists that disrupt RAR signaling may cause detrimental effects in humans and wild animals.

In recent years, RAR agonist contamination has been detected in aquatic environments. Gardiner et al. [10] first detected RAR agonistic activity in a Minnesota lake and a California pond in North America, where deformed frogs were frequently discovered. We have recently detected RAR agonistic activity in wastewater treatment plants (WWTPs) and rivers in Beijing, China [11] and in the Kinki region of Japan [12–14]. Despite the detection of RAR agonist contamination in various environments, there has been little available information regarding the causative contaminants. Recently, the contaminants responsible for the majority of the RAR agonistic activity in sewage in Beijing, China were identified as 4-oxo-atRA and 4-oxo-13-cis RA (4-oxo-13cRA), which are oxidative metabolites of RAs [11]. However, whether 4-oxo-RAs are typically the major RAR agonists in the environment or if other RAR agonists exist is unknown. These questions should be clarified for detailed assessment of possible risks of RAR agonist contamination.

In the present study, we examined the RAR agonistic activity in several municipal WWTPs in Osaka, Japan, using the RARα yeast two-hybrid assay. Because significant activity was commonly detected in WWTP influents, we identified the major causative compounds by applying bioassay-directed high-performance liquid chromatography (HPLC) fractionations followed by liquid chromatography ion trap time-of-flight mass spectrometry (LC/MS-IT-TOF) analysis. The identified compounds were then monitored in municipal WWTPs to determine their contribution to the total RAR agonistic activity in municipal wastewater and their fates during the activated sludge treatment process.

13.2 MATERIALS AND METHODS

13.2.1 CHEMICALS

All-trans RA (atRA), 9cRA, and 13cRA were purchased from Sigma-Aldrich. Ammonium acetate, dimethylsulfoxide (DMSO), and formic acid were purchased from Wako Pure Chemical Industries. Acetonitrile was purchased from Kanto Chemical. Methanol (MeOH) was purchased from Sigma-Aldrich or Kanto Chemical. 4-Hydroxy-atRA, 4-oxo-atRA, 4-oxo-9cRA, and 4-oxo-13cRA were purchased from Toronto Research Chemicals. Methanol, acetonitrile, and ammonium acetate were of HPLC grade and formic acid was of liquid chromatography/mass spectrometry (LC/MS) grade, while the other chemicals were of the highest grade commercially available.

13.2.2 WASTEWATER SAMPLES

Grab samples of influent (after the primary settling tank) and effluent (after the final settling tank) were collected from six municipal WWTPs (WWTPs-A, -B, -C, -D, -E, and -F) in Osaka, Japan during the period from December 2008 to March 2010. WWTPs-A, -B, -C, and -D employ conventional activated sludge process for biological treatment, while WWTP-E and WWTP-F utilize anaerobic-oxic and anaerobic-anoxic-oxic processes, respectively (Supplemental Data, Fig. S1). All the investigated WWTPs receive mainly domestic wastewater and small amounts of industrial discharges. All samples were collected in the morning or early in the afternoon (between 10 am and 2 pm), transported to the laboratory on ice, and subjected to solid phase extraction (SPE) within 6 h. The quality of influent and effluent samples was as follows: pH, 6.8 to 7.8 and 6.2 to 7.1; dissolved organic carbon, 13.2 to 88.0 mg/L and 3.5 to 21.6 mg/L; total nitrogen, 22.5 to 43.7 mg-N/L and 4.9 to 21.6 mg-N/L, respectively (see Supplemental Data, Table S1 for detailed water quality data).

13.2.3 SAMPLE PREPARATION

Samples used in the present study were extracted and concentrated by SPE as previously described by Inoue et al. [13] with minor modifications. Briefly, filtered samples were passed through the Oasis HLB cartridges (6 ml/500 mg, Waters), which were preconditioned with 6 ml of MeOH and 6 ml of ultrapure water (UPW). After sample loadings, the cartridges were washed with 6 ml of UPW (for samples used in the yeast two-hybrid assays and identification of RAR agonists) or 6 ml of 40% acetonitrile (for samples used in the LC/MS analyses of identified RAR agonists) and dried. Subsequently, the absorbed substances were eluted with 6 ml of MeOH. After evaporation under a gentle nitrogen stream, the dried residues were dissolved in DMSO (for yeast two-hybrid assays) or MeOH (for chemical analyses), resulting in concentration factors of 10,000 and 5,000-fold, respectively, when compared with the original sample. The concentrated samples were stored in the dark at −20°C until use.

13.2.4 YEAST TWO-HYBRID ASSAY

Retinoic acid receptor agonistic activities of the wastewater samples and identified compounds were measured by a yeast two-hybrid assay employing the recombinant yeast, *Saccharomyces cerevisiae* Y190, which contained human RARα and the coactivator transcriptional intermediary factor 2 15. Assays were conducted as previously described 13. The final concentration factors of the wastewater samples applied to the assay system were set at 0.1 to 100 by appropriately diluting 10,000-fold concentrated samples, while the concentrations of identified compounds ranged from 10 pM to 10 μM. Negative control experiments were performed with 1% DMSO without any sample. Positive controls with atRA dissolved in DMSO at varying concentrations (10 pM to 1.0 μM) were included in the assays of wastewater samples. All assays were conducted in triplicate. The relative RAR agonistic activity (%) of the wastewater sample was calculated by setting the maximum mean β-galactosidase activity of atRA to 100% and the mean activity of the negative control to 0%. For the determination of RAR agonistic potency of identified RAR agonists, their maximum β-galactosidase activities were set to 100%. The concentration factor of wastewater sample and concentration of identified RAR agonist that gave 50% relative RAR agonistic activity (EC50) were calculated by the Prism 5J for Windows program (GraphPad Software). The atRA equivalents (atRA-EQ_{bio}) of wastewater samples and atRA equivalency factors of identified RAR agonists were estimated by comparing their EC50 values with that of atRA.

13.2.5 REVERSE-PHASE HPLC FRACTIONATION

Reverse-phase HPLC fractionation of the samples was performed in three steps. First fractionation was performed using a Shimadzu LC-10Avp HPLC system as previously described by Inoue et al. [13], while the second and third fractionations were performed using a Shimadzu Prominence ultra-fast LC system composed of an SIL-20A automatic sampler, two LC-20AD solvent delivery units, a DGU-20A$_3$ degasser, a CTO-20A column oven, an SPD-20A ultraviolet (UV)/visible (VIS) detector, and a CBM-20A communications bus module (Shimadzu). The chromatographic separations were performed with a Shim-Pack VP-ODS column (250 × 4.6 mm inner diameter [i.d.]; particle size, 5 μm; Shimadzu) in the first and second fractionations and with a Shim-Pack VP-ODS column (150 × 4.6 mm i.d.; particle size, 5 μm; Shimadzu) in the third fractionation, both of which were maintained at 40°C.

In the first HPLC fractionation, 100 μl of the extracted sample in MeOH was injected into the column and eluted over 45 min at a flow rate of 1.0 ml/min, with UV detection at 254 nm. Acetonitrile/UPW was used as the mobile phase with the following gradient of acetonitrile: 0 to 3 min, 20%; 3 to 30 min, a linear gradient from 20 to 100%; 30 to 40 min, 100%; 40 to 45 min, 20%. The discrete fractions were collected at 2-min intervals, dried, and then dissolved in DMSO (for yeast two-hybrid assay) or MeOH (for next HPLC fractionation). In the second HPLC fractionation, 50 μl of the bioactive fraction obtained in the first fractionation was injected and eluted over 45 min at a flow rate of 1.0 ml/min, with UV detection at 360 nm. Acetonitrile/1% formic acid was used as the mobile phase with the same gradient conditions of acetonitrile as those of the first HPLC fractionation. The discrete fractions were collected at 0.5-min intervals, dried, and dissolved in DMSO or MeOH. In the third HPLC fractionation, 10 μl of the bioactive fraction obtained in the second fractionation was injected and eluted over 30 min at a flow rate of 1.0 ml/min, with UV detection at 360 nm. The MeOH/100 mM ammonium acetate was used as the mobile phase with the following gradient of MeOH: 0 to 5 min, 70%; 5 to 26 min, a linear gradient from 70 to 97%; 26 to 28 min, 97%; 28 to 30 min, 70%. All of the detected peaks were isolated, dried, and dissolved in MeOH for LC/MS-IT-TOF analysis.

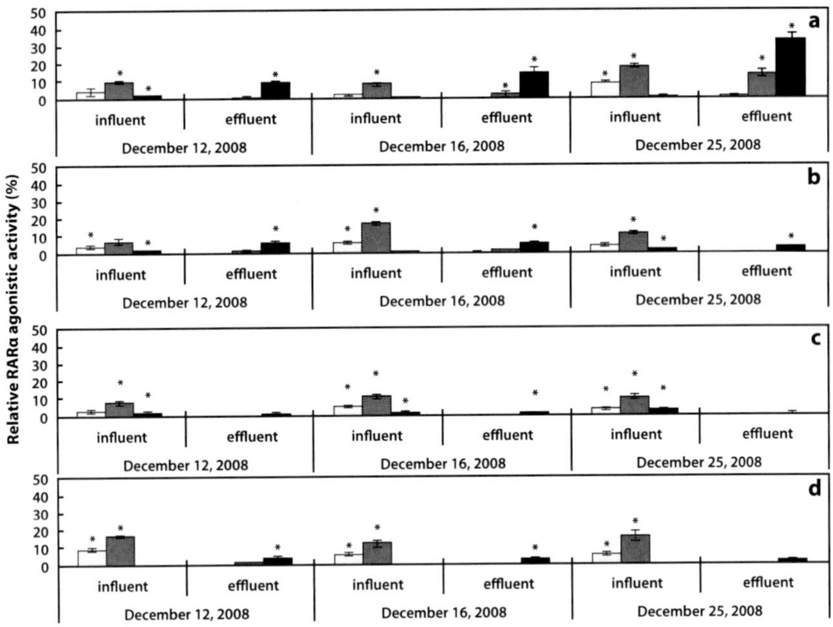

FIGURE 1: Retinoic acid receptor α (RARα) agonistic activity of influent and effluent samples collected from wastewater treatment plants-A (a), -C (b), -E (c), and -F (d) in December 2008. Data are shown as mean±standard deviation in triplicate measurements. Open, light-colored, and filled bars indicate the data at concentration factors of 1, 10, and 100, respectively. *p<0.05 versus the negative control.

13.2.6 LC/MS-IT-TOF ANALYSIS TO IDENTIFY RAR AGONISTS

The bioactive peaks obtained in the third HPLC fractionation were analyzed using a Shimadzu LCMS-IT-TOF system consisting of an SIL-20AC automatic sampler, two LC-20AD solvent delivery units, a DGU-20A3 degasser, a CTO-20AC column oven, an SPD-20AV UV/VIS detector, a CBM-20A communications bus module, and LC/MS-IT-TOF hybrid mass spectrometer with an electrospray ionization (ESI) interface (Shimadzu). A standard (2 µl) or sample (10 µl) dissolved in MeOH was injected into the LC/MS-IT-TOF system. Chromatographic separation was performed with a Shim-Pack VP-ODS column (150 × 2.0 mm i.d.; particle size, 5 µm; Shimadzu) maintained at 40°C. MeOH/10 mM ammonium acetate was used as the mobile phase with the following gradient of MeOH: 0 to 5 min, 70%; 5 to 20 min, a linear gradient from 70 to 100%; 20 to 30 min, 70%. The flow rate of the mobile phase was 0.2 ml/min. The mass spectrometer was operated in the scan and selected ion monitoring modes with the following conditions: ionization mode, positive ion ESI; nebulizer gas, nitrogen; nebulizer gas flow, 1.5 l/min; probe voltage, 4.5 kV; block heater temperature, 200°C; curved desolvation line temperature, 200°C; scan range, m/z 150 to 500; monitoring ion, m/z 301.22 (RAs) and 315.20 (4-oxo-RAs). All the analyses were entrusted to Shimadzu.

13.2.7 QUANTIFYING RAS AND 4-OXO-RAS BY LC/MS ANALYSIS

A quantitative method for RAs and 4-oxo-RAs in untreated and treated wastewater samples was established by the combination of SPE, HPLC fractionation, and LC/MS analysis. Details of establishing the quantitative method and the recovery experiments to determine the quantification limits and recoveries of RAs and 4-oxo-RAs are provided in the Supplemental Data. In the established method, LC/MS analysis was conducted using a Shimadzu LCMS-2010EV system consisting of a Shimadzu Prominence UFLC system and an LCMS2010EV single quadruple mass spectrometer with an ESI interface (Shimadzu). Aliquots (10 µl) of standard or sample dissolved in MeOH were injected into the LC/MS system. Conditions for the separation column and mobile phase were the same as those of the third HPLC fractionation described above. The mass spectrometer was operated in the selected ion monitoring mode with the following conditions: ionization mode, negative ion ESI; nebulizer gas, nitrogen; nebulizer gas flow, 1.5 l/min; probe voltage, −3.5 kV; block heater temperature, 200°C; curved desolvation line temperature, 250°C; monitoring ion, m/z 299 (at 20.1 and 19.0 min for atRA and 13cRA, respectively) and 313 (at 5.9 and 7.0 min for 4-oxo-atRA and 4-oxo-13cRA, respectively).

13.2.8 STATISTICAL ANALYSIS

Statistical analysis was conducted using SPSS software (Ver 15.0 for Windows). Comparisons of the RAR agonistic activity of the wastewater samples before HPLC fractionation with that of the negative control and between the activities of influent and effluent samples in a WWTP were performed using a one-way analysis of variance and then tested by Dunnett's post-hoc test. Differences were considered statistically significant at $p<0.05$.

13.3 RESULTS AND DISCUSSION

13.3.1 RAR AGONISTIC ACTIVITY IN WWTPS

To elucidate the occurrence of RAR agonists in WWTPs employing different biological treatment processes, the RARα agonistic activity of influent and effluent samples was examined in WWTPs-A, -C, -E, and -F on three different dates in December 2008 (Fig. 1). The concentration factors of influent and effluent samples in the assays were set at 1, 10, and 100-fold compared with the original samples.

Significant RARα agonistic activity ($p<0.05$) was detected in all influent samples, irrespective of the WWTP and sampling date. The relative RARα agonistic activities increased with the sample concentration factor from 1 to 10 in all tested influent samples; however, the activities in influent samples at a concentration factor of 100 were lower than those at a concentration factor of 10. Because optical density values at 620 nm in the assay system (an indicator of yeast cell density) for influent samples at a concentration factor of 100 were markedly lower than those at concentration factors of 1 and 10 (data not shown), the reduction of RARα agonistic activities at the concentration factor of 100 can be attributed to the toxic effect on the yeast cell by highly concentrated constituents in influent samples.

Comparison of the RARα agonistic activity of the influent and effluent samples at a concentration factor of 10 revealed that the activities were lowered by all the activated sludge treatments investigated, although the reductions were statistically significant ($p<0.05$) in 7 of the 12 cases. In effluent samples, significant RARα agonistic activity ($p<0.05$) was detected at concentration factors of 1 to 100 in all but three samples, which were obtained from WWTP-E on December 12 and 25, 2008 and from WWTP-F on December 25, 2008. RARα agonistic activity was especially high in WWTP-A on December 25, 2008 (Fig. 1a), with significant activity ($p<0.05$) detected even at a concentration factor of 10. Estimation of atRA-EQbio values based on the RARα agonistic activities revealed that the highest atRA-EQbio values of the influent and effluent samples were respectively 319 and 158 ng-atRA/L in WWTP-A on December 25, 2008.

To clarify the temporal variation of RARα agonist contamination in WWTP influent and effluent, investigations of the RARα agonistic activity were carried out on 10 distinct dates from December 2008 to March 2010 in WWTP-A (Fig. 2). Significant RARα agonistic activity ($p<0.05$) was detected in all influent and effluent samples, although the contamination level varied temporally. In addition, comparison of the activity of the influent and effluent samples at a concentration factor of 10 on the same sampling date showed that the activity was lowered significantly ($p<0.05$) during the biological treatment on all the sampling dates, except March 3 and November 20, 2009.

These results suggest that RARα agonists are consistently present in municipal wastewater in Japan and that they are partially removed by activated sludge treatment, although further confirmation with composite samples is required. Retinoic acid receptor α agonist levels observed in the present study were higher than those reported in sewage treatment plants in Beijing, China, where the highest atRA-EQ values in the influent and effluent were 13.2 ng-atRA/L and 3.4 ng-atRA/L, respectively [11], suggesting greater contamination in our survey area.

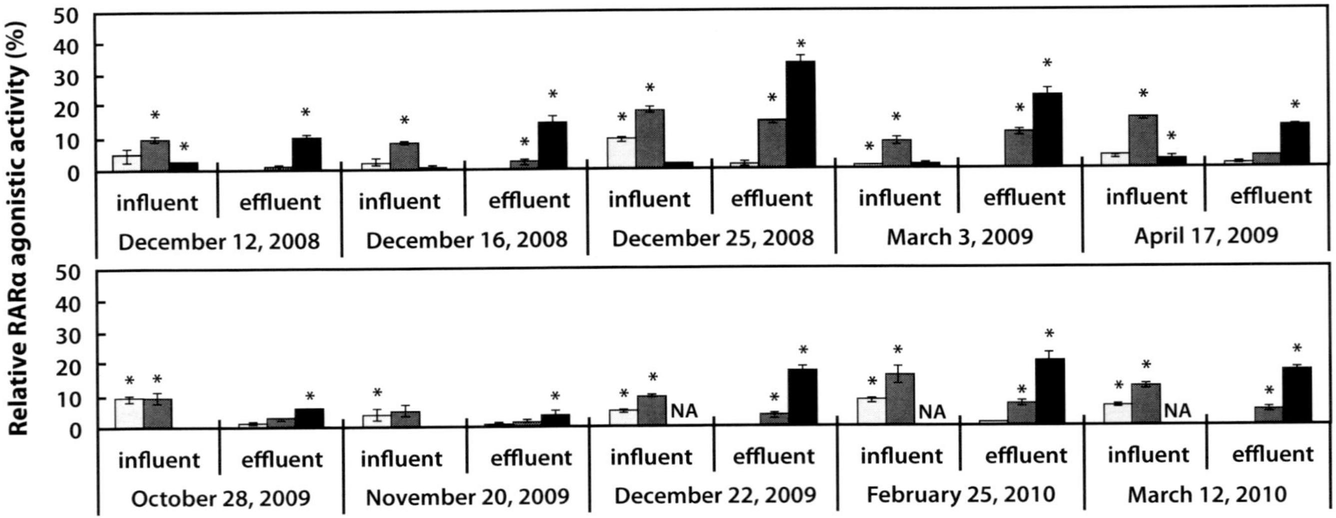

Figure 2. Temporal variations of retinoic acid receptor α (RARα) agonistic activity of influent and effluent samples collected from wastewater treatment plant-A. Data are shown as mean ± standard deviation in triplicate measurements. Open, light-colored, and filled bars indicate the data at concentration factors of 1, 10, and 100, respectively. *$p<0.05$ versus the negative control. NA = not analyzed.

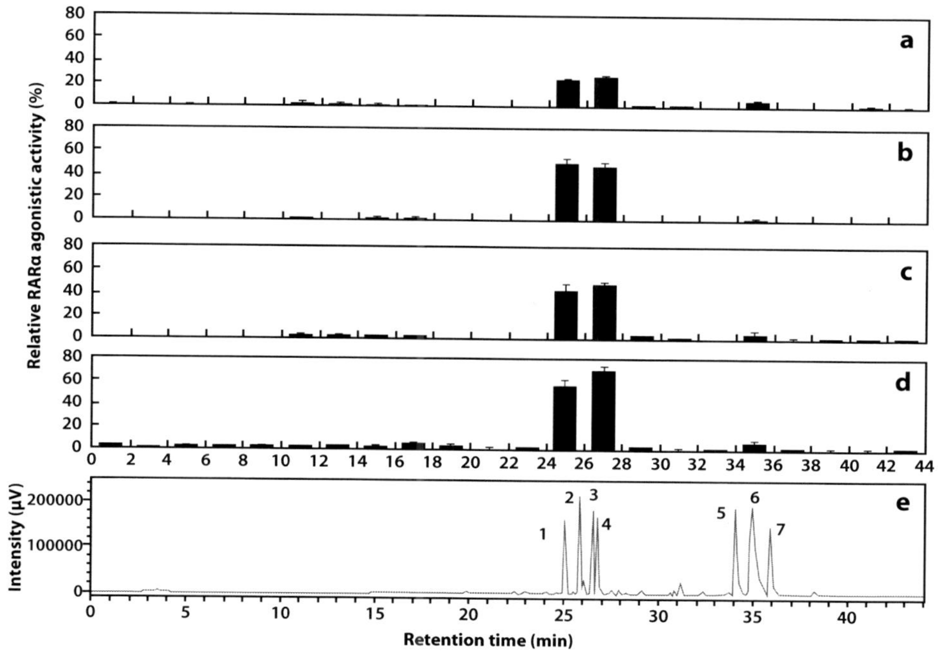

FIGURE 3: Retinoic acid receptor α (RARα) agonistic activity of high-performance liquid chromatography (HPLC) fractions in the first HPLC fractionation of influent samples collected from wastewater treatment plants-A (a), -C (b), -E (c), and -F (d) in June 2009. Results at a concentration factor of 100 are presented. Data are shown as mean ± standard deviation in triplicate measurements. Bottom panel (e) shows the HPLC chromatograms of 4-hydroxy-retinoic acid (4-hydroxy-RA, peak 1), 4-oxo-all-trans RA (4-oxo-atRA, peak 2), 4-oxo-13-cis RA (4-oxo-13cRA, peak 3), 4-oxo-9-cis RA (4-oxo-9cRA, peak 4), 13cRA (peak 5), atRA (peak 6), and 9cRA (peak 7). The detection wavelength was 360 nm

13.3.2 IDENTIFYING MAJOR RAR AGONISTS IN WASTEWATER

To characterize the RARα agonists present in municipal wastewater, influent samples collected from WWTPs-A, -C, -E, and -F in June 2009 were fractionated by reverse-phase HPLC, and fractions collected every 2 min from 0 to 44 min were subjected to the yeast assay. Irrespective of the WWTP, two consecutive fractions between 24 and 26 min and between 26 and 28 min (bioactive fraction I) showed the highest RARα agonistic activity (Fig. 3a–d). In addition, a fraction between 34 and 36 min (bioactive fraction II) also showed marked RARα agonistic activity in all investigated WWTPs. Similar results were observed in repeated investigations in WWTP-A in March, April, and July 2009 (Supplemental Data, Fig. S2). These results suggest the presence of multiple and common RARα agonists in the investigated WWTPs, which were detected in bioactive fractions I and II. Considering that the investigated WWTPs receive mainly domestic wastewater and the causative RARα agonists are commonly present in the wastewater, it is highly likely that the causative RARα agonists in wastewater are natural compounds in sewage. To identify the potential compounds, natural RAs and the metabolites that are known RAR agonists [16, 17] were analyzed by reverse-phase HPLC under the same conditions as those in the aforementioned HPLC fractionation. All-trans retinoic acid, 9cRA, 13cRA, 4-hydroxy-atRA, 4-oxo-atRA, 4-oxo-9cRA, and 4-oxo-13cRA were detected at retention times (RTs) of 35.0, 35.9, 34.1, 25.1, 25.9, 26.8, and 26.5 min, respectively (Fig. 3e). Thus, the peaks of 4-oxo-RAs and 4-hydroxy-atRA overlapped with the bioactive fraction I and those of the RAs overlapped with the bioactive fraction II (Fig. 3), suggesting the contribution of these compounds to the RARα agonistic activities in wastewater.

To identify the RARα agonists present in bioactive fractions I and II, these fractions were obtained from an influent sample collected from WWTP-A in July 2009 and subjected to further HPLC fractionations under distinct conditions. In the second HPLC fractionations, peaks were detected at RTs of 21 to 27 min and RTs of 31 to 34 min in the bioactive fractions I and II, respectively (Supplemental Data, Fig. S3a,c). Fractionations at 0.5-min intervals revealed a marked RARα agonistic activity in fractions between 22.5 to 25.0 min (bioactive fraction I′) for the bioactive fraction I and between 31.5 and 33.0 min (bioactive fraction II′) for the bioactive fraction II (Supplemental Data, Fig. S3b,d). When the bioactive fractions I′ and II′ were subjected to the third HPLC fractionation, four peaks (A, B, C, and D) and five peaks (E, F, G, H, and I) were detected, respectively (Fig. 4a,c). Among these peaks, five peaks A (RT 5.9 min), B (RT 7.0 min), C (RT 8.4 min), E (RT 18.9 min), and F (RT 19.9 min) exhibited a strong RARα agonistic activity (Fig. 4b,d). High-performance liquid chromatography analysis of RAs, 4-oxo-RAs, and 4-hydroxy-atRA under the same conditions as those in the third HPLC fractionation revealed that peaks of 4-oxo-atRA, 4-oxo-13cRA, 13cRA, and atRA overlapped with bioactive peaks A, B, E, and F, respectively, suggesting that these RAs and 4-oxo-RAs would be the causative RARα agonists in municipal wastewater. By contrast, a candidate compound for bioactive peak C was not revealed in the present study and must be investigated further.

Thus, peaks A, B, E, and F were subjected to LC/MS-IT-TOF analyses together with the candidate compounds (4-oxo-atRA, 4-oxo-13cRA, 13cRA, and atRA, respectively). Peak A showed a protonated molecular ion of m/z 315.20 at RT of 5.0 min similar to 4-oxo-atRA (Fig. 5a,b). Peak B showed a protonated molecular ion of m/z 315.20 at RT of 5.7 min similar to 4-oxo-13cRA (Fig. 5c,d). Peak E showed a protonated molecular ion of m/z 301.22 at RT of 17.9 min similar to 13cRA (Fig. 5e,f). Peak F showed a protonated molecular ion of m/z 301.22 at RT of 17.1 min similar to atRA (Fig. 5g,h). These results confirmed that bioactive peaks A, B, E, and F were 4-oxo-atRA, 4-oxo-13cRA, 13cRA, and atRA, respectively. Although the presence of 4-oxo-atRA and 4-oxo-13cRA has been reported previously in Beijing, China [11], the present study revealed that atRA and 13cRA, the parent compounds of 4-oxo-RAs, are also present in municipal wastewater.

Identified RARα agonists are likely to have originated from human urine. It has been shown that RAs and their oxidative metabolites are eliminated from human and animal bodies through urinary excretion, mainly as glucuronide conjugates (retinoyl-β-glucuronides) [18, 19], which do not have high binding affinity to RARs [20]. Because the glucuronide conjugates are deconjugated easily by microbial activity [21, 22], it is likely that RAs and 4-oxo-RAs with RAR agonistic activity can be produced through the deconjugation of their glucuronides in the sewage system and WWTPs.

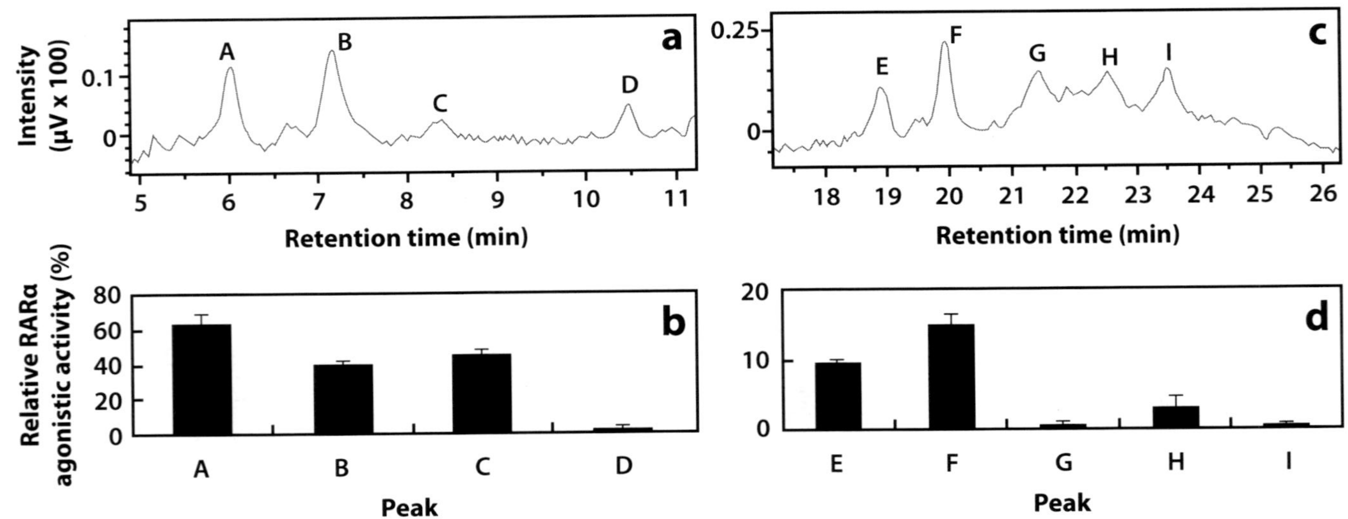

FIGURE 4: High-performance liquid chromatography (HPLC) chromatograms of bioactive fractions I′ (a) and II′ (c), which were obtained in the second HPLC fractionation of influent sample collected from wastewater treatment plant-A in July 2009, in the third HPLC fractionation and retinoic acid receptor α (RARα) agonistic activity of peaks isolated from bioactive fractions I′ (b) and II′ (d). Concentration factors of the samples in HPLC analyses and of RARα agonistic activity measurements were 5,000 and 100, respectively. The detection wavelength in HPLC analyses was 360 nm. Data for RARα agonistic activity are shown as mean ± standard deviation in triplicate measurements.

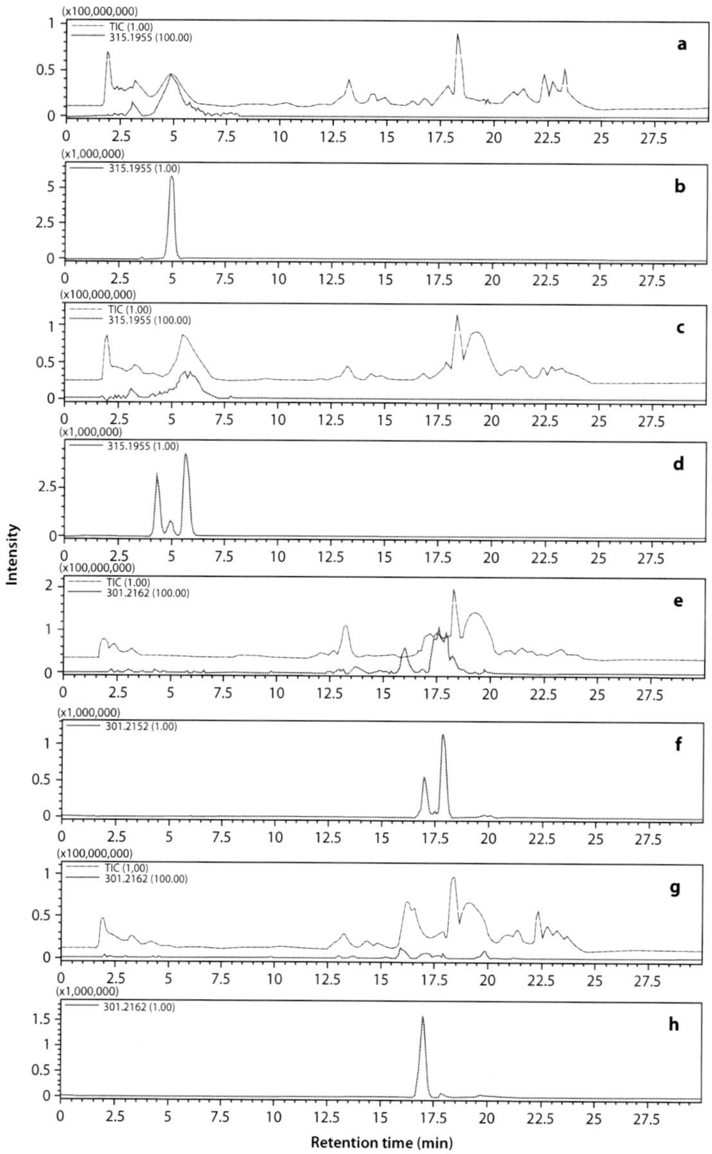

FIGURE 5: Mass chromatograms of peaks A (a), B (c), E (e), and F (g) obtained from third high-performance liquid chromatography fractionation and 4-oxo-all-trans retinoic acid (4-oxo-atRA [b]), 4-oxo-13-cis RA (4-oxo-13cRA [d]), 13cRA (f) and atRA (h) analyzed by liquid chromatography ion trap time-of-flight mass spectrometry system.

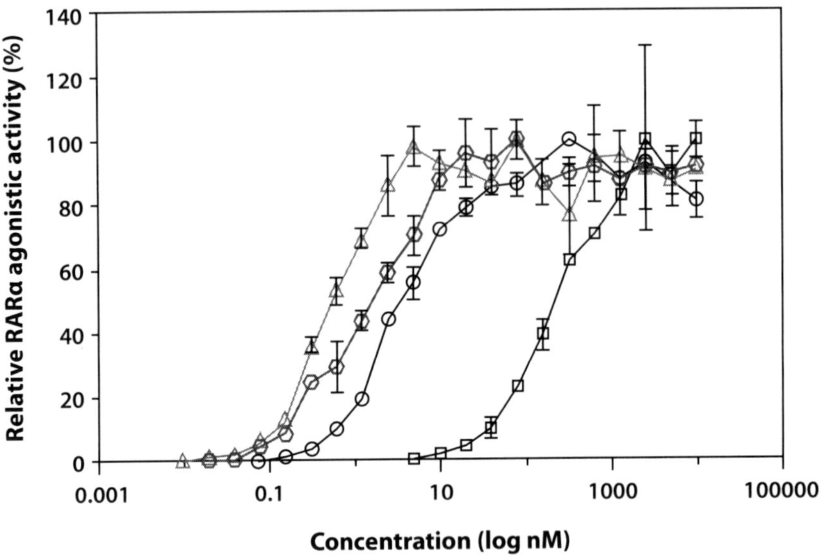

FIGURE 6: Dose–response curves of retinoic acid (RA) receptor α agonistic activity of all-trans RA (atRA, circle), 13-cis RA (13cRA, square), 4-oxo-atRA (triangle), and 4-oxo-13cRA (diamond). Data are shown as mean ± standard deviation in triplicate measurements.

13.3.3 RAR AGONISTIC ACTIVITIES OF IDENTIFIED RAR AGONISTS

Retinoic acid receptor α agonistic activity of four identified RARα agonists, namely, atRA, 13cRA, 4-oxo-atRA, and 4-oxo-13cRA, was evaluated by the yeast two-hybrid assay. Figure 6 shows the dose–response curves of RARα agonistic activity of the four compounds, where the highest mean activity of each compound was set at 100%. Based on the dose–response curves, the EC50 values for atRA, 13cRA, 4-oxo-atRA, and 4-oxo-13cRA were estimated to be 1.6 nM, 240 nM, 0.58 nM, and 4.2 nM, respectively. Thus, the atRA equivalency factors for atRA, 13cRA, 4-oxo-atRA, and 4-oxo-13cRA were calculated to be 1, 0.0067, 2.8, and 0.38, respectively.

The RAR agonistic potencies of RAs and 4-oxo-RAs determined here were roughly comparable to those obtained in the yeast two-hybrid assays of Zhen et al. [11] and Wu et al. [23]. In addition, based on the maximum mean β-galactosidase activities induced by four compounds, the maximum RARα agonistic activities of 13cRA, 4-oxo-atRA, and 4-oxo-13cRA were estimated to be 1.3 times lower, 1.3 times higher, and 1.2 times higher, respectively, than that of atRA (data not shown).

13.3.4 FATES OF RAS AND 4-OXO-RAS IN WWTPS

The fates of RAs, 4-oxo-RAs, and total RARα agonistic activity in six WWTPs (WWTPs-A to -F) were investigated in March 2010 using a newly established quantitative method for RAs and 4-oxo-RAs (Supplemental Data, Fig. S4) and yeast two-hybrid assay. Because grab samples of influent and effluent collected at similar time were applied in the present study, concentrations of RAs and 4-oxo-RAs and total RARα agonistic activity in the samples were not necessarily representative; thus, we tentatively determined the fates and removal efficiencies of RAs, 4-oxo-RAs, and total RARα agonistic activity in WWTPs based on the results of these investigations.

Concentrations of RAs and 4-oxo-RAs and atRA-EQchem values, which were estimated from measured concentrations and atRA equivalency factors of RAs and 4-oxo-RAs, are shown in Table 1 with and without correction applying the recoveries of RAs and 4-oxo-RAs given in the Supplemental Data. Retinoic acid all-trans equivalents values, which were estimated from the RARα agonistic activity detected with the yeast assay, are also presented in Table 1. In influent samples, atRA, 4-oxo-atRA, and 4-oxo-13cRA were detected, and their concentrations without recovery correction ranged from <1.0 to 5.0 ng/L, <0.5 to 5.3 ng/L, and 1.7 to 70.2 ng/L, respectively. Their concentrations were from <4.3 to 21.5 ng/L, <0.6 to 6.9 ng/L, and 2.3 to 94.3 ng/L with recovery correction, respectively. By contrast, concentrations of 13cRA were below the quantification limit (<5.0 ng/L and <13.6 ng/L without and with recovery correction, respectively) in all WWTPs although the compound was detected at concentrations below the quantification limit but above the detection limit in WWTPs-A (2.7 ng/L without recovery correction) and -E (1.5 ng/L without recovery correction). As a result, atRA-EQchem values in influent samples were estimated to be from 3.1 to 44.1 ng/L and 9.1 to 66.0 ng/L without and with recovery correction, respectively. Zhen et al. [11] reported that without recovery correction the concentrations of 4-oxo-atRA and 4-oxo-13cRA were from 4.7 to 10.4 ng/L and 2.3 to 7.1 ng/L, respectively, and atRA-EQchem were estimated to be from 18 to 41 ng/L in WWTP influents in Beijing, China. Thus, the contamination levels of our samples appeared to be comparable to previously reported contamination levels. The percentage of the atRA-EQchem value (without including the recoveries of RAs and 4-oxo-RAs) to the atRAEQbio value in influent samples ranged from 57 to 278%, except for WWTPs-A (16%) and -C (31%). When the recovery correction was performed, the atRA-EQchem value accounted for more than 93% of the atRA-EQbio value, except for WWTP-A (47%). These results suggested that four identified RAR agonists were responsible for the majority of RARα agonistic activity in WWTP influents. However, the results also suggested the possible occurrence of unidentified RAR agonists in WWTP influents, and further study is needed. In effluent samples, RAs and 4-oxo-RAs were undetectable and the atRA-EQchem value without including the recoveries of RAs and 4-oxo-RAs was estimated to be the lowest value of 0.7 ng/L in all WWTPs but WWTP-D, where 0.6 ng/L of 4-oxo-13cRA without recovery correction was detected. These results may suggest that RAs and 4-oxo-RAs were removed readily from the aquatic phase by the activated sludge treatment, regardless of the treatment process. The atRA-EQbio value was also reducedconsiderably (92–99% reduction) during the activated sludge treatment in most WWTPs. One exception was observed in an effluent sample from WWTP-D, in which the atRA-EQbio value (19.5 ng/L) was much higher than the atRA-EQchem value (0.8 and 1.6 ng/L without and with recovery correction, respectively). A similar result was observed in repeated investigations (data not shown). This suggests the occurrence of unidentified RAR agonist(s) within activated sludge treatment in WWTP-D, which must be studied further. RAR agonist(s) within activated sludge treatment in WWTP-D, which must be studied further.

TABLE 1: Concentrations of retinoic acids (RAs) and 4-oxo-RAs and all-trans RA (atRA) equivalents estimated from concentrations and atRA-equivalency factors of RAs and 4-oxo-RAs (atRA-EQchem) and from RA receptor α agonistic activity measured by yeast two-hybrid assay (atRA-EQbio) in wastewater treatment plant (WWTP) influents and effluents

Wastewater treatment plant	Sample	Concentration (ng/L)[a]				atRA-EQchem (ng/L)[a, b]	atRA-EQbio (ng/L)
		atRA	13cRA	4-oxo-atRA	4-oxo-13cRA		
A	Influent	1.7 (7.3)	<5.0 (<13.6)	<0.5 (<0.6)	1.7 (2.3)	3.1 (9.1)	19.2
	Effluent	<0.5 (<1.4)	<2.5 (<4.5)	<0.25 (<0.4)	<0.5 (<0.7)	0.7 (1.4)	1.6
B	Influent	2.1 (9.0)	<5.0 (<13.6)	2.5 (3.3)	70.2 (94.3)	35.9 (54.1)	12.9
	Effluent	<0.5 (<1.4)	<2.5 (<4.5)	<0.25 (<0.4)	<0.5 (<0.7)	0.7 (1.4)	0.5
C	Influent	5.0 (21.5)	<5.0 (<13.6)	<0.5 (<0.6)	8.6 (11.6)	9.0 (26.9)	29.0
	Effluent	<0.5 (<1.4)	<2.5 (<4.5)	<0.25 (<0.4)	<0.5 (<0.7)	0.7 (1.4)	1.4
D	Influent	<1.0 (<4.3)	<5.0 (<13.6)	1.6 (2.1)	32.0 (42.9)	17.2 (24.4)	17.8
	Effluent	<0.5 (<1.4)	<2.5 (<4.5)	<0.25 (<0.4)	0.6 (0.8)	0.8 (1.6)	19.5
E	Influent	2.6 (11.4)	<5.0 (<13.6)	<0.5 (<0.6)	12.0 (16.1)	7.9 (18.4)	13.9
	Effluent	<0.5 (<1.4)	<2.5 (<4.5)	<0.25 (<0.4)	<0.5 (<0.7)	0.7 (1.4)	0.2
F	Influent	2.5 (10.8)	<5.0 (<13.6)	5.3 (6.9)	70.0 (94.1)	44.1 (66.0)	38.7
	Effluent	<0.5 (<1.4)	<2.5 (<4.5)	<0.25 (<0.4)	<0.5 (<0.7)	0.7 (1.4)	0.5

[a]*Concentrations without and with recovery correction for RAs and 4-oxo-RAs were shown outside and in parentheses, respectively.* [b] *Concentrations of RAs and 4-oxo-RAs less than their respective quantification limit (QL) in our analytical method were assigned a proxy value of QL/2.*

13.3.5 PRELIMINARY RISK ASSESSMENT

The possibility of biological adverse effects on aquatic animals by RAR agonists in WWTP effluents was assessed by comparing the RAR agonist contamination level in WWTP effluent determined in the present study and the RAR agonist concentrations at which detrimental effects were observed in previous studies. In this assessment, we applied the RAR agonist contamination level that was corrected with the recoveries of RAs and 4-oxo-RAs. In WWTP effluents investigated in the present study, the concentration of 4-oxo-13cRA was 0.8 ng/L at highest, while atRA, 13cRA, and 4-oxo-atRA were under the quantification limit (1.4 ng/L, 4.5 ng/L, and 0.4 ng/L, respectively).

13.4 CONCLUSION

Retinoic acid receptor agonist contamination was investigated in municipal WWTPs in Osaka, Japan. Investigations with yeast assays revealed that municipal wastewaters in investigated WWTPs consistently contain RARα agonists. The major causative compounds of the contamination were identified as RAs (atRA and 13cRA) and 4-oxo-RAs (4-oxo-atRA and 4-oxo-13cRA), possibly originally from urine. These identified RAR agonists in municipal wastewater were readily removed by the activated sludge treatment, and concomitantly the RAR agonist contamination level was largely reduced in most cases. The measured level of RAR agonist contamination in WWTP effluent is not likely to cause any RAR-mediated deleterious biological effects. However, the results of the present study also revealed that unidentified RAR agonists are present during the activated sludge treatment. Thus, there is a need to identify unknown RAR agonists and thoroughly investigate the occurrence, fate, and ecotoxicity of RAR agonists to assess in detail the potential risks derived from RAR agonists in WWTP effluent. In addition, investigations applying composite samples are needed to determine the fates and removal efficiencies of RAR agonists during wastewater treatment processes in detail.

REFERENCES

1. Chambon P. 1996. A decade of molecular biology of retinoic acid receptors. FASEB J 10: 940–954.
2. Kastner P, Mark M, Chambon P. 1995. Nonsteroid nuclear receptors: What are genetic studies telling us about their role in real life? Cell 83: 859–869.
3. Herrmann K. 1995. Teratogenic effects of retinoic acid and related substances on the early development of the zebrafish (Brachydanio rerio) as assessed by novel scoring system. Toxicol In Vitro 9: 267–283.
4. Haga Y, Suzuki T, Takeuchi T. 2002. Retinoic acid isomers produce malformations in postembryonic development of the Japanese flounder, Paralichthys olivaceus. Zool Sci 19: 1105–1112.
5. Degitz SJ, Kosian PA, Makynen EA, Jensen KM, Ankley GT. 2000. Stage- and species-specific developmental toxicity of all-trans retinoic acid in four native North American ranids and Xenopus laevis. Toxicol Sci 57: 264–274.
6. Degitz SJ, Holcombe GW, Kosian PA, Tietge JE, Durhan EJ, Ankley GT. 2003. Comparing the effects of stage and duration of retinoic acid exposure on amphibian limb development: Chronic exposure results in mortality, not limb malformations. Toxicol Sci 74: 139–146.
7. Alsop DH, Brown SB, van der Kraak GJ. 2004. Dietary retinoic acid induces hindlimb and eye deformities in Xenopus laevis. Environ Sci Technol 38: 6290–6299.
8. Mulder GB, Manley N, Grant J, Schmidt K, Zeng W, Eckhoff C, Maio-Price L. 2000. Effects of excess vitamin A on development of cranial neural crest-derived structures: A neonatal and embryologic study. Teratology 62: 214–226.
9. Ritchie HE, Brown-Woodman PD, Korabelnikoff A. 2003. Effect of co-administration of retinoids on rat embryo development in vitro. Birth Defects Res A 67: 444–451.
10. Gardiner D, Ndayibagira A, Grün F, Blumberg B. 2003. Deformed frogs and environmental retinoids. Pure Appl Chem 75: 2263–2273.
11. Zhen H, Wu X, Hu J, Xiao Y, Yang M, Hirotsuji J, Nishikawa J, Nakanishi T, Ike M. 2009. Identification of retinoic acid receptor agonists in sewage treatment plants. Environ Sci Technol 43: 6611–6616.
12. Inoue D, Matsui H, Sei K, Hu J, Yang M, Aragane J, Hirotsuji J, Ike M. 2009. Evaluation of effectiveness of chemical and physical sewage treatment technologies for removal of retinoic acid receptor agonistic activity detected in sewage effluent. Water Sci Technol 59: 2447–2453.
13. Inoue D, Nakama K, Sawada K, Watanabe T, Takagi M, Sei K, Yang M, Hirotsuji J, Hu J, Nishikawa J, Nakanishi T, Ike M. 2010. Contamination with retinoic acid receptor agonists in two rivers in the Kinki region of Japan. Water Res 44: 2409–2418.
14. Inoue D, Nakama K, Sawada K, Watanabe T, Matsui H, Sei K, Nakanishi T, Ike M. 2011. Screening of agonistic activities against four nuclear receptors in wastewater treatment plants in Japan using a yeast two-hybrid assay. J Environ Sci 23: 125–132.

15. Nishikawa J, Saito K, Goto J, Dakeyama F, Matsuo M, Nishihara T. 1999. New screening methods for chemicals with hormonal activities using interaction of nuclear hormone receptor with coactivator. Toxicol Appl Pharmacol 154: 76–83.
16. Idres N, Marill J, Flexor MA, Chabot GG. 2002. Activation of retinoic acid receptor-dependent transcription by all-trans-retinoic acid metabolites and isomers. J Biol Chem 227: 31491–31498.
17. Pijnappel WW, Hendriks HF, Folkers GE, van den Brink CE, Dekker EJ, Edelenbosch C, van der Saag PT, Durston AJ. 1993. The retinoid ligand 4-oxo-retinoic acid is a highly active modulator of positional specification. Nature 366: 340–344.
18. Li S, Barua AB, Huselton CA. 1996. Quantification of retinoyl-β-glucuronides in rat urine by reversed-phase high-performance liquid chromatography with ultraviolet detection. J Chromatogr B Biomed Appl 683: 155–162.
19. Marill J, Idres N, Capron CC, Nguyen E, Chabot GG. 2003. Retinoic acid metabolism and mechanism of action: A review. Curr Drug Metab 4: 1–10.
20. Sani BP, Barua AB, Hill DL, Shih TW, Olson JA. 1992. Retinoyl β-glucuronide: Lack of binding to receptor proteins of retinoic acid as related to biological activity. Biochem Pharmacol 43: 919–922.
21. Ternes TA, Kreckel P, Mueller J. 1999. Behaviour and occurrence of estrogens in municipal sewage treatment plants—II. Aerobic batch experiments with activated sludge. Sci Total Environ 225: 91–99.
22. Gomes RL, Scrimshaw MD, Lester JN. 2009. Fate of conjugated natural and synthetic steroid estrogens in crude sewage and activated sludge batch studies. Environ Sci Technol 43: 3612–3618.
23. Wu X, Hu J, Jia A, Peng H, Wu S, Dong Z. 2010. Determination and occurrence of retinoic acids and their 4-oxo metabolites in Liaodong Bay, China, and its adjacent rivers. Environ Toxicol Chem 29: 2491–2497.

There are several supplemental files that are not available in this version of the article. To view this additional information, please use the citation on the first page of this chapter.

Author Notes

CHAPTER 1

Acknowledgments

The work was funded by a grant from the Consejo Estatal de Ciencia y Tecnología del estado de Jalisco (COECYTAL). Thanks to Noemy A. Hernández Razo and Rosa E. Lozano Mares for technical assistance.

Conflict of Interest

The authors declare no conflict of interest.

CHAPTER 2

Acknowledgments

The author would like to thank the people of the utilities departments and environmental regulatory offices in Kosrae, Pohnpei, Chuuk, and Yap for their professional assistance provided unreservedly during the fieldwork stage of this project.

Conflict of Interest

The author declares no conflicts of interest.

CHAPTER 3

Acknowledgments

This work was supported by the National Natural Science Foundation of China (1077029), the Funds for Creative Research Groups of China (Grant no. 51121062), the National Natural Science Foundation of China (51078105), and the Open Project of the State Key Laboratory of Urban Water Resource and Environment, Harbin Institute of Technology (no. QA201019).

CHAPTER 5

Conflict of Interest

The authors have not declared any conflict of interests.

Acknowledgments

The authors would like to thank the National institute for Occupational Health and the University of Johannesburg for funding the project and the Tshwane University of Technology for the bursary. They also want to acknowledge the Young Water Professional, South Africa, who through their writing workshop contributed to the submission of this work.

CHAPTER 6

Acknowledgments

The authors would like to acknowledge BRF - Brasil Foods for providing the infrastructure and financial support, as well as the National Council for Scientific and Technological Development (CNPq) and the Brazilian Federal Agency for Support and Evaluation of Graduate Education (CAPES) for the grants supporting this study.

CHAPTER 7

Author Contributions

Conceived and designed the experiments: ACS JDJ JF HS. Performed the experiments: ACS JDJ RG GAK GF JF RHL MJB BO HS. Analyzed the data: ACS HS. Contributed reagents/materials/analysis tools: ACS JF MJB HS. Contributed to the writing of the manuscript: ACS HS.

CHAPTER 8

Acknowledgments

Part of this work was presented in the IWA Regional Conference on Wastewater Purification & Reuse. Iraklion, Hellas, 28–30 March 2012.

Author Contributions

Andreas Ilias coordinated the preparation of the manuscript, collected and organized the data, and was the corresponding author, Athanasios Panoras analyzed and codified the data, and contributed to manuscript preparation and Andreas Angelakis had the original idea, supervised the research and review it.

Conflict of Interest

The authors declare no conflict of interest.

CHAPTER 9

Acknowledgments

The authors would like to thank all participants of the INTAFERE project (http://www.intafere.de/english/). Madeleine Payne (Smart Water Research Centre, Griffith University, Gold Coast, Australia) is greatly acknowledged for reviewing and editing the manuscript. Matthias Oetken and Maren Heß are acknowledged for fruitful discussions and helpful advices. Furthermore, the authors thank Heike Heidenreich and Gerlinde Liepelt from the International Institute of Higher Education (Zittau, Germany) for their analyses of sediment-bound pollutants.

Author Contributions

Conceived and designed the experiments: DS AM KQ AB JO. Performed the experiments: DS AM KQ AB JO. Analyzed the data: DS AM KQ AB JO. Contributed reagents/materials/analysis tools: DS AM KQ AB JO. Wrote the paper: DS AM KQ AB JO.

CHAPTER 10

Competing Interests

The authors have no competing interests to declare.

Author Contributions

JCA completed data analysis and drafted the manuscript. JCC collected samples using SPE and POCIS and analyzed samples by LC/MS. JKC and JEL collected samples and monitoring data and assisted in preparation of the manuscript. CWK analyzed the ARG samples. CSW and MLH devised the study, secured funding, and supervised personnel during sampling, analysis, and manuscript preparation. All authors read and approved the final manuscript.

Acknowledgments

The authors thank Hilary Bews, Teresa Senderewich, and Shira Joudan for assistance with sampling and fieldwork, Pascal Cardinal for assistance with sample analyses and Weldon Hiebert for his mapping assistance. We would also like to thank the municipality of St. Clements, Manitoba. Funding was provided by the Riddell Endowment Fund, Manitoba Conservation and Water Stewardship, the Natural Sciences and Engineering Research Council of Canada, and the Canada Research Chairs Program.

CHAPTER 11

Acknowledgment

Authors thank New Mexico State University Agricultural Experiment Station, Water Resources Research Institute, and City of Las Cruces for help and support to this study.

CHAPTER 12

Acknowledgments

The authors would like to thank the North Carolina Water Resources Research Institute and the East Carolina University Coastal Water Resources Center for partial funding of this research. The authors would also like to acknowledge the assistance that John Woods, Jim Watson, Sarah Hardison, Matt Smith, Eliot Anderson-Evans, Caitlin van Dodewaard, and Amberlynne VanDusen provided with field and/or lab work and data compilation.

Author Contributions

All authors contributed to the conception, field and/or lab work, and development of this manuscript.

Conflict of Interest

The authors declare no conflict of interest.

CHAPTER 13

Acknowledgments

We thank J. Nishikawa from the School of Pharmacy and Pharmaceutical Sciences, Mukogawa Women's University, Japan, for kindly providing the recombinant yeasts for the yeast two-hybrid assays. The present study was supported in part by the Environment Research and Technology Development Fund (C-0802) of the Ministry of the Environment, Japan, and the Grant-in-Aid for Young Scientists (B) 20760362 from the Ministry of Education, Culture, Sports, Science and Technology, Japan.

Index

S

T

U

V

W

Y

Z